Methods in Molecular Biology

Series Editor
John M. Walker
School of Life and Medical Sciences
University of Hertfordshire
Hatfield, Hertfordshire, AL10 9AB, UK

For further volumes:
http://www.springer.com/series/7651

Decellularized Scaffolds and Organogenesis

Methods and Protocols

Edited by

Kursad Turksen

Ottawa Hospital Research Institute, Ottawa, ON, Canada

Humana Press

Editor
Kursad Turksen
Ottawa Hospital Research Institute
Ottawa, ON, Canada

ISSN 1064-3745 ISSN 1940-6029 (electronic)
Methods in Molecular Biology
ISBN 978-1-4939-7655-3 ISBN 978-1-4939-7656-0 (eBook)
https://doi.org/10.1007/978-1-4939-7656-0

Library of Congress Control Number: 2018950491

Cover Illustration Caption: Image provided by Pedro M. Baptista.

This Humana Press imprint is published by the registered company Springer Science+Business Media, LLC part of Springer Nature.
The registered company address is: 233 Spring Street, New York, NY 10013, U.S.A.

Preface

While our understanding of how to grow and differentiate cells in culture has dramatically improved over time, one of the challenges remains how to instruct particular cells of interest to recognize and respond to their environment so as to proliferate, differentiate, and function for restoration of original tissue and organ form and function. Among recent developments is the use of decellularized tissues and organs as biologically active scaffolds. In this volume, I have attempted to provide an overview of these approaches by attracting many experts in the field to summarize their recently developed methods in this area. I am grateful for their willingness to share novel approaches and insights that will provide new opportunities for those already in the field or moving to enter the field.

I would like to acknowledge Dr. John Walker, the Editor-in-Chief of the *Methods in Molecular Biology* series. I thank him for supporting the idea of this volume from its beginning and his ongoing support in related projects.

I am grateful to Patrick Marton, Executive Editor of the *Methods in Molecular Biology* series and Springer Protocols, for always being there to see projects to completion. It continues to be a pleasure to work with him.

David Casey, Editor of the *Methods in Molecular Biology* series and Springer Protocols, has also been a pleasure to work with. David has been there throughout the finalization stages of the volume, never missing a detail and continuously providing support and encouragement. I thank him for his support and friendship.

Ottawa, ON, Canada ***Kursad Turksen***

Contents

Contributors

FABIO ACOCELLA • *Department of Health, Animal Science and Food Safety, Università degli Studi di Milano, Milan, Italy*
RANDA AKOURI • *Laboratory for Transplantation and Regenerative Medicine, Department of Obstetrics and Gynecology, Sahlgrenska Academy, University of Gothenburg, Gothenburg, Sweden*
OMAR S. ALJITAWI • *University of Rochester Medical Center, Rochester, NY, USA*
MANUEL ALMEIDA • *Instituto de Investigación Sanitaria de Aragón (IIS Aragón), Zaragoza, Spain*
AHMED BAKER ALSHAIKH • *Laboratory for Transplantation and Regenerative Medicine, Department of Obstetrics and Gynecology, Sahlgrenska Academy, University of Gothenburg, Gothenburg, Sweden*
BRIAN G. AMSDEN • *Department of Chemical Engineering, Queen's University, Kingston, ON, Canada*
YONI BAERT • *Biology of the Testis, Research Laboratory for Reproduction, Genetics and Regenerative Medicine, Vrije Universiteit Brussel (VUB), Brussels, Belgium*
LIANHUA BAI • *Hepatobiliary Institute, Southwest Hospital, Third Military Medical University, Chongqing, China*
PEDRO M. BAPTISTA • *Instituto de Investigación Sanitaria de Aragón (IIS Aragón), Zaragoza, Spain; Center for Biomedical Research Network Liver and Digestive Diseases (CIBERehd), Zaragoza, Spain; Instituto de Investigación Sanitaria de la Fundación Jimé nez Díaz, Madrid, Spain; Biomedical and Aerospace Engineering Department, Universidad Carlos III de Madrid, Madrid, Spain*
CATHERINE A. BAUTISTA • *University of California, San Diego, CA, USA*
BAHAR BILGEN • *Providence VA Medical Center, Providence, RI, USA; Department of Orthopaedics, The Warren Alpert Medical School of Brown University, Rhode Island Hospital, Providence, RI, USA*
GIANNI BISOGNO • *Department of Women's and Children's Health, University of Padova, Padova, Italy*
VALENTINO BONTEMPO • *Department of Health, Animal Science and Food Safety, Università degli Studi di Milano, Milan, Italy*
SZANDRA BRAMBILLA • *University of Pavia, Pavia, Italy*
TIZIANA A.L. BREVINI • *Department of Health, Animal Science and Food Safety, Università degli Studi di Milano, Milan, Italy; Laboratory of Biomedical Embryology, Centre for Stem Cell Research, Università degli Studi di Milano, Milan, Italy*
STEFANO BRIZZOLA • *Department of Health, Animal Science and Food Safety, Università degli Studi di Milano, Milan, Italy*
JANINA BURK • *Faculty of Veterinary Medicine, Equine Clinic and Hospital Leipzig, Universität Leipzig, Leipzig, Germany*
ERIC E. BUSE • *Children's Mercy-Kansas City, Rochester, NY, USA*
FRANCESCO CALABRESE • *Department of General Surgery, IRCCS Policlinico San Matteo, Pavia, Italy*
GUOBAO CHEN • *Key Laboratory of Biorheological Science and Technology (Chongqing University), Ministry of Education, Bioengineering College, Chongqing University,*

Chongqing, People's Republic of China; Mechanobiology and Regenerative Medicine Laboratory, Bioengineering College, Chongqing University, Chongqing, People's Republic of China

GUOPING CHEN • *International Center for Materials Nanoarchitectonics, National Institute for Materials Science, Ibaraki, Japan*

YIBIN CHEN • *Center for Engineering in Medicine, Massachusetts General Hospital, Harvard Medical School, Boston, MA, USA; Shriners Hospitals for Children in Boston, Boston, MA, USA*

ANNE GEORGE CHERIAN • *Department of Obstetrics and Gynaecology, Christian Medical College Vellore Hospital, Vellore, Tamil Nadu, India*

ANTONIO CITRO • *San Raffaele Diabetes Research Institute, IRCCS San Raffaele Scientific Institute, Milan, Italy*

LORENZO COBIANCHI • *Department of Clinical, Surgical, Diagnostic and Paediatric Sciences, University of Pavia, Pavia, Italy; Department of General Surgery, IRCCS Policlinico San Matteo, Pavia, Italy*

GABRIEL L. CONVERSE • *Children's Mercy-Kansas City, Kansas City, MO, USA*

ALONZO D. COOK • *Brigham Young University, Provo, UT, USA*

VALERIA CORRADETTI • *IRCCS Policlinico San Matteo, Unit of Nephrology, Dialysis and Transplantation, Pavia, Italy*

ARCHNA DHASMANA • *Department of Polymer and Process Engineering, Indian Institute of Technology, Roorkee, India*

TOMMASO DOMINIONI • *Department of General Surgery, IRCCS Policlinico San Matteo, Pavia, Italy*

STEFANIA D'AGOSTINO • *Fondazione Istituto di Ricerca Pediatrica Città della Speranza, Padova, Italy*

AYŞE ESER ELÇIN • *Tissue Engineering, Biomaterials and Nanobiotechnology Laboratory, Ankara University Faculty of Science, and Ankara University Stem Cell Institute, Ankara, Turkey*

YAŞAR MURAT ELÇIN • *Biovalda Health Technologies, Inc., Ankara, Turkey; Tissue Engineering, Biomaterials and Nanobiotechnology Laboratory, Ankara University Faculty of Science, Ankara, Turkey*

INA ERBE • *Faculty of Veterinary Medicine, Equine Clinic and Hospital Leipzig, Universität Leipzig, Leipzig, Germany*

LAUREN E. FLYNN • *Department of Chemical and Biochemical Engineering, The University of Western Ontario, London, ON, Canada; Department of Anatomy and Cell Biology, Schulich School of Medicine and Dentistry, The University of Western Ontario, London, ON, Canada*

CHIARA FRANZIN • *Fondazione Istituto di Ricerca Pediatrica Città della Speranza, Padova, Italy*

ZACHARY GALLIGER • *Department of Pediatrics, Blood and Marrow Transplant Program, University of Minnesota, Minneapolis, MN, USA*

PIERGIORGIO GAMBA • *Department of Women's and Children's Health, University of Padova, Padova, Italy*

FULVIO GANDOLFI • *Laboratory of Biomedical Embryology, Centre for Stem Cell Research, Università degli Studi di Milano, Milan, Italy; Department of Agricultural and Environmental Science—Production, Landscape, Agroenergy, Università degli Studi di Milano, Milan, Italy*

SHARON GEERTS • *Center for Engineering in Medicine, Massachusetts General Hospital, Harvard Medical School, Boston, MA, USA; Shriners Hospitals for Children in Boston, Boston, MA, USA*

MATTEO GHIRINGHELLI • *Department of Health, Animal Science and Food Safety, Università degli Studi di Milano, Milan, Italy*

ELIZABETH GILLIES • *Department of Chemical and Biochemical Engineering, The University of Western Ontario, London, ON, Canada; Department of Chemistry, The University of Western Ontario, London, ON, Canada*

ELLEN GOOSSENS • *Biology of the Testis, Research Laboratory for Reproduction, Genetics and Regenerative Medicine, Vrije Universiteit Brussel (VUB), Brussels, Belgium*

WARREN L. GRAYSON • *Translational TE Center, Johns Hopkins University School of Medicine, Baltimore, MD, USA; Department of Biomedical Engineering, Johns Hopkins University School of Medicine, Baltimore, MD, USA; Department of Materials Science and Engineering, Johns Hopkins University, Baltimore, MD, USA; Institute for NanoBioTechnology, Johns Hopkins University, Baltimore, MD, USA; Johns Hopkins University, Translational Tissue Engineering Center, Baltimore, MD, USA*

SWETA K. GUPTA • *Department of Polymer and Process Engineering, Indian Institute of Technology, Roorkee, India; Department of Chemical Engineering, University of Rhode Island, Kingston, RI, USA*

CLEMENS GÖGELE • *Institute of Anatomy, Paracelsus Medical University, Salzburg and Nuremberg, Nuremberg, Germany*

NIELS HAMMER • *Department of Anatomy, Otago School of Biomedical Sciences, University of Otago, Dunedin, New Zealand*

MATS HELLSTRÖM • *Laboratory for Transplantation and Regenerative Medicine, Department of Obstetrics and Gynecology, Sahlgrenska Academy, University of Gothenburg, Gothenburg, Sweden*

MATTHEW J. HODGSON • *Brigham Young University, Provo, UT, USA*

RICHARD A. HOPKINS • *Children's Mercy-Kansas City, Rochester, NY, USA*

TAKASHI HOSHIBA • *Frontier Center for Organic Materials, Yamagata University, Yamagata, Japan; Innovative Flex Course for Frontier Organic Materials Systems, Yamagata University, Yamagata, Japan; International Center for Materials Nanoarchitectonics, National Institute for Materials Science, Ibaraki, Japan*

HIROYUKI IJIMA • *Department of Chemical Engineering, Faculty of Engineering, Kyushu University, Fukuoka, Japan*

MARIA JARAMILLO • *Center for Engineering in Medicine, Massachusetts General Hospital, Harvard Medical School, Boston, MA, USA; Shriners Hospitals for Children in Boston, Boston, MA, USA*

NAOKI KAWAZOE • *International Center for Materials Nanoarchitectonics, National Institute for Materials Science, Ibaraki, Japan*

YUKO KITAGAWA • *Department of Surgery, Keio University School of Medicine, Tokyo, Japan*

CHRISTOPHER C. KNUTSON • *Brigham Young University, Provo, UT, USA*

SANJAY KUMAR • *Center for Stem Cell Research, A Unit of inStem Bengaluru, Christian Medical College, CMC Rehab Campus Bagayam, Vellore, Tamil Nadu, India*

DANDAN LI • *University of Rochester Medical Center, Rochester, NY, USA*

JINGTING LI • *Stem Cell and Tissue Engineering Laboratory, Department of Orthopaedics, Division of Exercise Physiology, and Mechanical and Aerospace Engineering, West Virginia University, Morgantown, WV, USA*

Li Liu • *Department of Plastic Surgery, Affiliated Hospital of Zunyi Medical College, Zunyi, China*

Yonggang Lv • *Key Laboratory of Biorheological Science and Technology (Chongqing University), Ministry of Education, Bioengineering College, Chongqing University, Chongqing, People's Republic of China; Mechanobiology and Regenerative Medicine Laboratory, Bioengineering College, Chongqing University, Chongqing, People's Republic of China*

Marcello Maestri • *Department of General Surgery, IRCCS Policlinico San Matteo, Pavia, Italy*

Edoardo Maghin • *Fondazione Istituto di Ricerca Pediatrica Città della Speranza, Padova, Italy*

Jin Mei • *Anatomy Department, Wenzhou Medical University, Wenzhou, China*

Narayan C. Mishra • *Department of Polymer and Process Engineering, Indian Institute of Technology, Roorkee, India*

Nima Momtahan • *Brigham Young University, Provo, UT, USA*

Sara Morini • *Instituto de Investigación Sanitaria de Aragón (IIS Aragón), Zaragoza, Spain; Department of Bioengineering and iBB, Institute for Bioengineering and Biosciences, Instituto Superior Técnico, Universidade de Lisboa, Lisboa, Portugal*

Maurizio Muraca • *Department of Women's and Children's Health, University of Padova, Padova, Italy*

Ethan Nyberg • *Translational TE Center, Johns Hopkins University School of Medicine, Baltimore, MD, USA; Department of Biomedical Engineering, Johns Hopkins University School of Medicine, Baltimore, MD, USA*

Kutluk Oktay • *Innovation Institute for Fertility Preservation and In Vitro Fertilization, New York, NY, USA; Laboratory of Molecular Reproduction and Fertility Preservation, Department of Obstetrics and Gynecology, New York Medical College, Valhalla, NY, USA*

Graziano Oldani • *Divisions of Abdominal and Transplantation Surgery, Hepato-Pancreato-Biliary Centre, Department of Surgery, Faculty of Medicine, University Hospital of Geneva, University of Geneva, Geneva, Switzerland*

Benjamin Ondruschka • *Institute of Legal Medicine, University of Leipzig, Leipzig, Germany*

Arvind Manikantan Padma • *Laboratory for Transplantation and Regenerative Medicine, Department of Obstetrics and Gynecology, Sahlgrenska Academy, University of Gothenburg, Gothenburg, Sweden*

Angela Panoskaltsis-Mortari • *Department of Pediatrics, Blood and Marrow Transplant Program, University of Minnesota, Minneapolis, MN, USA*

Mahmut Parmaksiz • *Tissue Engineering, Biomaterials and Nanobiotechnology Laboratory, Ankara University Faculty of Science, and Ankara University Stem Cell Institute, Ankara, Turkey*

Ming Pei • *Stem Cell and Tissue Engineering Laboratory, Department of Orthopaedics, West Virginia University, Morgantown, WV, USA*

Andrea Peloso • *Department of Clinical, Surgical, Diagnostic and Paediatric Sciences, University of Pavia, Pavia, Italy; Department of General Surgery, IRCCS Policlinico San Matteo, Pavia, Italy; Divisions of Abdominal and Transplantation Surgery, Hepato-Pancreato-Biliary Centre, Department of Surgery, Faculty of Medicine, University Hospital of Geneva, University of Geneva, Geneva, Switzerland*

Martina Piccoli • *Fondazione Istituto di Ricerca Pediatrica Città della Speranza, Padova, Italy*

IRIS PLA-PALACÍN • *Instituto de Investigación Sanitaria de Aragón (IIS Aragón), Zaragoza, Spain*

MICHELA POZZOBON • *Fondazione Istituto di Ricerca Pediatrica Città della Speranza, Padova, Italy; Department of Women and Children Health, University of Padova, Padova, Italy; Stem Cells and Regenerative Medicine Lab, Istituto di Ricerca Pediatrica Città della Speranza, Padova, Italy*

ALEXANDRA N. RINDONE • *Translational TE Center, Johns Hopkins University School of Medicine, Baltimore, MD, USA; Department of Biomedical Engineering, Johns Hopkins University School of Medicine, Baltimore, MD, USA*

SANDRO ROSSI • *Medicina Interna e Gastroenterologia, Università di Pavia, Fondazione IRCCS Policlinico S. Matteo, Pavia, Italy*

SUSANNE PAULINE ROTH • *Faculty of Veterinary Medicine, Equine Clinic and Hospital Leipzig, Universität Leipzig, Leipzig, Germany; Saxonian Incubator for Clinical Translation, Universität Leipzig, Leipzig, Germany*

MATTIA SAGGIORO • *Department of Women's and Children's Health, University of Padova, Padova, Italy; Fondazione Istituto di Ricerca Pediatrica Città della Speranza, Padova, Italy*

PILAR SAINZ-ARNAL • *Instituto de Investigación Sanitaria de Aragón (IIS Aragón), Zaragoza, Spain; Instituto Aragonés de Ciencias de la Salud (IACS), Zaragoza, Spain*

GUNDULA SCHULZE-TANZIL • *Institute of Anatomy, Paracelsus Medical University, Salzburg and Nuremberg, Nuremberg, Germany*

SILKE SCHWARZ • *Institute of Anatomy, Paracelsus Medical University, Salzburg and Nuremberg, Nuremberg, Germany*

NANA SHIRAKIGAWA • *Department of Chemical Engineering, Faculty of Engineering, Kyushu University, Fukuoka, Japan*

ARTHI SHRIDHAR • *Department of Chemical and Biochemical Engineering, The University of Western Ontario, London, ON, Canada*

STEVEN SKOLASINSKI • *Department of Medicine, Division of Pulmonary, Allergy, Critical Care, and Sleep, 350 Variety Heart and Research Center, University of Minnesota, Minneapolis, MN, USA*

MIN JONG SONG • *Laboratory for Transplantation and Regenerative Medicine, Department of Obstetrics and Gynecology, Sahlgrenska Academy, University of Gothenburg, Gothenburg, Sweden; Division of Gynecologic Oncology, Department of Obstetrics and Gynecology, Daejeon St. Mary's Hospital, The Catholic University of Korea, Seoul, South Korea*

BALASUBRAMANIAN SUNDARAM • *Center for Stem Cell Research, A Unit of inStem Bengaluru, Christian Medical College, CMC Rehab Campus Bagayam, Vellore, Tamil Nadu, India*

KAZUKI TAJIMA • *Department of Surgery, Keio University School of Medicine, Tokyo, Japan*

ENES TAYLAN • *Innovation Institute for Fertility Preservation and In Vitro Fertilization, New York, NY, USA*

TOM TRISTAN TIEMANN • *Laboratory for Transplantation and Regenerative Medicine, Department of Obstetrics and Gynecology, Sahlgrenska Academy, University of Gothenburg, Gothenburg, Sweden; Department of Obstetrics and Gynecology, University Hospital Heidelberg, Heidelberg, Germany*

CATERINA TREVISAN • *Fondazione Istituto di Ricerca Pediatrica Città della Speranza, Padova, Italy; Department of Women and Children Health, University of Padova, Padova, Italy*

BASAK E. UYGUN • *Center for Engineering in Medicine, Massachusetts General Hospital, Harvard Medical School, Boston, MA, USA; Shriners Hospitals for Children in Boston, Boston, MA, USA*

ZAIRONG WEI • *Department of Plastic Surgery, Affiliated Hospital of Zunyi Medical College, Zunyi, China*

RENPEI XIA • *Hepatobiliary Institute, Southwest Hospital, Third Military Medical University, Chongqing, China*

HIROSHI YAGI • *Department of Surgery, Keio University School of Medicine, Tokyo, Japan*

WEI YANG • *Hepatobiliary Institute, Southwest Hospital, Third Military Medical University, Chongqing, China*

YALING YU • *Department of Orthopedic Surgery, Shanghai Jiao Tong University Affiliated Sixth People's Hospital, Shanghai, China*

ALESSANDRO ZENOBI • *Department of Health, Animal Science and Food Safety, Università degli Studi di Milano, Milan, Italy; Laboratory of Biomedical Embryology, Centre for Stem Cell Research, Università degli Studi di Milano, Milan, Italy*

HONGYU ZHANG • *Hepatobiliary Institute, Southwest Hospital, Third Military Medical University, Chongqing, China*

JIANSE ZHANG • *Anatomy Department, Wenzhou Medical University, Wenzhou, China*

YUJUN ZHANG • *Hepatobiliary Institute, Southwest Hospital, Third Military Medical University, Chongqing, China*

Methods in Molecular Biology (2018) 1577: 1–10
DOI 10.1007/7651_2017_34

Published online: 27 May 2017

Decellularization Methods for Scaffold Fabrication

Sweta K. Gupta, Narayan C. Mishra, and Archna Dhasmana

Abstract

Decellularization is the process of removal of native cells from tissue, leaving behind a three-dimensional (3D) ultrastructure of extracellular matrix (ECM) proteins while preserving the bioactivity and mechanics of the tissue. It offers a unique top-down approach for fabricating ECM based natural scaffold for tissue engineering application. Herein, this chapter presents the fabrication of decellularized scaffold employing different methods: whole organ perfusion, immersion and agitation, pressure gradient, and supercritical fluid. The decellularized scaffold aims to exploit the nature-designed 3D architecture, a successful platform technology, for creating scaffolding materials for tissue engineering and regenerative medicine.

Keywords Decellularization, Scaffold, Cells, Extracellular matrix, Tissue engineering

1 Introduction

The term "decellularization" means removal of cells, and their lipid and nuclear membranes (major immunogenic components) from the tissues [1], while retaining the composition, biological, and mechanical properties of protein-rich three-dimensional (3D) extracellular matrix (ECM). The ECM is predominantly composed of complex network of structural proteins (collagens and elastin), specialized proteins (fibrillin, fibronectin, and laminin), proteoglycans (heparin sulphate, chondroitin sulphate, keratin sulphate, and GAGs), and growth factors [2]. The decellularized/acellular matrix does not induce antigenicity, inflammatory, and immune response responsible for tissue rejection [3, 4], and the success of these decellularized scaffolds upon implantation is due to the molecular cues provided by the natural ECM components. Intact decellularized ECM scaffold acts as the reservoir of numerous molecular constituents (cell adhesive proteins) and the growth factors found in the native tissue [5]. Upon implantation, the injected cells repopulate the matrix surface, and eventually the matrices degrade with time and are replaced and remodeled by ECM proteins

The original version of this chapter was revised. An erratum to this chapter can be found at DOI 10.1007/7651_2017_110

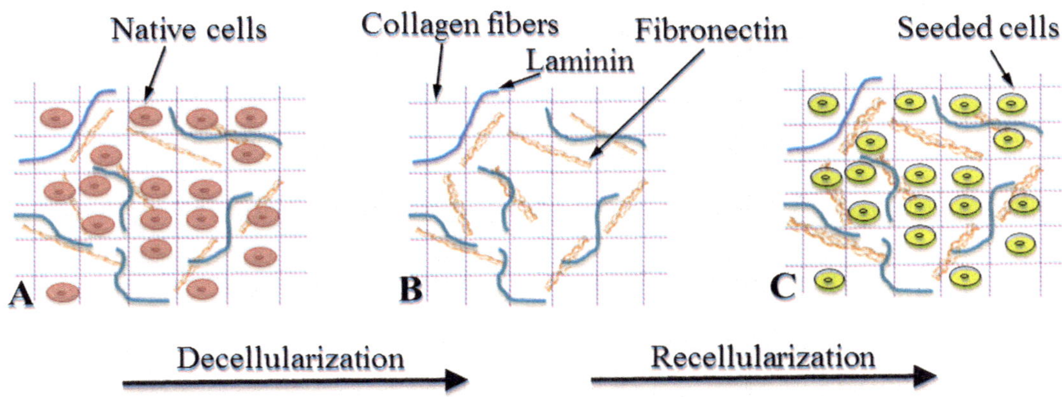

Fig. 1 Decellularization and recellularization of an organ: (**a**) native tissue, (**b**) decellularized tissue, and (**c**) recellularized tissue

secreted by transplanted or ingrowing cells. The native 3D architecture of decellularized ECM provides a unique niche of protein composition, mechanical properties, and bioactive molecules that opens a huge potential to substitute the artificial scaffolds and use the decellularized scaffold for tissue regeneration (Fig. 1). The primary goals of regenerative medicine are the healthy remodeling of the organs instead of the formation of scar tissue. Decellularized ECMs promote constructive host tissue remodeling instead of the formation of scar tissue, which is of great benefit, and avoid long-term complications and the overall costs [6–8].

2 Materials

There are different strategies (chemical, physical, and enzymatic) to decellularize a particular organ or tissues, and sometimes these strategies are used in combinations to enhance the effectiveness of decellularization process.

2.1 Chemical Agents

(1) Acids: Acids catalyze hydrolytic degradation of biomolecules. The most common acid used for decellularization is acetic acid (0.5–1 M); however, it damages ECM structure by removing/denaturing collagen thereby, reducing the overall ECM strength. Peracetic acids (0.1 %) remove nucleic acid and have minimum effect on ECM structure and composition.

(2) Bases: Bases are commonly used to remove hairs from dermal tissue. The common bases used for decellularization are ammonium hydroxide, calcium hydroxide, and sodium hydroxide. If exposed for long time, bases can remove growth factors from ECM and decrease mechanical properties of the decellularized matrix by disrupting collagens.

(3) Organic solvents: Organic solvents tend to dehydrate the cells and solubilize their cell membrane, which causes cell lysis and

removal. Alcohols, acetones, and 1 % tributyl phosphate (TBP) are common organic solvents used for decellularizing dense tissues. Acetone destroys the ECM ultrastructure as it is a tissue-fixative, whereas TBP has been reported to have viricidal (inactivate viruses in blood) activities without compromising the coagulation factor [9].

(4) Hypotonic/hypertonic solutions: Hypotonic solutions rupture cell membrane and cause cell lysis by osmotic shock. This method, combined with chemical treatment, causes minimal change in ECM composition. Hypertonic solution has been reported to dissociate DNA from proteins.

(5) Ionic detergents: An ionic detergent used in tissue decellularization includes sodium dodecyl sulphate (SDS), Triton X-200, sodium dodecyl cholate (SDC), and sodium hypochlorite. 0.1 % SDS is the most common treatment method; however, some studies have used a series of SDS concentrations (0.01, 0.025, 0.05, 0.075, and 0.1 %) for decellularization, which helps in removing cells while maintaining the overall tissue architecture and composition [10].

(6) Nonionic detergents: Nonionic detergents are mild and disrupt lipid–lipid and lipid–protein interactions, while leaving protein–protein interactions intact. 0.05 % Triton X-100 is the common nonionic detergent used for decellularization [11].

(7) Zwitterionic detergents: Zwitterionic detergents exhibit the properties of both nonionic and ionic detergents and are used for thinner tissues, which require mild treatment. The common zwitterionic detergents used for decellularization include Sulfobetaine-10 (SB10), Sulfobetaine-16 (SB16), and 3-[(3-cholamidopropyl) dimethylammonio]-1-propanesulfonate (CHAPS) [1].

2.2 Physical Agents

(1) Freezing and Thawing: Multiple freeze–thaw cycles disrupt cell membrane by formation of intracellular ice crystals; however, this method has not been reported complete removal of nuclear materials. This method is usually combined with ionic/nonionic detergents and nucleases for effective decellularization.

(2) Force/Pressure: Pressure can be used to lyse cells in the tissues that are not densely populated by cells (e.g., lung and liver). Force can be applied to delaminate the tissue layers from small intestine and urinary bladder, which are characterized by natural planes of dissection.

(3) Electroporation: Electric field oscillation disrupts the cell membrane and causes cell lysis.

(4) Agitation: Agitation is the most common method for decelluarizing thin tissues, which do not have proper vasculature. It facilitates chemical exposure and disrupts cells.

(5) Sonication: Ultrasonic sound waves are applied to rupture the cell membrane.

2.3 Enzymatic Agents

(1) Trypsin: Cleaves peptide bonds on the carboxyl-side of arginine and lysine. 0.5 % trypsin (48 h exposure time) has been reported to cause damage to aortic valve ECM [12] and 1 % trypsin (24 h exposure time) extensive collagen damage in porcine dermis [13]. 0.02 % trypsin (1 h exposure time) suggests negligible change in porcine adipose tissue ECM after decellularization [14].

(2) Exo/Endo nucleases: Catalyze the hydrolysis of terminal/interior bonds of ribonucleotides and deoxyribonucleotides. Treatment with DNase (0.2–0.5 mg/mL) and RNase (0.2–50 μg/mL) is used in combination with other decellularization methods which facilitate the complete removal of residual DNA/RNA in the decellularized tissues [15].

(3) Dispase: Cleaves specific peptides, mainly collagen IV and fibronectin, in the basement membrane and separates epithelial sheets from the substrate. Dispase II (4 mg/mL, 45 min) has been used to decellularize porcine cornea followed by hypertonic/detergent treatment (1.5 M NaCl/0.1 % SDS at 200 rpm for 12 h) [16] whereas, to decellularize porcine skin, it was first treated with hypertonic/alkaline treatment (20 % sodium sulfide, 1 M NaCl, 85 % glycerol, 2/1 (v/v) chloroform/methanol) to remove hair, fat, and epidermis, and then exposed to Dispase II (0.24 mg/mL, 3 h) [13].

(4) Phospholipase A2: hydrolyzes phospholipid component and is used in combination with other decellularizing agents. It helps in maintaining the collagen and proteoglycans in the tissue but might affect the composition of GAGs in the tissue [17].

3 Methods

The effectiveness of cell removal from the tissues depends on the source, composition, and density of the tissue, and the decellularization protocol needs to be optimized depending upon the tissue types. The various different type of methods used for decellularization include whole organ perfusion method for thick 3D tissues or complete organs, immersion, and agitation method for thin tissues, pressure gradient based decellularization for hollow tissues, and use of supercritical fluid to facilitate long-term storage of the decellularized tissue.

3.1 Whole Organ Perfusion Method

To achieve the effective decellularization of whole organs and vascularized tissues, antegrade or retrograde perfusion technique via perfusion bioreactor is generally used. This process utilizes the organ's natural plumbing to remove the cells rapidly from inside out of the organ, which helps in preserving the 3D ultrastructure of tissue ECM. Due to the innate vascular networks present within the organs, the diffusion distance of oxygen to the cells is minimized. This technique involves cannulating the vascular network and applying physiological perfusion pressure which helps in effective permeation of the decellularizing agent through the organs native vasculature (arteries, arterioles, and capillaries) and helps in the removal of all the cellular components/debris through the venous system. As the organs are dense with vascular capillaries and most cells are located in close proximity to the capillaries, the effective surface area of the decellularizing agent to reach and remove the cellular components is increased, without exposing them to the scaffold. By applying antegrade perfusion of detergents and subsequent washes to arterial vasculature at low physiological pressures, successful decellularization of complex organs (i.e., laryngeal muscle [18], rat heart [19], lung [20–23], kidney [24], and liver [25, 26]) has been performed. The resultant acellular matrix/scaffold obtained after the completion of the vascular perfusion techniques retains the intact 3D ultrastructure, appropriate microenvironment, mechanical properties, and vascular conduits of the organs, which enables the recellularization and growth of the new functional organ of clinically significant thickness [21].

3.2 Immersion and Agitation

Although the whole organ perfusion decellularization technique is more effective, by exploiting the tissue's vascular network, but all tissues do not have innate vasculature that will helps in straightforward, discrete, and complete access [27]. The common approach of decellularization for such tissues that do not possess vascular network is immersion and agitation of tissue in decellularizing agent. This technique depends on the intensity of agitation, detergent used, tissue density, and thickness [28]. Thin tissue (urinary bladder and small intestine submucosa) can be decellularized relatively in small time (1–2 h) as compared to dense tissue (dermis, tendon, and trachea), which takes longer exposure time for decellularization (12–72 h). This process can be applied to only small sections of tissues and never whole organ. In this process, an organ is decellularized by immersing in a reservoir of decellularizing solution where the decellularization process is diffusion limited because the detergent migrates slowly from the outer surface to the inner side and then again to the outer side after dissolving the cellular materials. This process also requires the use of endonucleases (DNase/RNase) to remove residual nuclear components. The limitation behind the process is that the cells start breaking before being exposed to the detergent and release various proteases, which degrade the surrounding ECM. The resultant scaffold, so formed, is partially degraded with compromised mechanical properties. This method has been used to decellularize variety of tissues, which includes urinary bladder, dermis, esophagus, trachea, cartilage, spinal cord, peripheral nerve, skeletal muscles, blood vessels, and heart valves [12, 29–33].

3.3 Pressure Gradient

Pressure gradient is used in case of hollow tissues (intestine, blood vessels, umbilical vein, and bladder) to accelerate and augment the delivery of decellularizing agents into the tissues and forcing the residual cellular components out of the tissues. During decellularization, pressure gradient can also be induced along with the enzyme treatment of the tissue, which will result in better maintenance of tissue composition and ultrastructure [13]. Pressure gradient (flow rate 50 mL/min and pulse frequency 2 Hz) has been used to decellularize umbilical vein [32] and bladder tissue by combining cyclic intravesical pressurization and submersion [34].

3.4 Supercritical Fluid

Supercritical fluid employs the use of inert substances (e.g., carbon dioxide), which has low viscosity and high transport properties. Cellular components are removed from the tissues when supercritical fluids are passed at a controlled rate similar to critical point drying and result in minimal disruption of tissue ECM mechanical properties. The obtained decellularized tissue is dry which is helpful for long-term storage, avoiding the need for lyophilization. The use of carbon dioxide as supercritical fluid (pressure 7.4 MPa and temperature 32 °C) has demonstrated the removal of cells from aortic tissue and an ethanol entrainer [35].

4 Notes

1. Effective decellularization of a tissue is an appropriate balance between cell removal and preservation of an overall ECM integrity. If incompletely decellularized, cellular immunogenic constituents present in the tissue may elicit an adverse host response [28].
2. If the tissue is not properly handled and/or if chemicals are overdosed, then there are chances of loss of original ECM structure and also there could be substantial loss of ECM proteins present in the tissue.
3. Due to "sticky" nature of DNA, sometimes it tends to adhere to ECM proteins strongly and makes the removal of DNA difficult. Therefore, in case of hypotonic/hypertonic treatment for tissue decellularization, additional step of enzymatic or chemical treatment is necessary to remove the residual cellular/nuclear components.
4. When the cells get disrupted during decellularization, they release proteases, which can cause damage to the innate ECM structure. Therefore, to inhibit the activity of proteases, it is desirable to include either protease inhibitor (leupeptin, aprotinin, and phenylmethylsulfonylfluoride (PMSF)) or buffered solution (pH 7–8). Inhibition of protease activity is also controlled by controlling temperature and time of decellularization.

5. Scaffold sterilization: Sometimes bacteria contaminate the remaining ECM components during decellularization process. Therefore, it is desirable to include 1 % antibiotic–antimycotic solutions (PenStrep and amphotericin B) during each decellularization step. To further disinfect the scaffold and prevent contamination, sterilizing agents (ethanol, ethylene oxide, peracetic acid, and gamma and UV radiations) should be used before the recellularization of the scaffolds.
6. Although there is no "gold standard" method for decellularization, because the removal of cells depends on the type of the tissue, e.g., in case of anatomically intact organ, whole organ perfusion decellularization is used, if the organ does not have proper vasculature and is few millimeter thick, then mechanical agitation in detergent/chemical agents is used, and if the organs are hollow, pressure gradients are used for decellularization.
7. Scaffold preservation/storage: Sometimes storing the decellularized matrices in 1× PBS in freezer (−20 or −80 °C) causes a significant change in the ultrastructure of ECM after thawing it. However, short-term storage of decellularized tissues could be done at 4 °C in 0.01–0.1 % peracetic acid solution. The decellularized organ/tissue scaffold can be cryopreserved at subzero temperature (−196 °C) in liquid nitrogen or could be freeze-dried and preserved for years at room temperature.
8. To optimize whole organ perfusion based decellularization, there are three things to control: perfusate temperature, perfusion flow, and control of pressure during decellularization.
9. Pressure-controlled perfusion decellularization, with appropriate modifications, can be achieved in small-animal experimental models (rat organs, 4–5 days) and scaled to clinically relevant models (porcine and human organs, 12–14 days).
10. To enhance the attachment of cells, infiltration into the scaffolds, and growth and proliferation, decellularized tissues are often coated with natural polymers (gelatin, chitosan, collagen, and RGD peptides) to provide motifs, which increase the attachment between the cells and the surface of the decellularized tissue.
11. After decellularization, it is essential to check whether the necessary components of the tissue (actin, vimentin, fibronectin, laminin, growth factors, elastic fibers, collagens, etc.) are still retained within the decellularized ECM or not, and various techniques, as mentioned in the following, help to check this:
    (a) Hematoxylin and eosin staining: Hematoxylin stains the ECM pink and eosin stains the cell nuclei blue in color. It shows the presence of cellular components in the ECM [36].

(b) Movat'spentachrome: stains glycosaminoglycans (GAGs) blue, cell nuclei and elastic fibers black, and collagen yellow [37].

(c) Masson's trichrome/Picrosirius red: stains ECM protein and shows that the structure of fibrin and collagen is maintained intact in ECM [38].

(d) Verhoeff van Gieson: stains elastin protein present in the ECM [39].

(e) Alcian blue/PAS: stains proteoglycans in the ECM [40].

(f) Safranin O stains: stains GAGs present in ECM [41].

(g) Oil Red and Sudan: stains the lipid molecules in the tissue sections [42].

(h) DAPI or Hoechst staining: shows the presence of DNA in the decellularized tissue by binding to the AT clusters in the minor groove of DNA [43, 44].

(i) Propidium Iodide and PicoGreen: shows the presence of DNA within the decellularized tissue [45].

(j) DNA probe techniques: used to verify the presence of DNA in the decellularized tissue [46].

Acknowledgements

The authors are thankful to the Council of Scientific and Industrial Research (Grant No. 27(0222)/10/EM R-II dated 31.05.10), India, for the financial support of this work.

References

1. Gilbert TW, Sellaro TL, Badylak SF (2006) Decellularization of tissues and organs. Biomaterials 27:3675–3683
2. Frantz C, Stewart KM, Weaver VM (2010) The extracellular matrix at a glance. J Cell Sci 123:4195–4200
3. Ketchedjian A, Jones AL, Krueger P et al (2005) Recellularization of decellularized allograft scaffolds in ovine great vessel reconstructions. Ann Thorac Surg 79:888–896, discussion 896
4. Grauss RW, Hazekamp MG, van Vliet S et al (2003) Decellularization of rat aortic valve allografts reduces leaflet destruction and extracellular matrix remodeling. J Thorac Cardiovasc Surg 126:2003–2010
5. Voytik-Harbin SL, Brightman AO, Kraine MR et al (1997) Identification of extractable growth factors from small intestinal submucosa. J Cell Biochem 67:478–491
6. Xu CC, Chan RW, Weinberger DG et al (2010) A bovine acellular scaffold for vocal fold reconstruction in a rat model. J Biomed Mater Res A 92:18–32
7. Parekh A, Mantle B, Banks J et al (2009) Repair of the tympanic membrane with urinary bladder matrix. Laryngoscope 119:1206–1213
8. Valentin JE, Turner NJ, Gilbert TW et al (2010) Functional skeletal muscle formation with a biologic scaffold. Biomaterials 31:7475–7484
9. Cartmell JS, Dunn MG (2000) Effect of chemical treatments on tendon cellularity and mechanical properties. J Biomed Mater Res 49:134–140
10. Schaner PJ, Martin ND, Tulenko TN et al (2004) Decellularized vein as a potential scaffold for vascular tissue engineering. J Vasc Surg 40:146–153

11. Gonzalez-Andrades M, Carriel V, Rivera-Izquierdo M et al (2015) Effects of detergent-based protocols on decellularization of corneas with sclerocorneal limbus. Evaluation of regional differences. Transl Vis Sci Technol 4:13
12. Meyer SR, Chiu B, Churchill TA et al (2006) Comparison of aortic valve allograft decellularization techniques in the rat. J Biomed Mater Res A 79:254–262
13. Prasertsung I, Kanokpanont S, Bunaprasert T et al (2008) Development of acellular dermis from porcine skin using periodic pressurized technique. J Biomed Mater Res B Appl Biomater 85:210–219
14. Brown BN, Freund JM, Han L et al (2011) Comparison of three methods for the derivation of a biologic scaffold composed of adipose tissue extracellular matrix. Tissue Eng Part C Methods 17:411–421
15. Grauss RW, Hazekamp MG, Oppenhuizen F et al (2005) Histological evaluation of decellularised porcine aortic valves: matrix changes due to different decellularisation methods. Eur J Cardiothorac Surg 27:566–571
16. Gonzalez-Andrades M, de la Cruz Cardona J, Ionescu AM et al (2011) Generation of bioengineered corneas with decellularized xenografts and human keratocytes. Invest Ophthalmol Vis Sci 52:215–222
17. Huang M, Li N, Wu Z et al (2011) Using acellular porcine limbal stroma for rabbit limbal stem cell microenvironment reconstruction. Biomaterials 32:7812–7821
18. Hou N, Cui P, Luo J et al (2011) Tissue-engineered larynx using perfusion-decellularized technique and mesenchymal stem cells in a rabbit model. Acta Otolaryngol 131:645–652
19. Ott HC, Matthiesen TS, Goh SK et al (2008) Perfusion-decellularized matrix: using nature's platform to engineer a bioartificial heart. Nat Med 14:213–221
20. Petersen TH, Calle EA, Zhao L et al (2010) Tissue-engineered lungs for in vivo implantation. Science 329:538–541
21. Ott HC, Clippinger B, Conrad C et al (2010) Regeneration and orthotopic transplantation of a bioartificial lung. Nat Med 16:927–933
22. Cortiella J, Niles J, Cantu A et al (2010) Influence of acellular natural lung matrix on murine embryonic stem cell differentiation and tissue formation. Tissue Eng Part A 16:2565–2580
23. Price AP, England KA, Matson AM et al (2010) Development of a decellularized lung bioreactor system for bioengineering the lung: the matrix reloaded. Tissue Eng Part A 16:2581–2591
24. Caralt M, Uzarski JS, Iacob S et al (2015) Optimization and critical evaluation of decellularization strategies to develop renal extracellular matrix scaffolds as biological templates for organ engineering and transplantation. Am J Transplant 15:64–75
25. Shupe T, Williams M, Brown A et al (2010) Method for the decellularization of intact rat liver. Organogenesis 6:134–136
26. Uygun BE, Soto-Gutierrez A, Yagi H et al (2010) Organ reengineering through development of a transplantable recellularized liver graft using decellularized liver matrix. Nat Med 16:814–820
27. Henderson PW, Nagineni VV, Harper A et al (2010) Development of an acellular bioengineered matrix with a dominant vascular pedicle. J Surg Res 164:1–5
28. Crapo PM, Gilbert TW, Badylak SF (2011) An overview of tissue and whole organ decellularization processes. Biomaterials 32:3233–3243
29. Tudorache I, Cebotari S, Sturz G et al (2007) Tissue engineering of heart valves: biomechanical and morphological properties of decellularized heart valves. J Heart Valve Dis 16:567–573, discussion 574
30. Cebotari S, Tudorache I, Jaekel T et al (2010) Detergent decellularization of heart valves for tissue engineering: toxicological effects of residual detergents on human endothelial cells. Artif Organs 34:206–210
31. Gui L, Chan SA, Breuer CK et al (2010) Novel utilization of serum in tissue decellularization. Tissue Eng Part C Methods 16:173–184
32. Montoya CV, McFetridge PS (2009) Preparation of ex vivo-based biomaterials using convective flow decellularization. Tissue Eng Part C Methods 15:191–200
33. Lehr EJ, Rayat GR, Chiu B et al (2011) Decellularization reduces immunogenicity of sheep pulmonary artery vascular patches. J Thorac Cardiovasc Surg 141:1056–1062
34. Bolland F, Korossis S, Wilshaw SP et al (2007) Development and characterisation of a full-thickness acellular porcine bladder matrix for tissue engineering. Biomaterials 28:1061–1070
35. Sawada K, Terada D, Yamaoka T et al (2008) Cell removal with supercritical carbon dioxide for acellular artificial tissue. J Chem Technol Biotechnol 83:943–949
36. Keane TJ, Londono R, Turner NJ et al (2012) Consequences of ineffective decellularization of biologic scaffolds on the host response. Biomaterials 33:1771–1781
37. Zou Y, Zhang Y (2012) Mechanical evaluation of decellularized porcine thoracic aorta. J Surg Res 175:359–368
38. Totonelli G, Maghsoudlou P, Garriboli M et al (2012) A rat decellularized small bowel scaffold

that preserves villus-crypt architecture for intestinal regeneration. Biomaterials 33:3401–3410

39. Wallis JM, Borg ZD, Daly AB et al (2012) Comparative assessment of detergent-based protocols for mouse lung de-cellularization and re-cellularization. Tissue Eng Part C Methods 18:420–432
40. Struecker B, Hillebrandt KH, Voitl R et al (2014) Porcine liver decellularization under oscillating pressure conditions—a technical refinement to improve the homogeneity of the decellularization process. Tissue Eng Part C Methods 21(3):303–313
41. Azhim A, Ono T, Fukui Y et al (2013) Preparation of decellularized meniscal scaffolds using sonication treatment for tissue engineering. Conf Proc IEEE Eng Med Biol Soc 2013:6953–6956
42. Porzionato A, Sfriso MM, Macchi V et al (2013) Decellularized omentum as novel biologic scaffold for reconstructive surgery and regenerative medicine. Eur J Histochem e4:57
43. Latt SA, Wohlleb JC (1975) Optical studies of the interaction of 33258 Hoechst with DNA, chromatin, and metaphase chromosomes. Chromosoma 52:297–316
44. Kakkar R, Suruchi GR (2005) Theoretical study of molecular recognition by Hoechst 33258 derivatives. J Biomol Struct Dyn 23:37–47
45. Kral T, Widerak K, Langner M et al (2005) Propidium iodide and PicoGreen as dyes for the DNA fluorescence correlation spectroscopy measurements. J Fluoresc 15:179–183
46. Jackson DW, Simon TM (2002) Donor cell survival and repopulation after intraarticular transplantation of tendon and ligament allografts. Microsc Res Tech 58:25–33

Methods in Molecular Biology (2018) 1577: 11–23
DOI 10.1007/7651_2017_53

Published online: 31 August 2017

Decellularized Liver Scaffold for Liver Regeneration

Wei Yang, Renpei Xia, Yujun Zhang, Hongyu Zhang, and Lianhua Bai

Abstract

After being initially hailed as the ultimate solution to end-stage organ failure, such as end-stage liver disease (ESLD), engineering of vascularized tissues has stalled because of the need for a well-structured circulatory system that can maintain the cells to be seeded inside the construct.

In the field of regenerative medicine, decellularized scaffolds, derived mainly from various non-autologous whole organs, have become an emerging treatment technique to overcome this obstacle. As a result of significant progress made in recent years, organogenesis through whole-organ decellularization scaffolds may now become more feasible than ever before. In this chapter, we describe in detail the necessary steps for liver organogenesis using a decellularized acellular scaffold (DAS), seed cell isolation, and recellularization in a bioreactor-like culture system. This new technique to re-engineer organs may have major implications for the fields of drug discovery, organ transplantation, and ultimately regenerative medicine.

Keywords Bioreactor, Decellularization, Organogenesis, Recellularization, Regenerative medicine, Stem cells, Tissue engineering

1 Introduction

Allogeneic organ transplantation remains the ultimate solution for end-stage organ failure like end-stage liver disease (ESLD); however, shortage of donor organs has resulted in extending transplantation waiting lists. Thus constructing a portable organ by tissue engineering in vitro might be a better choice at present.

Cellular components can be well removed from whole organs by detergent perfusion to produce a "rejectless" acellular extracellular matrix (ECM) that retains most of the ECM components and vascular and microcirculatory structures, which can be anastomosed with the recipient circulation; thus tissue engineering by decellularization/recellularization for whole organogenesis now appears more feasible than ever before [1–3]. Progress is also rapidly being made as researchers address several key challenges, for example, ensuring correct cell distribution, seeding, donor/recipient blood compatibility, angiogenesis, immunological concerns, cell sources, and matrices for whole-organ tissue engineering.

Over the past few years, some of the techniques used have been optimized to a point where the decellularization of whole organs is now possible to generate a decellularized acellular scaffold (DAS) for organ bioengineering [4–6]. The method of decellularization by perfusion actively "pushes" a detergent solution into the vasculature of an organ with a pump, allowing for decellularization of organs, which was previously unattainable with passive diffusion of detergents. The DAS from whole organs prepared in this fashion can then be readily recellularized with seed cells such as hepatocytes, endothelial cells, and stem/progenitor cells for organogenesis (Fig. 1).

Classical collagenase digestion of the tissues [7] and a novel "Percoll-plate-wait" procedure for adult liver hepatic stem/progenitor cell isolation [8] can provide a highly functional cell source with great potential for cell adhesion, proliferation, differentiation, and organogenesis [3]. However, their delivery into the DAS requires development and use of a different seeding procedure for effective recellularization. Once cell seeding is completed, a day–month maintenance period with continuous culture media distribution into the recellularized DAS and a bioreactor [9]/bioreactor-like culture system (BLCS) [3] is required to effectively proliferate/differentiate the seeding cells for organ formation and maturation. The proliferated/differentiated seeding cells from the stem/

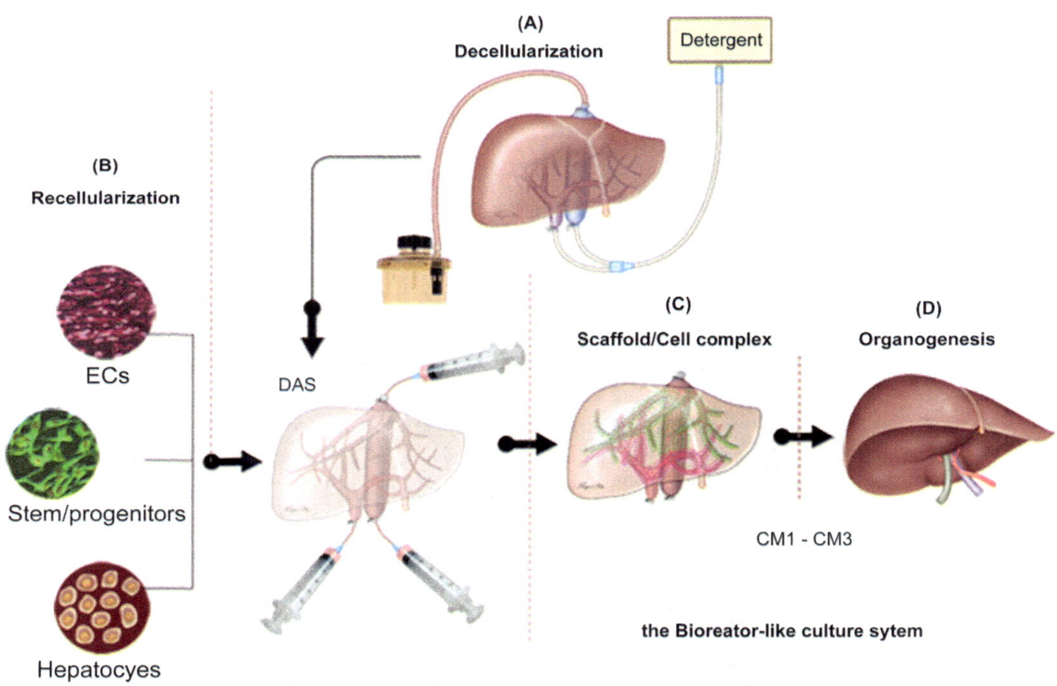

Fig. 1 Strategy of liver organogenesis. (**a**) Generation of liver decellularized acellular scaffold (DAS) by detergent perfusion, with entry of fluid into the organ via portal vein cannulation. (**b**) Seeding of cells into the DAS. (**c**) Culture DAS–cell complex in the bioreactor-like culture system (BLCS). (**d**) Induction of liver organ formation with conditioned media (CMs) in the BLCS

progenitor cells will then exhibit typical functions, such as synthesis of the liver functional protein albumin, low-density lipoprotein (LDL), and urea, as well as diverse phenotypic markers of biliary cholangiocytes (CK19) and hepatocytes from engineered liver organs [3, 10].

2 Materials

2.1 Decellularization

All experimental solutions are made with ultrapure water (prepared by purifying deionized water) and use analytical grade reagents. All prepared reagents are stored at room temperature (unless indicated otherwise). All waste disposal regulations are diligently followed when waste materials are disposed of. The following materials and equipment are used:

1. 0.9% sterile saline containing heparin lithium salt (100 U/mg; Southwest Hospital, Chongqing, China).
2. Sodium dodecyl sulfate (SDS) detergent with low-digestive trypsin–EDTA (ethylene diamine tetraacetic acid) solution (SDS/trypsin–EDTA solution): 0.1% SDS (Sigma, St. Louis, MO, USA), 0.005% trypsin (Hyclone; Thermo Fisher Scientific, Waltham, MA, USA), and 0.002% EDTA (Amresco Inc., Solon, OH, USA) in distilled water.
3. Peristaltic pump (BT101F; LeadFluid Technology Co., Baoding, China).
4. Silicone tubing (16#, size for mouse; LeadFluid, USA).

2.2 Seed Cell Isolation and Recellularization

All experimental animal protocols are approved by the Animal Ethics Committee of the Third Military Medical University (#SYXK-PLA-20120031). Animals used for decellularization are anesthetized with an intraperitoneal injection of 1% sodium pentobarbital 60 mg/kg) (Merck Millipore, Darmstadt, Germany), and surgical procedures are performed in a fully equipped animal research laboratory. The following materials and equipment are used:

1. Inbred C57BL/6 mice are used as donor livers.
2. Tissue medium: Minimum essential medium without calcium and magnesium (wMEM) for cell digestion, and phosphate buffer saline without calcium and magnesium (wPBS) (Hyclone, USA) for cell washing.
3. Digestive solution: 0.005% trypsin, 0.002% EDTA, and 900 U deoxyribonuclease (DNase I) (Roche Diagnostics, Inc., Indianapolis, IN, USA) in 5 mL Dulbecco's modified Eagle's medium (Hyclone; Thermo Fisher Scientific, Waltham, MA,

USA) (DMEM)/F12 medium (0.07 mg/mL, AMPD1; Sigma, USA) for cell digestion.

4. Complete culture medium: DMEM/F12 containing 10% (w/v) fetal bovine serum (FBS) (ZSGB, China), 200 mM (18.25 ng/mL) glutamine (Invitrogen Inc., Carlsbad, CA, USA) (Hyclone, USA), and 1% antibiotic–antimycotic solution (Hyclone SV30010; USA).
5. Cell washing medium: DMEM/F12 medium and phosphate buffer saline (PBS).
6. 40 and 70 μm cell strainers (Falcon, NJ, USA).
7. Endothelial cell (EC) culture medium (EC-CM = CM1): DMEM containing 2 mM L-glutamine, 1× non-essential amino acids (100×), 1× sodium pyruvate (100×), 25 mM HEPES (pH 7.0–7.6) (1 M; Gibco, Invitrogen Inc., Carlsbad, CA, USA), 100 μg/mL of endothelial cell growth supplements (ECGS) (100×; Sciencell Research Laboratories, Carlsbad, CA, USA), and 20% FBS.
8. Trophic factors: Endothelial growth factor (EGF) (Gibco, USA), bovine insulin, human transferrin, L-thyroxine, sodium selenite, humic amine, progesterone (Sigma), hepatic growth factor (HGF) (BioVision, San Francisco, CA, USA), basic fibroblast growth factor (bFGF) (PeproTech, Rocky Hill, NJ, USA), tumor suppressor M, and dexamethasone (Sigma-Aldrich, USA).
9. Anti-CD31 (clone #390; Biolegend, CA, USA), anti–von Willebrand factor (anti-vWf) (clone #C12; Santa Cruz Biotechnology, Santa Cruz, CA USA), and anti-neural/glial antigen 2 (anti-NG2) (clone #L20; Santa Cruz Biotechnology) antibodies.
10. Control seeding media (Ctrl-CM): DMEM/F12 medium.
11. Conditioned media (CMs): Three conditioned media (CM1, CM2, and CM3) with various additives. All CMs contain one portion of developmental liver homogenate (postnatal p0) and three portions of DMEM/F12 plus 10 ng/mL HGF, 5 ng/mL bFGF, 0.5 mg/mL human insulin, 0.5 mg/mL human transferrin, 40 μg/mL L-thyroxine, 34 μg/mL human transferrin, 0.5 μg/mL humic amine and 6 μg/mL progesterone (CM2), 1 mL 100 ng/mL HGF, 50 ng/mL bFGF, 20 ng/mL tumor suppressor M and 0.1 μM dexamethasone (CM3), and CM1 (=EC-CM).
12. Vascular cannulas: 20–24G (for mouse) (Becton Dickinson, Medical Devices Co., Jiangsu, China).
13. 50 mL conical centrifuge tubes (Biologix, China).
14. 26G syringe needle (Shandong Weigao Group Medical Polymer Co., Ltd., China).

15. Scalpel and scissors.
16. Several different forceps such as fixation, mosquito, Adson's, Kelly's, and Jeweler's forceps, etc.
17. Retractors and towel clamps
18. Surgical suture: Silk suture 11-0, including thin silk thread, monofilament nylon suture, monofilament polypropylene sutures, and absorbable thread (Ningbo Medical Needle Co., Ltd.).
19. 75 cm^2 culture flasks (Thermo) coated with poly-L-lysine (PLL; 10.67 $\mu g/cm^2$) (Sigma).
20. Several sizes of micro-clips (Shanghai Apparatus & Instrument Co., Ltd., XEC290, China).
21. Betadine disinfection (Sichuan Huatian Profession Industry Co., Ltd., China) for surgical area sterile conditions.

All instruments are kept on a cork board during surgery to prevent damage to their fine tips.

2.3 Tissue Engineering Processing in the BLCS for Liver Organogenesis [3]

The following materials and equipment are used:

1. A centrifuge (Thermo ST16R, USA).
2. 6- and 12-well plates for cultures (Biologix, China).
3. A shaker (TS-2000A; Haimen Kylin-Bell Lab Instruments Co., Jiangsu, China) with an agitation rate setting of 40–42 min/min.
4. An incubator (Thermo, MA, USA).

The shaker is placed in the incubator, where dissolved oxygen is maintained at 40% or 60% with air, oxygen, or nitrogen; CO_2 is added to maintain pH at 7.2, and the temperature is set at 37 °C throughout the entire process.

3 Methods

3.1 Operation for Harvesting Liver Organs

All procedures are performed under aseptic conditions:

1. A longitudinal abdominal incision is made to visualize the liver, lower abdominal cavity, and rib cage.
2. The supra hepatic vena cava is transected as close to the atrium as possible, along with the falciform and cardiac ligaments.
3. The diaphragm is dissected carefully around the esophagus in order to separate it from the liver and diaphragm, and the common bile duct is transected as close to the duodenum as possible.

4. The adipose tissue layers surrounding the portal vein (PV) are dissected carefully in order to visualize the vein and its branches.
5. The lateral branches are ligated with silk suture (4-0) and cut closer to the intestines (the distal end regarding the liver).
6. The PV is transected about 1.5–2 cm away from the liver.
7. The infra hepatic vena cava located under the right lobe of the liver is dissected carefully and transected without damaging the liver lobe.
8. The intact liver is removed gently by holding it by the diaphragm.

3.2 DAS Preparation and Sterilization

1. The cannula is attached in the PV to a peristaltic pump by using a 20–24G catheter (for mouse) depending on the diameter of the vein.
2. 0.5 L of 0.9% saline containing heparin is perfused through the abdominal aorta at the rate of 5 mL/min (mouse).
3. 4 L of detergent solution (SDS/trypsin–EDTA) is perfused, following an initial wash with 1 L of distilled water solution.
4. 3 L of PBS is perfused through the liver to remove all of the decellularization detergent present (Fig. 2).
5. The DAS is put into a 50 mL conical tube with its cannula in deionized water after decellularization.
6. The DAS is sterilized in 1.5 Mrad of gamma radiation with a cobalt 60 gamma irradiator (see **Note 1**) or under ultraviolet (UV) light, ready for recellularization.
7. The sterilized DAS is placed in deionized water and stored at 4 °C until use.

3.3 Isolation of Seeding Cells

3.3.1 Novel Hepatic Stem/Progenitor Cell Isolation from the Adult Liver Periportal Vascular Region ($NG2^+$ HPCs) [8]

Step 1: Percoll Gradient Isolation

1. The liver periportal vascular region is cut carefully (two livers at a time) into small fragments with scissors in a petri dish with 200 μL cold wMEM solution, and the small tissue chunks are homogenized by a scalpel.
2. The small tissue chunks are digested in 2 mL of digestive solution and placed in a 37 °C 5% CO_2 incubator for 10 min (see **Note 2**).
3. 3 mL of complete medium is added to stop the digestion, following 300 μL (7 μg/mL) DNase I (Sigma, AMPD1)

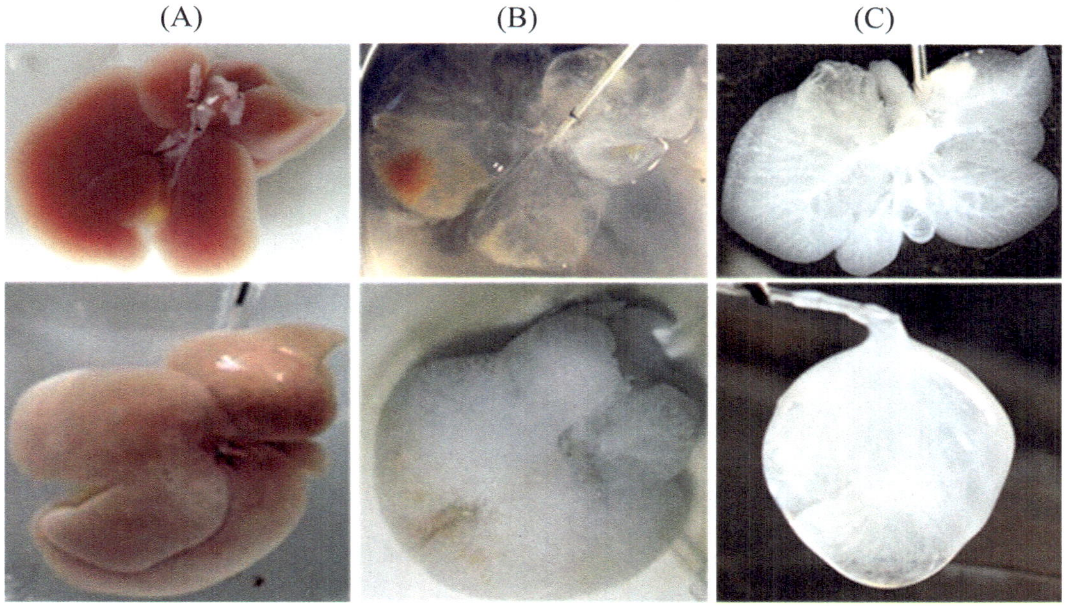

Fig. 2 Generation of decellularized acellular scaffold (DAS) by using sodium dodecyl sulfate (SDS) with a low concentration of trypsin–EDTA solution. (**a**) Appearance of mouse liver immediately after isolation. (**b**) After 2 h of decellularization. (**c**) Post-detergent perfusion

made with complete medium, and placed in a 50 mL conical tube.

4. The heterogenous suspension is triturated gently with a glass pipet to break the remaining tissue pieces, and 10 mL DMEM/F12 wash medium is added.
5. Centrifugation at 250 × *g* is performed for 10 min.
6. The supernatant is aspirated, and the pellet is suspended in 10 mL of DMEM/F12 medium and centrifuged again at 250 × *g* for 10 min.
7. The supernatant is aspirated and the resuspension is passed through a 70 μm cell strainer, followed by a 40 μm cell strainer.
8. The resulting suspension is centrifuged at 250 × *g* for 10 min.
9. The supernatant is aspirated and the resuspension is washed twice with DMEM/F12 medium before layering onto a Percoll gradient (GE Healthcare, 10055500) (see **Note 3**).
10. 7.5 mL of homogenized cells is layered carefully in 30% stock isotonic Percoll (SIP) over 7.5 mL 70% SIP in one tube to form a uniform layer (see **Note 4**).
11. Centrifugation at 700 × *g* is performed at 25 °C for 30 min for breakdown.

12. Cells are collected from the 70–30% interface and the collected fraction is put into a 50 mL conical tube in 10 mL DMEN/F12 medium.
13. Centrifugation at 250 × *g* is performed for 10 min.
14. The supernatant is aspirated and 1 mL of sterile ice-cold distilled water is added for 6 s to remove erythrocytes, followed immediately by 1 mL of 2× PBS with 4% (vol/vol) FBS to quench the reaction.
15. Centrifugation at 250 × *g* is performed for 10 min.

Step 2: Plating of Cells

1. The supernatant is aspirated and the pellet is suspended in 3 mL of complete medium.
2. The cells are counted and plated (1×10^6 cells) on PLL-coated 75 cm^2 flasks, and the cells are cultured at 37 °C in 5% CO_2 for 20 min.
3. An additional 22 mL of complete medium is added to start the primary culture.

Step 3: Waiting Procedure

1. 1 mL of complete medium is added every 7 days and the cells are allowed to grow for 21 days during the primary culture period.
2. 1 mL of complete medium is added every 3–5 days until 6 weeks when cell colonies appear (~21 days). Of these cells, 95–98% express NG2 and exhibit hepatic progenitor cell properties, so these cells are assigned as the primary culture (passage 0, p0), namely these cell populations are the $NG2^+$ hematopoietic progenitor cells (HPCs) (Fig. 3) (see **Note 5**).

3.3.2 EC and Hepatocyte Isolation

1. Classical methods are used to isolated ECs [11].
2. EC colonies usually appear at between 3 and 5 days during primary culture.

3.3.3 Subculture

A regular subculture procedure is performed with trypsin/EDTA/DNase solution when the cells grow to more than 80% confluence.

3.4 BLCS Preparation and Sterilization

The BLCS provides the in vitro portable environment that is necessary for appropriate tissue bioengineering. All illustrated BLCS are configured according to Fig. 4a, using 6- or 12-well plates, a shaker, and a tissue culture biosafety cabinet (a normal incubator).

1. The 6- or 12-well plate is placed on a shaker at 40–42 rpm (see **Note 6**).

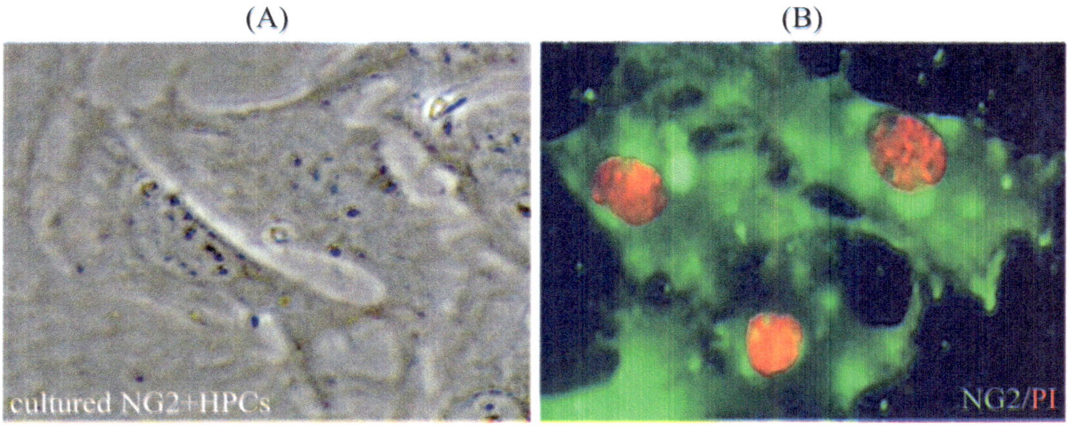

Fig. 3 Isolation of adult hepatic stem/progenitor cells. (**a**) Cultured hepatic stem/progenitor cells isolated from the adult mouse liver periportal vascular region. (**b**) The isolated cells express neural/glial antigen 2 (NG2), namely $NG2^+$ hematopoietic progenitor cells (HPCs)

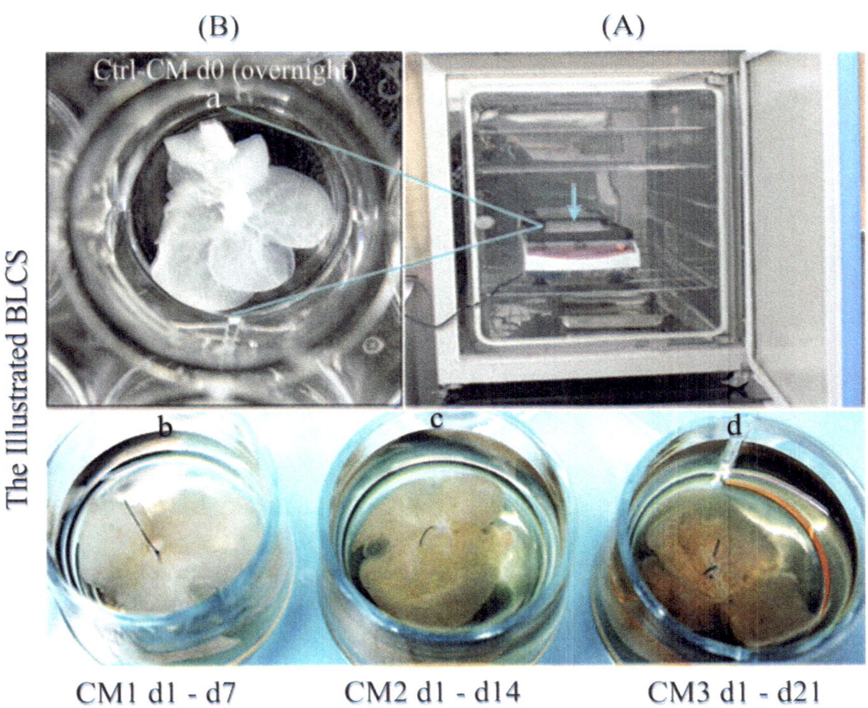

Fig. 4 Schematic diagrams of the recellularized DAS (decellularized acellular scaffold) in the bioreactor-like culture system (BLCS) for liver organogenesis. (**A**) DAS–cell complex in the BLCS. (**Ba**) DAS–cell complex cultures in the BLCS on day 0 in Dulbecco's modified Eagle's medium (DMEM)/F12 medium. (**Bb–d**) DAS–cell complex cultures in the BLCS from day 1 to day 21 in different conditioned media (CMs: CM1–CM3)

2. The shaker is placed in the biosafety incubator cabinet using sterile gloves to reduce the likelihood of contamination.
3. The culture is maintained in the incubator at 37 °C in 5% CO_2 overnight prior to seeding of cells (see **Note 7**).

3.5 Recellularization

3.5.1 Digestion of Cultured Cells

1. Culture medium is aspirated from cultured cells, which are then washed once with wash medium (DMEM/F12 or PBS).
2. The wash medium is aspirated and replaced with 3 mL of 0.05% trypsin/0.02% EDTA solution, followed by incubation at 37 °C for 5 min.
3. DMEM/F12/10% FBS medium is added, followed by pipetting up and down carefully to stop digestion, and then the cells are transferred to a 50 mL conical tube.
4. The dish is rinsed with 5 mL of DMEM/10% FBS for any leftover or unattached cells remaining on the culture flask or dish.
5. Centrifugation at 250 × *g* is performed for 5 min. The supernatant is aspirated and the pellet is reconstituted with 10 mL of DMEM/F12 medium.
6. The cells are strained through 70 and 40 μm cell strainers to remove cell aggregates.
7. Centrifugation at 250 × *g* is performed for 5 min.
8. The supernatant is aspirated, followed by centrifugation at 250 × *g* for 5 min.
9. The supernatant is aspirated, and the cells are resuspended and counted (see **Note 8**).
10. Centrifugation at 250 × *g* is performed for 5 min.
11. The supernatant is aspirated, keeping the cells in Ctrl-CM, and the tube is placed on ice until ready for injection.

3.5.2 Injection of Seed Cells into the Prepared DAS

1. Preparation of ECs: The stem/progenitor cell–derived EC isolation uses either fluorescence-activated cell sorting (FACS) and magnetic beads or an aortic EC isolation method [11].
2. ECs are seeded (0.5–1 × 10^6) for 7 days, and then hepatocytes or stem/progenitor cells are co-seeded (1–5 × 10^6), such as $NG2^+$ HPCs or mesenchymal stem cells [12], into scaffolds through both the PV and the inferior vena cava (IVC) by a needle injection method (see **Note 9**).
3. The seed cells are co-infused every 4 h, with a total of two repetitions (8 h).

4. The cell–scaffold complex is cultured in the BLCS.

3.6 Organogenesis

The liver organogenesis process consists of four phases: (1) seeding of cells; (2) maintaining the seeded cells in DAS (described above); and (3) culture of the DAS–cell complex in the BLCS with constant agitation (40–42 rpm) to allow the cell–scaffold complex to form a tissue-like construct or organogenesis in the system for about 21 days [3].

1. After 1 h of cell seeding, some of the DAS–cell complex is analyzed immediately for viability and retention; some is left in the culture with CM1 (=EC-CM) and cultured in 24-well plates in the BLCS at 37 °C in 5% CO_2 for 7 days to promote EC differentiation (CM1 d1–d7) (Fig. 4Ba, b) (see **Note 10**).
2. CM1 is replaced with CM2 for the next 7 days to maintain stem/progenitor cell self-renewal (second week in CM2 d1–d14) (Fig. 4Bc).
3. Subsequently CM2 is replaced with CM3 in the third week until day 21 (for the third week in CM2 d1–d21) (Fig. 4Bd) (see **Note 11**) to promote liver organ formation (organogenesis) (Fig. 5).
4. A small amount of some of the complex is collected every week for viability, retention, and DNA extraction for measurement. Some of the remaining complex is fixed in a 4% paraformaldehyde solution for paraffin embedding, and the rest of the complex continues to culture until 3 weeks to induce organogenesis.

4 Notes

1. The number of tubes depends on the size/volume of the pellet obtained.
2. DAS sterilization with gamma irradiation is sometimes not available. However, this method is strongly recommended because use of UV light and chemical disinfectants sometimes has the undesirable effect of changing the biomaterial mechanical properties.
3. Tissue chunks in digestive solution (containing 2 mL 0.05% trypsin/0.02% EDTA solution, 0.3 mL DNase I (70 μg/mL) in 5 mL solution) require gentle shaking.
4. The Percoll gradient is prepared from SIP at 70% in 1× wPBS (white color) and 30% in 1× wMEM (red color).
5. There is no contamination of fibroblast-like cells by using the “Percoll-plate-wait” procedure for $NG2^+$ HPC cell isolation.

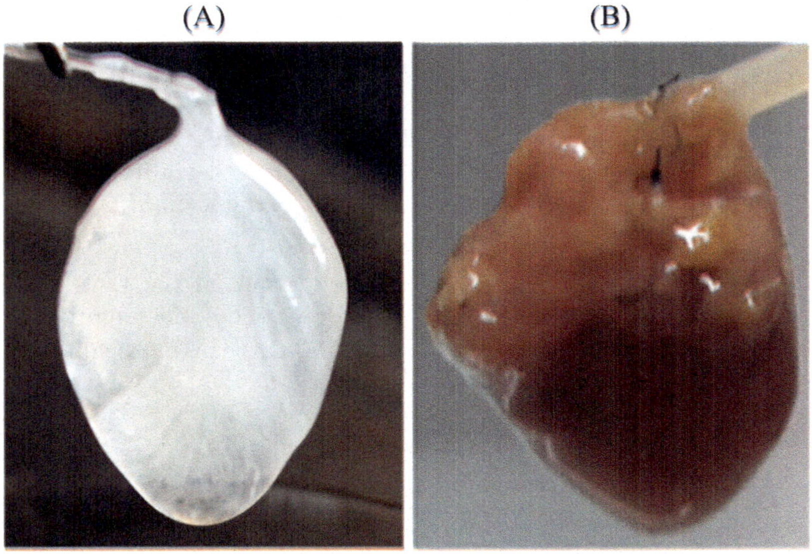

Fig. 5 Liver organogenesis induced by conditioned media (CMs). During CM changes every week, the bioengineered liver is gradually formed on the decellularized acellular scaffold (DAS) (**a**) and has formed a liver-like tissue or organ (**b**) at day 21

6. The DAS–cell complex is placed in 6- or 12-well plates, and the plates are then put on a shaker (at 40–42 rpm) and cultured in an incubator-like cabinet in three different CMs (CM1–CM3) at 37 °C in 5% CO_2.
7. The gas/media exchange occurs through the DAS/ECM–cell complex to construct liver organs.
8. Straining the cells prior to injection is essential to remove cell aggregates that might clump in the DAS small vasculature network.
9. When the cells are injected into the DAS during cell seeding, the outside ports of the PV and IVC are wiped with alcohol before and after cell injection to reduce the likelihood of contamination. One should be absolutely certain that the cells are injected into the DAS through the PV and IVC ports and eventually recirculate into the whole DAS.
10. For the first week, the medium change procedure is not performed during the DAS–cell complex culture in CM1.
11. CM2 and CM3 media are changed at 2-day intervals to ensure adequate oxygen and nutrient delivery to the cells.

Acknowledgments

This work was supported by the National Natural Science Foundation of China (NSFC) grant 81570573 to L.H.B. and University Southwestern Hospital grant SHW2014 LC01 to L.H.B.

References

1. Guyette JP, Charest JM, Mills RW et al (2016) Bioengineering human myocardium on native extracellular matrix. Circ Res 118:56–72
2. Uygun BE, Soto-Gutierrez A, Yagi H et al (2010) Organ reengineering through development of a transplantable recellularized liver graft using decellularized liver matrix. Nat Med 16:814–820
3. Zhang H, Siegel CT, Li J et al (2016) Functional liver tissue engineering by an adult mouse liver-derived neuro-glia antigen 2-expressing stem/progenitor population. J Tissue Eng Regen Med. https://doi.org/10.1002/term.2311
4. Baptista PM, Moran EC, Vyas D et al (2016) Fluid flow regulation of revascularization and cellular organization in a bioengineered liver platform. Tissue Eng Part C Methods 22:199–207
5. Guyette JP, Gilpin SE, Charest JM et al (2014) Perfusion decellularization of whole organs. Nat Protoc 9:1451–1468
6. Song JJ, Guyette JP, Gilpin SE et al (2013) Regeneration and experimental orthotopic transplantation of a bioengineered kidney. Nat Med 19:646–651
7. Bartlett DC, Hodson J, Bhogal RH et al (2014) Combined use of N-acetylcysteine and liberase improves the viability and metabolic function of human hepatocytes isolated from human liver. Cytotherapy 16:800–809
8. Zhang H, Siegel CT, Shuai L et al (2016) Repair of liver mediated by adult mouse liver neuro-glia antigen 2-positive progenitor cell transplantation in a mouse model of cirrhosis. Sci Rep 6:21783–21797
9. Caralt M, Velasco E, Lanas A et al (2014) Liver bioengineering: from the stage of liver decellularized matrix to the multiple cellular actors and bioreactor special effects. Organogenesis 10:250–259
10. Sabetkish S, Kajbafzadeh AM, Sabetkish N et al (2015) Whole-organ tissue engineering: decellularization and recellularization of three-dimensional matrix liver scaffolds. J Biomed Mater Res A 103:1498–1508
11. Kobayashi M, Inoue K, Warabi E et al (2005) A simple method of isolating mouse aortic endothelial cells. J Atheroscler Thromb 12:138–142
12. Watt SM, Gullo F, van der Garde M et al (2013) The angiogenic properties of mesenchymal stem/stromal cells and their therapeutic potential. Br Med Bull 108:25–53

Methods in Molecular Biology (2018) 1577: 25–33
DOI 10.1007/7651_2017_61

Published online: 08 August 2017

Wharton's Jelly Matrix Decellularization for Tissue Engineering Applications

Gabriel L. Converse, Dandan Li, Eric E. Buse, Richard A. Hopkins, and Omar S. Aljitawi

Abstract

Scaffolds, both natural and synthetic, used in tissue engineering provide mechanical support to cells. Tissue decellularization has been used to provide natural extracellular matrix scaffolds for tissue engineering purposes. In this chapter we focus on describing the methodology used to decellularize Wharton's jelly matrix, the mucous connective tissue that surrounds umbilical cord vessels, to obtain decellularized Wharton's jelly matrix (DWJM); an extracellular matrix that can be used for tissue engineering purposes. We also, briefly, describe our experience with processing DWJM for cell seeding and recellularization.

Keywords Cell seeding, Decellularized Wharton's jelly matrix, Natural scaffolds, Recellularization, Tissue decellularization, Wharton's jelly matrix

1 Introduction

Successful tissue engineering requires a combination of four elements including cells, growth factors, scaffolds, and the mechanical environment [1]. Scaffolds provide mechanical support to the cells, which actively influence cellular responses including cell attachment and proliferation [2]. Several types of scaffolds are available for tissue engineering, which include natural or synthetic scaffolds. Given the many favorable characteristics of natural scaffolds, there has been significant interest in tissue decellularization as a method to provide natural tissue that could replace the tissue of interest [3–6].

Wharton's jelly is a mucous connective tissue that surrounds the umbilical cord vessels and is covered by a layer of simple amniotic epithelium [7]. This tissue, with the support of other umbilical cord tissues, must work to maintain blood flow during fetal movements, during labor and during cord abnormalities like knots. To perform in this capacity, this matrix has several unique biomechanical characteristics related to its biochemical structure. This unique tissue is made of mesenchymal cells immersed in ground substance that is rich in collagen and hyaluronan and contains sulfated glycosaminoglycans [7]. The firmness of the intact cord is related to the fibrous nature of Wharton's jelly matrix, which is attributable to collagen and elastin fibers. On the other hand, the matrix is porous with canalicular structures that house the proteoglycans and hyaluronic acid. Interaction with water forms a highly viscous fluid within the pore space of the matrix [8]. The viscous nature of this fluid and the presence of these canalicular structures allow the fluid to move when compressed. This poroelastic behavior of Wharton's jelly allows the umbilical cord to resist compressive loading [8].

Wharton's jelly matrix has several characteristics desired in a scaffold for tissue engineering. First, it hosts and supports Wharton's jelly mesenchymal stem cells, which makes this matrix potentially capable of supporting other populations of undifferentiated mesenchymal cells. The 3 dimensional (3D) environment provided by this matrix could potentially even enhance mesenchymal stem cell culture in vitro as seen previously in 3D systems [9]. Hyaluronan also widely expresses the archetypal hyaluronan receptor, CD 44, which is expressed on chondrocytes, hematopoietic marrow cells, and osteocytes [10], which allows these cells to attach to the matrix. CD 44 is expressed on mesenchymal stem cells, which explains their attachment to the matrix [11]. Additionally, because of its poroelastic behavior, we expect a scaffold made of Wharton's jelly matrix to withstand compressive forces. Finally, Wharton's jelly is a rich source of peptide growth factors, notably insulin like growth factor-1 (IGF-1), but also platelet derived growth factor (PDGF) [12]. These growth factors are found to be associated with extracellular matrix proteins and in turn they control cell proliferation, differentiation and synthesis, and remodeling of the extracellular matrix [13]. These growth factors might help in controlling the cellular processes described above and potentially positively impact the growth and differentiation of transplanted mesenchymal cells.

Given the many favorable features of Wharton's jelly as a matrix for tissue engineering, we attempted to decellularize Wharton's jelly matrix to obtain a 3D matrix structure that can be used for tissue engineering [1]. Our working hypothesis was that we can apply decellularization techniques to Wharton's jelly matrix to obtain an acellular scaffold for tissue engineering purposes. Herein, we describe our experience with decellularization of Wharton's jelly matrix. We also describe the process of making decellularized Wharton's jelly matrix (DWJM) wafers that can be used for cell seeding. Finally, we briefly describe our experience with recellularization of DWJM.

2 Materials

2.1 Decellularization Solutions

1. *Transport Solution*: Add 20 mL of Amphotericin B (250 μg/mL) and 20 mL of Penicillin–Streptomycin (10,000 U Penicillin/10 mg Streptomycin per mL) to 460 mL Lactated Ringer's in the BSC, maintaining sterility of the solution. Filter sterilize the solution if sterility is in question. Mix thoroughly by repeated pipetting.
2. *0.05% Triton X-100 Solution*: Add 0.5 mL Triton X-100 to 999.5 mL of deionized water in a glass beaker. Add a magnetic stir bar and stir until the Triton X-100 is fully dispersed and no longer visible in the solution (see notes 1 and 2).
3. *Hypertonic Salt Solution*: Add 9 g NaCl, 0.476 g $MgCl_2$, 37.3 g KCl, and 125 g to 1 L of normal (0.9%) saline in a glass beaker. Add a magnetic stir bar and stir until all components are fully dissolved and the solution is clear (see notes 1 and 2).
4. *Enzymatic Solution*: Add 67 mg $MgCl_2$ and 550 μL Benzonase (250 U/μL) to 350 mL of deionized water in a glass beaker. Add a magnetic stir bar and stir until all components are dispersed and no longer visible in the solution. Adjust pH to 8.5 using NH_4OH (see notes 1 and 3).
5. *N-Lauroylsarcosine Solution*: Add 19.4 g *N*-Lauroylsarcosine to 1 L of deionized water in a glass beaker. Add a magnetic stir bar and stir until the *N*-Lauroylsarcosine is fully dispersed and no longer visible in the solution (see notes 1 and 2).
6. *Ethanol Solution*: Add 400 mL of 200 proof Ethanol to 600 mL of deionized water in a glass beaker. Add a magnetic stir bar and stir until ethanol is fully dispersed and no longer visible in the solution (see notes 1 and 2).
7. *Saline Mannitol Solution*: Add 18 g NaCl and 125 g D-mannitol to 1 L of normal (0.9%) saline in a glass beaker. Add a magnetic stir bar and stir until all components are fully dissolved and the solution is clear (see notes 1 and 2).
8. *Cryopreservation Media*: Add 100 mL dimethyl sulfoxide and 100 mL fetal bovine serum to 800 mL RPMI 1640. Mix thoroughly by repeated pipetting (see note 4).

2.2 Other Decellularization Materials

1. Deionized Water.
2. TMD-8 Ion Exchange Resin Beads.
3. XAD-16 Amberlite Ion Exchange Resin Beads.
4. Dowex Monosphere 550A UPW Ion Exchange Resin Beads.
5. Dialysis Tubing.
6. Surgical Staples.

7. Surgical Instruments.
8. Centrifuge.
9. Shaker Plate.
10. Magnetic Stir Plate.

2.3 Wafer Cutting Materials

1. Lactated Ringer's Solution.
2. Liquid Nitrogen.
3. Wafer Cutter (Fig. 1).
4. Silicone Mat.
5. OCT Freeze Medium.
6. Molds for Cryoembedding.
7. Forceps.
8. Scissors.
9. 0.9% Saline.
10. Bowl.
11. 2 mL Tube.
12. Cryopreservation Media (as described above).

Fig. 1 Custom-built wafer cutter. The frozen OCT block is inserted into the opening (indicated by *arrow*). The position of the block and the wafer thickness is adjusted using the screw opposite to the opening

2.4 Recellularization of DWJM Wafers

Suggested culture media based on seeded cells:

1. Purified umbilical cord blood (UCB) hematopoietic stem and progenitor cells (CD34+ HSPCs): StemSpan™ SFEM (Catalog # 09650 Stemcell Technologies) supplemented with StemSpan™ CC110 (Catalog # 02697 Stemcell Technologies).
2. Leukemia cell lines HL 60, Kasumi I, and MV 411: RPMI 1640 supplemented with 10% FBS and 1% pen/strep.
3. Primary bone marrow stromal cells (MSCs): DMEM supplemented with 20% FBS and 1% pen/strep.
4. Phosphate-buffered saline (PBS).
5. Collagenase II.
6. 24-Well non-tissue culture treated plates.

3 Methods

3.1 Decellularization of Wharton's Jelly Matrix

1. Place umbilical cord in Transport Solution immediately following harvest and store at 4 °C for up to 72 h.
2. Dissect Wharton's jelly from the umbilical cord, removing the umbilical arteries and vein, in the biosafety cabinet using sterile technique (see note 5).
3. Place Wharton's jelly into a sterile 50 mL centrifuge tube. The tissue should occupy between 5 and 10 mL of the tube. Fill tube to the 45 mL mark with Hypertonic Salt Solution and treat for 1 h (see notes 6 and 7).
4. Transfer the Wharton's jelly to a new, sterile 50 mL centrifuge tube and fill to the 45 mL mark with Triton-X 100 solution. Treat for 1 h (see notes 6 and 7).
5. Transfer the Wharton's jelly to a new, sterile 50 mL centrifuge tube and fill to the 45 mL mark with Hypertonic Salt Solution. Treat for 1 h (see notes 6 and 7).
6. Transfer the Wharton's jelly to a new, sterile 50 mL centrifuge tube and fill to the 45 mL mark with sterile, Deionized Water. Treat for 1 h (see notes 6 and 7).
7. Transfer the Wharton's jelly to a new, sterile 50 mL centrifuge tube and fill to the 45 mL mark with Enzymatic Solution. Treat 12–16 h (see notes 7–9).
8. Transfer the Wharton's jelly to a new, sterile 50 mL centrifuge tube and fill to the 45 mL mark with *N*-Lauroylsarcosine Solution. Treat for 2 h (see notes 6 and 7).
9. Transfer the Wharton's jelly to a new, sterile 50 mL centrifuge tube and fill to the 45 mL mark with Ethanol Solution. Treat for 10 min (see notes 6 and 7).

10. Add 200 mL of sterile, Deionized Water to a sterile glass jar. Seal 1.5 g each of TMD-8, XAD-16 Amberlite and Dowex Monosphere 550A UPW Ion Exchange Resin Beads within dialysis tubing and place in the jar (see note 10).
11. Seal the Wharton's jelly within dialysis tubing (see note 10).
12. Place the Wharton's jelly packets in the jar containing the ion exchange resin beads. Treat for 6 h on a shaker plate set to 150 RPM at 4 °C.
13. Remove the Wharton's jelly from the dialysis tubing packets and transfer to a new, sterile 50 mL centrifuge tube. Fill to the 45 mL mark with Saline Mannitol Solution and treat for 1 h (see notes 6 and 7).
14. Transfer the decellularized Wharton's jelly to a new, sterile 50 mL centrifuge tube and fill to the 45 mL mark with Cryo-preservation Media (see notes 7 and 11).
15. Store at −80 °C until use.

3.2 Preparation of DWJM Wafers

1. To thaw previously frozen DWJM, remove the container from the freezer and allow it to sit at room temperature for 7 min, place in a water bath at 37 °C for 7 min.
2. Once thawed, remove DWJM from container and rinse in Lactated Ringer's Solution for 7 min.
3. Fill an appropriate container with liquid nitrogen (see note 12).
4. Cut the DWJM with sterile scissors to a size that fits within a mold for cryoembedding (see note 13).
5. Cool the wafer cutter by immersing it into the large container of liquid nitrogen (see note 12).
6. Remove cutter from liquid nitrogen and place on silicone mat (see note 12).
7. Fill cryoembedding mold with OCT freeze medium so that the Wharton's jelly is completely covered and the OCT medium is leveled with the top of the mold.
8. With forceps, hold the loaded mold over the surface of the liquid nitrogen and carefully watch it as the OCT freezes (see notes 12 and 14).
9. Once the OCT is frozen, remove the block from the cassette, and allow the frozen block to sit on the silicone mat for ~30 s to warm slightly to ensure that it is not too brittle for cutting.
10. Load the frozen block into the cooled wafer cutter.
11. Cut wafers to the desired thickness (see note 13).
12. With sterile forceps pick up cut wafers and place in a large bowl of sterile saline to wash out OCT medium.

13. Remove the cut wafers from sterile saline with sterile forceps and place in 2 mL tube.
14. Fill tubes with 1.5 mL of tissue freeze media and place in freezer (−20 °C).

3.3 Recellularization of DWJM Wafers

1. Preparing DWJM wafers for cell seeding: Thaw cryopreserved DWJM wafers, typically stored at −80 °C, in 37 °C water bath. Wash the thawed wafers (Fig. 2) (in PBS 3 times followed by immersing them in culture media overnight at 37 °C with 5% CO_2.
2. Centrifuge the cells of interest at 350 × *g* for 5 min and resuspend cells in culture media to a density of 1–3 × 10^6/mL.
3. Prior to cell seeding, transfer DWJM wafers to non-tissue culture treated plates to optimize the chance of cells attaching to the DWJM wafers. The cell seeding density varies from one type of cells to another. For leukemia cell lines, we typically seed 2–5 × 10^5 cells (see note 15) in 2-mL culture media into each well of a 24-well plate that is 70% covered by DWJM wafers. Incubate seeded DWJM wafers at 37 °C with 5% CO_2. The frequency of media changes and the volume added/changed also varies depending on the seeded cell type.
4. Seeded DWJM evaluation/processing: Seeded DWJM are washed in PBS three times followed by fixing in 4% paraformaldehyde at 4 °C overnight and paraffin embedded for histology. To isolate cells seeded adherent to or embedded in DWJM

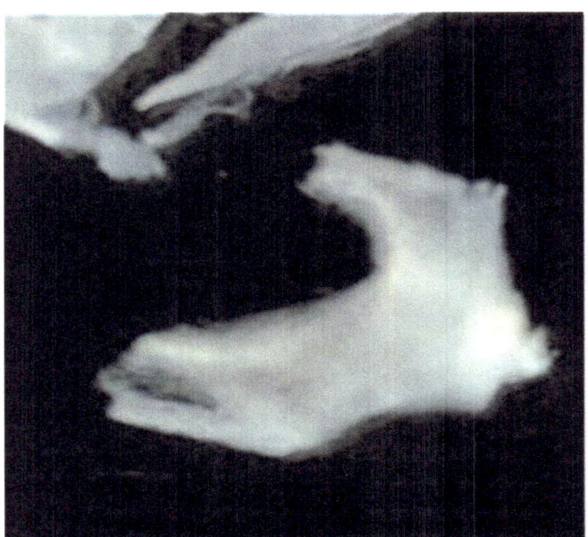

Fig. 2 Decellularized Wharton's Jelly matrix wafers. Wafers have uniform thickness but variable shapes otherwise

wafers, we use collagenase II to digest DWJM (0.05 g collagenase II in 50 mL DMEM) for 2 h.

4 Notes

1. Sterile filter solution prior to use.
2. Solution has a shelf life of 6 weeks.
3. Solution should be made day of use.
4. Depending on the desired end use of the decellularized Wharton's jelly, alternative protein sources (e.g., recombinant human albumin, lamb serum, etc.) may be substituted.
5. Use sterile surgical gloves and instruments during this process. It is useful to create a sterile field within the biosafety cabinet using a sterile drape. It is also useful to have additional personnel available for assistance as needed, for example, to assist in opening instrument packaging.
6. The Wharton's jelly floats in the decellularization solutions. This step is performed in the centrifuge at 4 °C at 500 rpm to ensure full immersion of the Wharton's jelly in the solution.
7. All transfers of the tissue between containers should be performed in the BSC. Use sterile instruments to perform each transfer. Alternatively, instruments can be stored 70% ethanol between uses.
8. This step is performed on a shaker plate 230 rpm at room temperature
9. This is an overnight step. Within the specified time range, there is flexibility to adjust the length of this step to accommodate personnel schedules.
10. Cut a section of dialysis tubing (approximately 10 cm). Wet the tubing with sterile, deionized water. Make an approximately 5 mm fold along one end of the tubing and crease. Repeat this twice more on the same end of the tubing. Staple the fold with approximately 3–4 surgical staples. Insert the Resin Beads or Wharton's jelly into the open end of the tubing. Fold and seal the open end of the tubing as described above, removing as much air as possible prior to sealing. Adding a small amount of deionized water to the tubing can help with this.
11. If desired, proceed immediately to cutting without freezing the tissue.
12. Follow appropriate safety procedures when handling liquid nitrogen.

13. The custom wafer cutter in our laboratory is designed to accommodate a $25 \times 20 \times 5$ mm block. A screw is used to advance the block during cutting to achieve the desired wafer thickness.
14. The OCT turns from clear to white as it freezes. Do not immerse and hold the cassette in the liquid nitrogen as this will cause the OCT medium to crack, which will also break the tissue into pieces.
15. For umbilical cord blood CD34$^+$ cells, we seed 3×10^4 cells in 0.5-mL culture media. For leukemia patient samples, we seed 3×10^5 cells in 2-mL culture media. For bone marrow MSCs, we seed 5×10^4 cells in 0.5-mL culture media.

References

1. Jadalannagari S, Converse G, McFall C, Buse E, Filla M, Villar MT, Artigues A, Mellot AJ, Wang J, Detamore MS, Hopkins RA, Aljitawi OS (2017) Decellularized Wharton's Jelly from human umbilical cord as a novel 3D scaffolding material for tissue engineering applications. PLoS One 12(2):e0172098
2. Zippel N, Schulze M, Tobiasch E (2009) Biomaterials and Mesenchymal stem cells for regenerative medicine. Recent Pat Biotechnol. doi: BIOT-ABSTRACT 08 [pii]
3. Hellstrom M, El-Akouri RR, Sihlbom C, Olsson BM, Lengqvist J, Backdahl H, Johansson BR, Olausson M, Sumitran-Holgersson S, Brannstrom M (2014) Towards the development of a bioengineered uterus: comparison of different protocols for rat uterus decellularization. Acta Biomater 10(12):5034–5042
4. Costa F, Dohmen P, Vieira E, Lopes SV, Colatusso C, Pereira EWL, Matsuda CN, Cauduro S (2007) Ross operation with decellularized pulmonary allografts: medium-term results. Rev Bras Cir Cardiovasc 22(4):454–462
5. Kiyotake EA, Beck EC, Detamore MS (2016) Cartilage extracellular matrix as a biomaterial for cartilage regeneration. Ann N Y Acad Sci 1383(1):139–159
6. Law JX, Liau LL, Aminuddin BS, Ruszymah BHI (2016) Tissue-engineered trachea: a review. Int J Pediatr Otorhinolaryngol 91:55–63
7. Franc S, Rousseau JC, Garrone R, van der Rest M, Moradi-Ameli M (1998) Microfibrillar composition of umbilical cord matrix: characterization of fibrillin, collagen VI and intact collagen V. Placenta 19(1):95–104
8. Ferguson VL, Dodson RB (2009) Bioengineering aspects of the umbilical cord. Eur J Obstet Gynecol Reprod Biol 144(Suppl 1): S108–S113. doi: S0301-2115(09)00133-X [pii]10.1016/j.ejogrb.2009.02.024
9. Grayson WL, Zhao F, Izadpanah R, Bunnell B, Ma T (2006) Effects of hypoxia on human mesenchymal stem cell expansion and plasticity in 3D constructs. J Cell Physiol 207 (2):331–339. doi:10.1002/jcp.20571
10. Bastow ER, Byers S, Golub SB, Clarkin CE, Pitsillides AA, Fosang AJ (2008) Hyaluronan synthesis and degradation in cartilage and bone. Cell Mol Life Sci 65(3):395–413. doi:10.1007/s00018-007-7360-z
11. Ishige I, Nagamura-Inoue T, Honda MJ, Harnprasopwat R, Kido M, Sugimoto M, Nakauchi H, Tojo A (2009) Comparison of mesenchymal stem cells derived from arterial, venous, and Wharton's Jelly explants of human umbilical cord. Int J Hematol 90(2):261–269. doi:10.1007/s12185-009-0377-3
12. Sobolewski K, Malkowski A, Bankowski E, Jaworski S (2005) Wharton's Jelly as a reservoir of peptide growth factors. Placenta 26 (10):747–752. doi:S0143-4004(04)00260-7 [pii] 10.1016/j.placenta.2004.10.008
13. Taipale J, Keski-Oja J (1997) Growth factors in the extracellular matrix. FASEB J 11(1):51–59

Methods in Molecular Biology (2018) 1577: 35–53
DOI 10.1007/7651_2017_71

Published online: 26 September 2017

3D Decellularized Native Extracellular Matrix Scaffold for In Vitro Culture Expansion of Human Wharton's Jelly-Derived Mesenchymal Stem Cells (hWJ MSCs)

Balasubramanian Sundaram, Anne George Cherian, and Sanjay Kumar

Abstract

Mesenchymal stem cells (MSCs) are derived from Wharton's jelly tissue of the human umbilical cord. Given appropriate culture conditions, these cells can self-renew and differentiate into multiple cell types across the lineages. Among the properties exhibited by these cells, immunomodulation through secretion of trophic factors has been widely exploited in a broad spectrum of preclinical/clinical regenerative applications. Moreover, the extracellular matrix is found to play a major role apart from niche cells in determining stem cell fate including that of MSCs. Therefore, the currently employed technique of two-dimensional culture expansion can alter the inherent properties of naïve MSCs originally residing within the three-dimensional space. This limitation can be overcome to some extent by using native extracellular matrix scaffold culture system which mimics the in situ microenvironment. In this chapter, we have elucidated the protocol for the preparation of a native extracellular matrix scaffold by decellularization of the MSC sheet and thereof culture expansion and characterization of human Wharton's jelly-derived MSCs.

Keywords Decellularization, Extracellular matrix, Mesenchymal stem cells, Scaffold, Wharton's jelly, Stem cell niche

Abbreviations

αMEM	Alpha modification of minimum essential medium
APC	Allophycocyanin
CD	Cluster of differentiation
DAPI	4′,6-Diamidino-2-phenylindole
DMSO	Dimethyl sulfoxide
DNase	Deoxyribonuclease
DPX	Distyrene-plasticizer-xylene
ECM	Extracellular matrix
EDTA	Ethylenediaminetetraacetic acid
FACS	Fluorescence-activated cell sorting
FITC	Fluorescein isothiocyanate
hWJ MSCs	Human Wharton's jelly-derived MSCs
IgG	Immunoglobulin G
ITS	Insulin-transferrin-selenium
MSCs	Mesenchymal stem cells

μl	Microliters
μM	Micromoles
ml	Milliliters
mM	Millimoles
PBS	Phosphate buffered saline
PBST	Phosphate buffered saline with Tween 20
PE	Phycoerythrin
Pen-Strep	Penicillin-Streptomycin
2D	Two dimensional
3D	Three dimensional

1 Introduction

Mesenchymal stem cells (MSCs) are multipotent cells, which has enormous scope in regenerative medicine as evidenced by the surge in a number of ongoing/completed MSC-related clinical trials [1]. Recent reports reiterate the presence and successful isolation of MSCs from almost all tissues ranging from fetal to adult sources. Fetal MSCs, especially human Wharton's jelly-derived MSCs (hWJ MSCs), have an advantage over the adult MSCs due to their (1) less ethical issues, (2) ease of isolation, (3) enhanced proliferation, and (4) immunomodulatory potential [2]. Recent developments emphasize the role of secretory factors and extracellular vesicles as a therapeutic mechanism of MSCs, rather than homing and differentiation into a reparative cell type [3]. MSCs, as either cell-based or cell-free products or the combination of both, hold immense therapeutic potential across a broad range of diseases.

In tissues, MSCs exist within the three-dimensional microenvironment composed of niche cells, growth factors, signaling molecules and extracellular matrix (ECM) [4]. The ECM is part of the stem cell niche and plays a major role in regulating the stem cell fate decisions such as their proliferation, differentiation and self-renewal. It is a complex and dynamic structure comprised of mechanical, biophysical and biochemical properties. Each tissue has a different ECM which differs in its components to impart a specific function to the constituent cells [5]. ECM exerts the regulatory role of stem cells by (1) direct interaction with the cells through integrin receptors, (2) secreted factors and (3) - ECM-bound growth factors. Integrin receptors communicate through the extracellular microenvironment to the intracellular cytoskeleton networks of cells, thus by modulating the cellular migration, cells survival, cellular differentiation and cell proliferation [6]. Moreover, ECM also acts as storage for insoluble, non-bioactive growth factors, which will be available to the cells upon enzymatic cleavage under specific physiological conditions.

Further, the biophysical characters of ECM such as stiffness also regulate the fate of cells through mechanotransduction [7].

As with other cell types, MSCs are also highly regulated by the surrounding ECM [8]. Geometric features such as cell shape influence the MSCs into preferential differentiation toward a specific cell type. Adherent cells which are more flattened differentiate into osteocytes, while the remaining rounded cells underwent adipogenic differentiation [9]. This preference of MSCs to differentiate into specific lineages is due to the stiffness of the substrate, which in turn modulates the cytoskeleton tension within the cells. Increased contractility induced by culturing cells in rigid substrate leads to osteocyte differentiation. On the other hand, less contractility produced by softer substrates induced MSCs into adipocytes [10]. Moreover, Nam et al. reported that electrospun scaffolds with same surface chemistry and microstructure but different mechanical properties such as soft and stiffer fibers favored chondrogenesis and osteogenesis, respectively [11]. Changes in the topography of nano-tubular titanium oxide alone induced MSCs into osteogenesis even without osteocyte differentiation media, further affirming the role of dimensions of surface structure in stem cell fate [12]. Thus, ECM plays a major role in biology, cell fate, and cell characteristics of MSCs, which in turn will be reflected on the therapeutic potential. However, the current technique of two-dimensional expansion of MSCs in tissue culture flask does not recapitulate the in vivo tissue microenvironment and thus, does not guarantee the preservation of properties of naïve MSCs, which exist within the tissues. Therefore, improving the MSCs' culture expansion technique will allow us to harness the actual potential of MSCs in different clinical settings.

In some extent, native decellularized ECM derived from MSCs mimics the properties of in vivo microenvironment as they are (1) three-dimensional nanofibrous structure and (2) composed of fibronectin, collagen1, collagen 3, and laminin ECM proteins [13]. There have been several studies reported that MSCs cultured on decellularized ECM had important traits required for a therapeutic application like enhanced proliferation and antioxidant properties than cells cultured on 2D tissue culture flask [14, 15]. These reports further prove that native decellularized ECM can be used as a scaffold for MSC in vitro culture expansion, which will serve as a platform to access the untapped therapeutic potential of MSCs.

In this article, we have described a step-by-step protocol for successful derivation and characterization of the native decellularized ECM for the expression of fibronectin protein (Fig. 1) and in vitro culture expansion of hWJ MSCs in ECM (Fig. 2), followed by surface marker expression analysis by FACS (Fig. 3) and in vitro tri-lineage differentiation (Fig. 4).

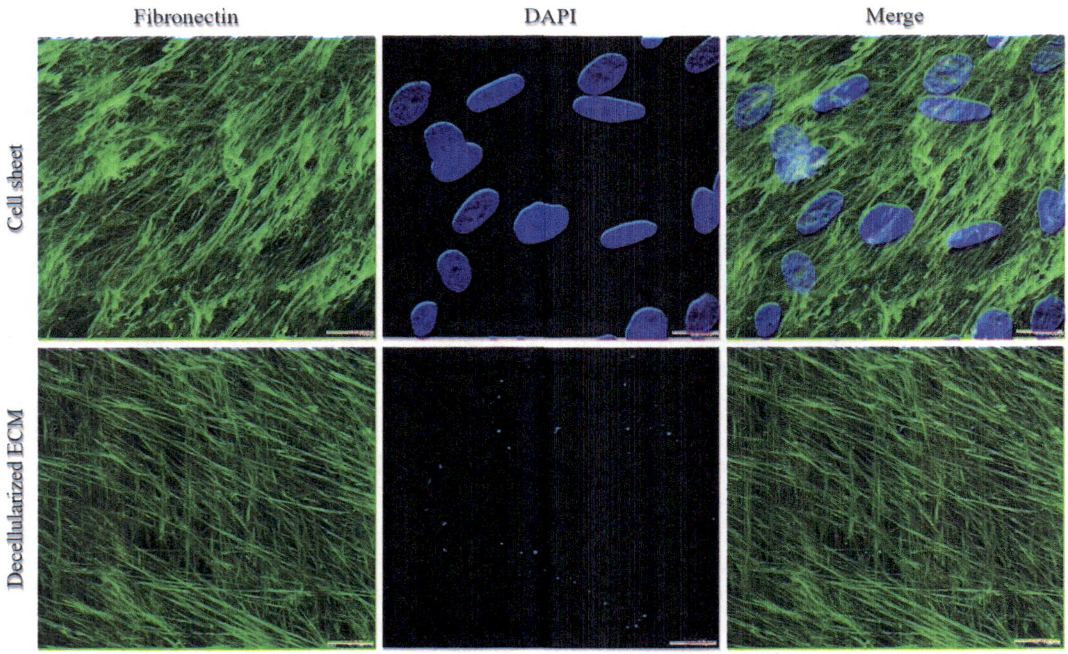

Fig. 1 Decellularization of cell sheet and characterization of ECM. Immunofluorescence staining of fibronectin before and after decellularization of cell sheet. Scale bar = 100 μm

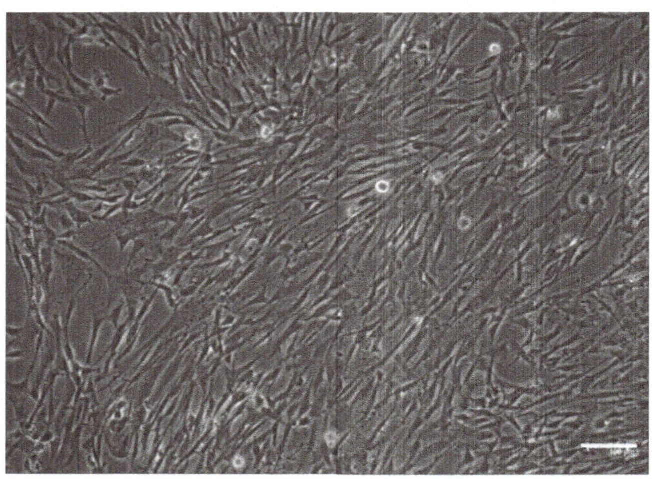

Fig. 2 In vitro culture expansion of hWJ MSCs. Phase-contrast microscopic images of spindle-shaped hWJ MSC cultured on native decellularized ECM. Scale bar = 100 μm

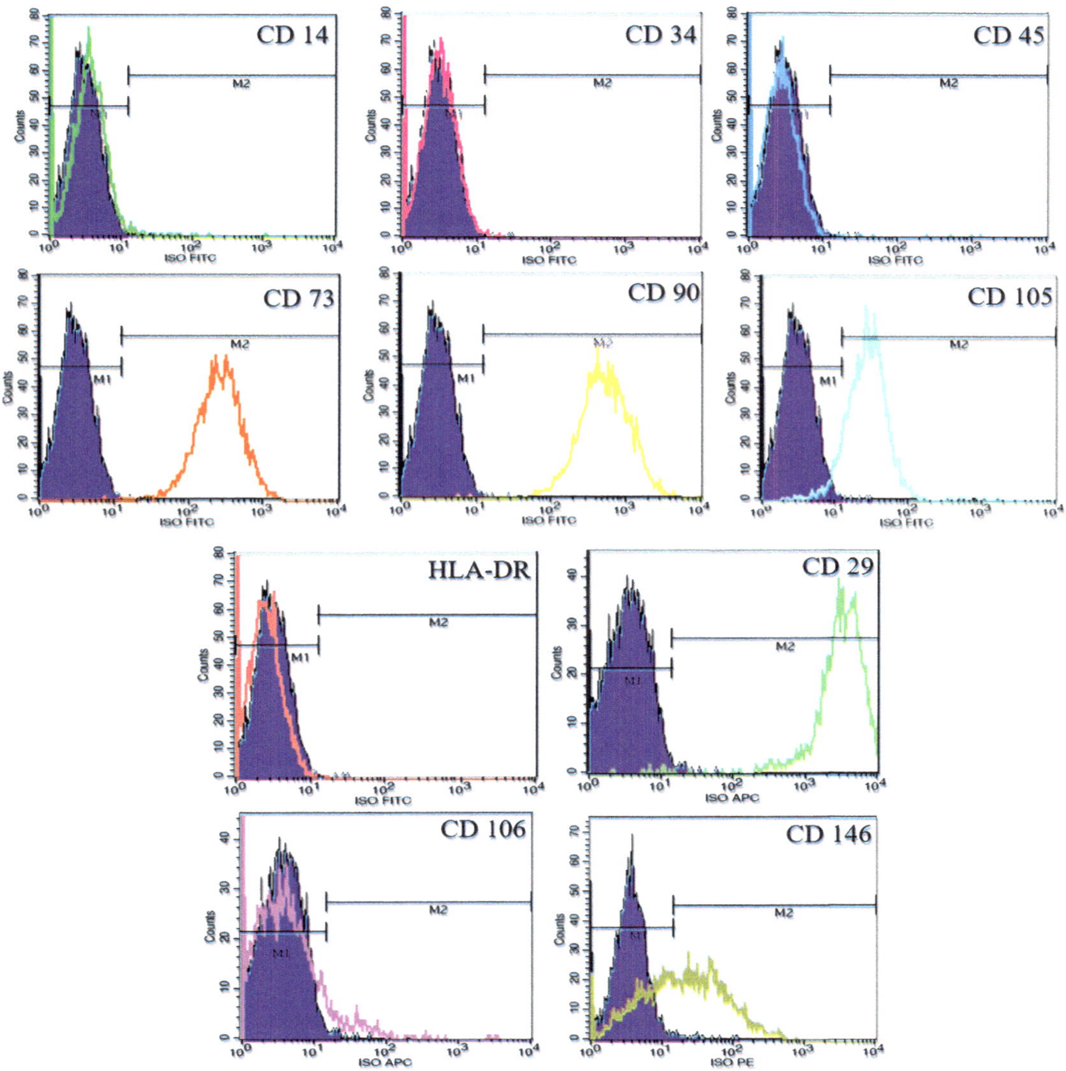

Fig. 3 Immunophenotypic analysis of hWJ MSCs. Flow cytometry analysis for surface marker expression by hWJ MSC. Results analyzed by CellQuest software. *FITC* fluorescein isothiocyanate, *APC* allophycocyanin, and *PE* phycoerythrin

2 Materials

2.1 Isolation of MSCs from Human Wharton's Jelly Tissue

1. Biosafety cabinet class 2.
2. Sterile tissue collection/transport container.
3. Tissue processing container.
4. Surgical scissors.
5. Scalpel blade.
6. Scalpel blade holder.

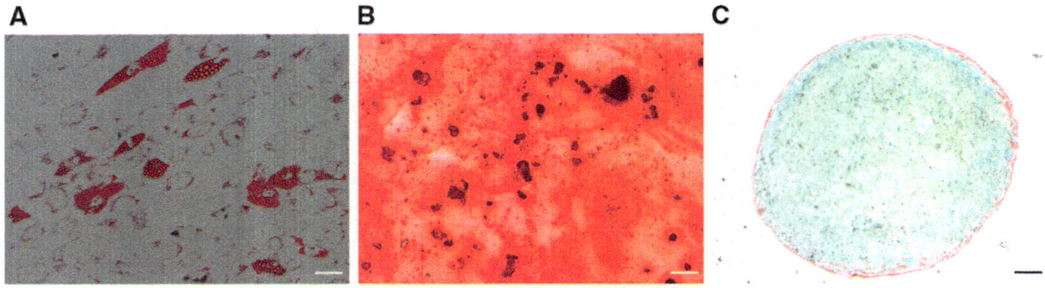

Fig. 4 In vitro tri-lineage differentiation of hWJ MSCs. (**a**) Oil Red O staining of lipid droplets of adipocytes, (**b**) alizarin red staining of calcium deposition by osteocytes, and (**c**) Alcian blue staining of chondrocytes. Scale bar = 100 μm

7. Tissue culture treated dishes (100 and 150 mm).
8. Tissue culture treated flasks (75 and 175 cm^2).
9. Circular water bath (37 °C).
10. Centrifuge instrument.
11. Centrifuge tubes (15 and 50 ml).
12. Pipette controller.
13. Serological pipettes (5, 10, and 25 ml).
14. Pipette (1–10, 20–200, and 100–1000 μl).
15. Sterile pipette tips (10, 200, and 1000 μl).
16. Phosphate-buffered saline (without Mg2+ and Ca2+, pH 7.4).
17. Penicillin (100 U/ml).
18. Streptomycin (100 μg/ml).
19. Amphotericin B (2.5 μg/ml).
20. collagenase I enzyme (12.5 U/ml).
21. Dispase enzyme (2% w/v).
22. Sterile red blood cell lysis buffer.
23. Ice cubes.
24. Mesenchymal stem cell culture and expansion media: Minimum essential medium (alpha modification), fetal bovine serum (10%), L-glutamine (1%), and Pen-Strep (1%). All the components mixed and filtered through a 0.22 μM filter.
25. Mesenchymal stem cell freezing medium: Mesenchymal stem cell expansion media (90%) and 10% dimethyl sulfoxide (DMSO). Components mixed and filtered through a 0.22 μM filter.
26. Trypsin EDTA (0.05%).
27. Humidified tissue culture incubator (37 °C and 5% CO_2).
28. Liquid nitrogen tank.

29. Freezing container (Mr. Frosty).
30. Freezer (−80 °C).
31. Hemocytometer.

2.2 In Vitro Characterization of hWJ MSCs

2.2.1 Surface Marker Expression Analysis

1. Human Wharton's jelly-derived MSCs (hWJ MSCs).
2. Trypsin EDTA (0.05%).
3. Phosphate-buffered saline (without Mg^{2+} and Ca^{2+}, pH 7.4).
4. Pipette controller.
5. Serological pipettes (5, 10, and 25 ml).
6. Centrifuge instrument.
7. Sterile pipette tips (10, 200, and 1000 μl).
8. Pipette (1–10, 20–200, and 100–1000 μl).
9. Fluorochrome-conjugated antibodies (BD Pharmingen): Antihuman CD 14 FITC (BD Cat # 555397), antihuman CD 29 APC (BD Cat # 559883), antihuman CD 34 FITC (BD Cat # 555821), antihuman CD 45 FITC (BD Cat # 555482), antihuman CD 73 FITC (BD Cat # 561254), antihuman CD 90 FITC (BD Cat # 555595), antihuman CD 105 FITC (BD Cat # 561443), antihuman CD 106 APC (BD Cat # 551147), antihuman CD 146 PE (BD Cat # 550315), FITC mouse IgG k isotype control (BD Cat # 555748), PE mouse IgG k isotype control (BD Cat # 556650), APC mouse IgG k isotype control (BD Cat # 550854), and antihuman HLA-DR FITC (eBioscience-11-9956-73).
10. Sterile flow cytometer tubes.
11. FACSCalibur flow cytometer instrument with CellQuest software (Becton Dickinson, USA).
12. Label with marker pen.
13. Aluminum foil.

2.2.2 In Vitro Tri-lineage Differentiation

1. Adipocyte differentiation media: Minimum essential media (alpha modification), fetal bovine serum (10%), dexamethasone (1 μM), 3-isobutyl-1 methylxanthine (500 μM), human recombinant insulin (10 μg/ml), indomethacin (1 μM), L-glutamine (1%), and Pen-Strep (1%). All the components mixed and filtered through a 0.22 μM filter.
2. Osteocyte differentiation media: Minimum essential media (alpha modification), fetal bovine serum (10%), beta

glycerophosphate (10 mM), ascorbic acid (50 μg/ml), dexamethasone (10^{-8} M), L-glutamine (1%), and Pen-Strep (1%). All the components mixed and filtered through a 0.22 μM filter.

3. Chondrocyte differentiation media: Minimum essential media (alpha modification), fetal bovine serum (10%), dexamethasone (100 nM), sodium pyruvate (100 μg/ml), TGF-β 1 (10 ng/ml), ascorbic acid (10 μg/ml), insulin-transferrin-selenium (ITS) (5 μg/ml), and gentamycin (5 μg/ml). All the components mixed together and filtered through a 0.22 μM filter.
4. Human Wharton's jelly-derived MSCs (hWJ MSCs).
5. Tissue culture treated plates (24 well).
6. Biosafety cabinet class 2.
7. Trypsin EDTA (0.05%).
8. Pipette controller.
9. Serological pipettes (5, 10, and 25 ml).
10. Humidified tissue culture incubator (37 °C and 5% CO_2).
11. Pipette (1–10, 20–200, and 100–1000 μl).
12. Sterile pipette tips (10, 200, and 1000 μl).
13. Phosphate-buffered saline (without Mg^{2+} and Ca^{2+}, pH 7.4).
14. Centrifuge tube (15 ml).
15. Oil Red O stain (0.5%).
16. Alizarin red stain (2%).
17. Alcian blue stain (3%).
18. Refrigerator (4 °C).
19. Paraformaldehyde (4%).
20. Mounting media (DPX).
21. Ice-cold methanol.
22. Syringe water.
23. Isopropanol (60%).
24. Xylene.
25. Ethanol (60 and 70%).
26. Mercurochrome dye.
27. Inverted microscope with camera.

2.3 Induction of ECM Deposition in Cultured hWJ MSCs

1. Human Wharton's jelly-derived MSCs (hWJ MSCs).
2. Cell culture treated dish (100 mm).
3. Biosafety cabinet class 2.
4. Trypsin EDTA (0.05%).

5. Pipette (1–10, 20–200, and 100–1000 μl).
6. Sterile pipette tips (10, 200, and 1000 μl).
7. Humidified tissue culture incubator (37 °C and 5% CO_2).
8. ECM induction media: Minimum essential medium (alpha modification), ascorbic acid (100 μm/ml), FBS (10%), Pen-Strep (1%), and L-glutamine (1%). All the components mixed and filtered through a 0.22 μM filter.
9. Syringe filter (0.22 μM).
10. Serological pipettes (5, 10, and 25 ml).
11. Pipette controller.
12. Gelatin (from porcine skin) solution (0.2%). Gelatin added to the distilled water should be autoclaved and filtered through a 0.22 μM filter.
13. Tissue culture incubator (humidified, 5% CO_2 and 37 °C).

2.4 Preparation of ECM by Decellularization of Cell Sheet

1. ECM-induced hWJ MSC sheet in a dish.
2. Phosphate-buffered saline (PBS) (without Mg^{2+} and Ca^{2+}, pH 7.4).
3. ECM extraction buffer: Ammonium hydroxide (20 mM), Triton X-100 (0.5%), and PBS.
4. DNase I enzyme (100 μm/ml).
5. DNase I buffer: Tris–HCl (50 mM), $MgCl_2$ (10 mM), and PBS.
6. Orbital shaker.
7. Biosafety cabinet class 2.
8. Serological pipettes (5, 10, and 25 ml).
9. Pipette controller.
10. Pipette (1–10, 20–200, and 100–1000 μl).
11. Sterile pipette tips (10, 200, and 1000 μl).
12. Parafilm.
13. PBS and Pen-Strep (2%).

2.5 Characterization of Decellularized Native ECM

1. Decellularized ECM.
2. Cover glass bottom tissue culture chamber (ibidi).
3. Phase-contrast microscope (Leica).
4. Laser scanning confocal microscope (Olympus FV1000).
5. Paraformaldehyde (4%).
6. Universal blocking solution (buffered solution of casein with additives) (BioGenex Inc.).
7. Antibody diluent.

8. Antibodies: Antifibronectin (P1H11) primary antibody (Santa Cruz Biotechnology SC 18825) and anti-mouse IgG (H + L) Alexa Fluor 488 secondary antibody (Cell Signaling-4408).
9. PBS.
10. Tween 20.
11. PBST (0.05% Tween 20 in 1× PBS).
12. Dancing shaker.
13. DAPI (1 μg/ml).

2.6 Culture Expansion of hWJ MSCs on Decellularized ECM

1. Decellularized ECM.
2. Liquid nitrogen-stored hWJ MSCs.
3. Mesenchymal stem cell culture and expansion media: Minimum essential medium (alpha modification), fetal bovine serum (10%), L-glutamine (1%), and Pen-Strep (1%). All the components mixed and filtered through a 0.22 μM filter.
4. Biosafety cabinet class 2.
5. Serological pipettes (5, 10, and 25 ml).
6. Pipette controller.
7. Pipette (1–10, 20–200, and 100–1000 μl).
8. Sterile pipette tips (10, 200, and 1000 μl).
9. Tissue culture incubator (humidified, 5% CO_2 and 37 °C).

3 Methods

3.1 Isolation of MSCs from Human Wharton's Jelly Tissue

1. The ethical approval for the use of human umbilical cord in this study is sanctioned by the Institutional Review Board (IRB) of Christian Medical College, Vellore, India (*see* Note 1).
2. Patients undergoing full-term elective cesarean have been explained by a medical professional about the use of tissues, and consent obtained before the collection of human umbilical cord sample (*see* Note 2).
3. Transfer the umbilical cord tissue into PBS-containing collection container immediately after the cesarean procedure. Maintain the sterility throughout the transport and keep the container in ice until processing the tissue in the laboratory (*see* Note 3).
4. Transfer the umbilical cord tissue into sterile tissue processing trough.
5. Add PBS (1×) containing penicillin-streptomycin (2×) along with antifungal Amphotericin B (2.5 μg/ml) solution into the trough. Close the trough and move the trough horizontally

back and forth to wash the umbilical cord. Discard the PBS carefully (*see* Note 4).

6. Repeat step 5 to make sure the umbilical cord tissue is sterile before processing.
7. Transfer the tissue into a cell culture dish (150 mm). Using a scalpel, remove remaining blood clots in the tissue and wash with PBS containing the antibiotic and antimycotic solution.
8. Using sharp scissors, cut the umbilical cord tissue into two 5 cm pieces.
9. Remove the umbilical vessels from the tissue by making an incision parallel to the axis of the umbilical cord.
10. Transfer the tissue into a sterile 50 ml centrifuge tube.
11. Add 25 ml of serum-free antibiotic-free MEM and mince the tissue into small pieces with sharp scissors.
12. Centrifuge at 700 × *g* for 10 min and discard the supernatant leaving the tissue pieces in the tube.
13. Add 20 ml of serum-free antibiotic-free MEM containing 12.5 U/ml of collagenase I enzyme, subsequently seal the tube with parafilm, and incubate in shaking water bath at 37 °C for overnight (*see* Note 5).
14. Add serum-free antibiotic-free MEM containing dispase I enzyme (2 mg/ml) to the digested tissue and mix properly.
15. Seal the tube with parafilm and incubate in shaking water bath for 2 h with occasional mixing for every 10 min.
16. Centrifuge the tube at 700 × *g* for 10 min at 4 °C.
17. Remove the supernatant and add 25 ml of PBS to the digested tissue.
18. Mix properly and centrifuge the tube at 700 × *g* for 10 min at 4 °C.
19. Remove the supernatant and repeat steps 17 and 18.
20. To the pellet, add 10 ml of RBC lysis buffer. Mix properly and incubate on ice for 10 min.
21. Centrifuge at 700 × *g* for 10 min at 4 °C. Remove the supernatant and wash the pellet with PBS containing the antibiotic solution.
22. Seed the enzymatically digested tissue pieces into 150 mm tissue culture treated dish containing 25 ml of mesenchymal stem cell culture media and incubate in the humidified tissue culture incubator (5% CO_2 and 37 °C) (*see* Note 6).
23. Change the media once in a week.
24. MSCs will migrate from the processed tissue into the substratum of the dish within 7–21 days.

25. Upon reaching 70–80% confluence, remove the media and wash the cells twice with PBS.
26. Add 3 ml of trypsin EDTA solution to the dish and rotate to cover the entire surface and incubate in a 37 °C incubator for 3 min (*see* Note 7).
27. Observe under phase-contrast microscope for the dissociation of the monolayer into single cells.
28. Add 10 ml of MSC culture media to the cells to quench the trypsin activity.
29. Pipette up and down to disperse the cells. Transfer the medium-containing cells into the 15 ml centrifuge tube.
30. Centrifuge the tube at 350 × *g* for 5 min to pellet down the cells.
31. Remove the supernatant and resuspend the cells in 1 ml αMEM. Count the cells using a hemocytometer and seed appropriate cells in tissue culture flasks for the characterization experiments.
32. Add 1 ml of MSC freezing media to the remaining cells. Mix properly and transfer into labeled cryovials. Transfer the cryovials into the Mr. Frosty freezing container and then to the −80 °C freezer (*see* Note 8).
33. The next day, transfer the vials from the −80 °C freezer to the liquid nitrogen container until further use.

3.2 Characterization of hWJ MSCs for the Expression of Surface Markers by Flow Cytometry

1. Transfer the hWJ MSC cryopreserved vial into the water bath for a quick thaw and seed the cells into a T75 culture flask containing MSC culture media. Incubate the flask in the humidified 37 °C tissue culture incubator (5% CO_2) (*see* Note 9).
2. Next day, remove the media and add MSC culture media to the flask. Observe the adherent cells under phase-contrast microscope for the confluency (*see* Note 10).
3. Upon reaching 70–80% confluency, remove the media from the flask and wash twice with PBS.
4. Add 1 ml of trypsin EDTA and rotate the flask to spread throughout the surface and incubate in 37 °C incubator for 3 min.
5. Observe the cells under the microscope for detachment from the surface of the flask.
6. Add 10 ml of MSC culture media to quench the trypsin activity and collect the cells in a centrifuge tube.
7. Centrifuge at 350 × *g* for 5 min in 37 °C to pellet down the cells.

8. Remove the supernatant and resuspend the cells in 1 ml PBS and count the cells using a hemocytometer.
9. Transfer approximately 1×10^5 cells in 100 μl of PBS into each flow cytometry tubes.
10. Add appropriate concentration of fluorochrome-conjugated antibodies to respective tubes and mix well with a pipette. Incubate the flow cytometer tubes in the dark for 20 min (*see* Note 11).
11. Add 1 ml of PBS to each tube and mix well.
12. Centrifuge the flow cytometer tubes at $350 \times g$ for 5 min at room temperature.
13. Remove the supernatant and resuspend the cells in 400 μl of PBS.
14. Use appropriate IgG control and unstained control sample during acquisition in FACSCalibur flow cytometer.
15. Acquire 10,000 events for each of the tubes and analyze the result with CellQuest software.

3.3 Characterization of hWJ MSCs for the In Vitro Differentiation Ability

3.3.1 Adipocyte Differentiation

(a) Seed 5×10^4 hWJ MSCs on a 24-well plate 1 day before the differentiation.

(b) Remove the media and add adipocyte differentiation media and incubate the plate in the humidified 37 °C tissue culture incubator (5% CO_2) (*see* Note 12).

(c) Change the media every other day for 30 days.

(d) Induction of hWJ MSCs into adipocyte can be observed by the appearance of lipid droplets in the cells.

(e) Remove the media and add 1 ml of 4% paraformaldehyde and incubate at room temperature for 3 h.

(f) Remove the fixative and wash with sterilized water.

(g) Add 60% ethanol to the cells and incubate them at room temperature.

(h) Incubate the cells with 0.5% Oil Red O for 20 min at room temperature.

(i) Remove the stain and wash with 60% isopropanol.

(j) Wash once with sterile water.

(k) Observe under a microscope and visualize the cells for stained lipid droplets. Capture the image using a camera attached to the microscope with the help of the software.

3.3.2 Osteocyte Differentiation

(a) Seed 5×10^4 hWJ MSCs on a 24-well plate 1 day before the differentiation.

(b) Remove the media and add the osteocyte differentiation media and incubate the plates in the humidified 37 °C tissue culture incubator (5% CO_2).

(c) Change the osteocyte differentiation media every alternative day for 21 days (*see* Note 12).

(d) Deposition of extracellular calcium indicates the differentiation of hWJ MSCs into osteocytes.

(e) At the end of the differentiation process, remove the media from the dish.

(f) Add 1 ml of 1× PBS and swirl the dish to remove the remaining media.

(g) Wash one more time with 1× PBS.

(h) Add 3 ml of 70% ethanol and leave the dish in 4 °C to fix the cells.

(i) Remove the fixative and wash twice with distilled water.

(j) Add 2 ml of alizarin red stain (2%) into the dish and incubate at 25 °C for 10 min.

(k) Rinse the dish with distilled water to remove the excess stain.

(l) Add 1 ml of distilled water to the dish and observe under the microscope with appropriate settings. Use the camera attached to the microscope to capture the image with the help of software.

3.3.3 Chondrocyte Differentiation

(a) Pellet culture of hWJ MSCs can be utilized for chondrocyte differentiation.

(b) To obtain a micropellet of hWJ MSCs, transfer 1×10^6 cells into a 15 ml centrifuge tube and centrifuge at $350 \times g$ for 5 min.

(c) Add chondrocyte induction media through the sides of the tube without disturbing the pellet (*see* Note 12).

(d) Transfer the tube into the humidified 37 °C tissue culture incubator (5% CO_2) and incubate upright for 30 days (*see* Note 13).

(e) Carefully change the media every alternative day for 30 days.

(f) At the end of differentiation, remove the media without disturbing the pellet and process for Alcian blue (3%) staining for the confirmation of chondrocyte differentiation.

(g) Use 10% formalin to fix the cells and use a drop of mercurochrome dye before embedding in paraffin.

(h) Use 5 μm deparaffinized sections for Alcian blue staining.

(i) After the staining, remove the excess stain by rinsing in distilled water.

(j) Air dry and immerse the sections in xylene.

(k) Use DPX mounting media to mount the sections and observe under the microscope with appropriate settings. Capture the image with the camera attached to the microscope with the help of software.

3.4 Induction of Extracellular Matrix Deposition by hWJ MSCs

1. Add 6 ml of 0.2% gelatin into the 100 mm dish and swirl to cover the entire surface of the plate. Incubate the dish in the humidified 37 °C tissue culture incubator (5% CO_2) for 3 h (*see* Note 14).
2. Remove the excess gelatin solution and dry the dish by incubating in the humidified 37 °C tissue culture incubator (5% CO_2) for 30 min.
3. Seed 5×10^5 hWJ MSCs into the gelatin pre-coated dish containing 15 ml of MSC culture media and incubate the dish in the humidified 37 °C tissue culture incubator (5% CO_2).
4. Upon hWJ MSCs reaching 80–90% confluency, remove the media and add 12 ml ECM induction media and incubate in the humidified 37 °C tissue culture incubator (5% CO_2) for 8 days.
5. Once in every 3 days change the media with fresh ECM induction media (*see* Note 15).
6. The formation of a confluent cell sheet indicates the successful induction of ECM deposition.
7. Proceed the ECM-induced cell sheet-containing dish for decellularization process.

3.5 Decellularization of hWJ MSC Sheet

1. Remove the media from the dish and wash the cell sheet twice with 1× PBS.
2. Add 5 ml of ECM extraction buffer into the culture dish and incubate them in the 37 °C incubator for 5 min with occasional mixing (*see* Note 16).
3. Observe the decellularization process under the phase-contrast microscope. Cytoplasmic and nuclear content will be disintegrated leaving the 3D ECM structure (*see* Note 17).
4. Remove the ECM extraction buffer.
5. Wash the decellularized ECM twice with PBS (*see* Note 18).

6. Add 6 ml of DNase I buffer containing 100 μm/ml of DNase I enzyme and incubate in the 37 °C incubator for 1 h (*see* Note 19).
7. Remove the DNase I buffer and wash twice with PBS.
8. Add PBS containing a 2× concentration of antibiotics to the ECM and seal the dish with paraffin and store in 4 °C until further use.

3.6 Characterization of Decellularized ECM

1. Observe the ECM under phase-contrast microscope. The presence of nanofibrous structure of ECM scaffold confirms the successful decellularization process.
2. Immunostaining for the presence of fibronectin protein followed by confocal z-stack imaging will reveal the 3D architecture of ECM.
3. Prepare the ECM in a cover glass bottom confocal dish.
4. Fix the decellularized ECM with 4% paraformaldehyde for 3 h.
5. Remove the fixative and wash thrice with PBS.
6. Add 1 ml of universal blocking buffer (1×) to the ECM and leave the dish at room temperature for 30 min.
7. Remove the blocking buffer and add 1 ml of antibody diluent (1:100 ratio) containing antifibronectin primary antibody.
8. Cover the dish with aluminum foil and keep it in a dancing shaker with intermittent shaking overnight at 4 °C.
9. Remove the antibody diluent and wash thrice with 1× PBST.
10. Add 1 ml of antibody diluent (1:250 ratio) containing anti-mouse IgG (H + L) Alexa Fluor 488 secondary antibody into the dish and incubate for 2 h at room temperature.
11. Remove the antibody diluent and wash twice with 1× PBST followed by washing once with PBS.
12. Add 1 ml of PBS containing DAPI (1 μg/ml) and incubate the dish in the dark for 5 min.
13. Wash thrice with 1× PBS.
14. Add 1 ml of PBS and observe under the confocal microscope with appropriate settings. Capture and analyze the image with the help of the software.

3.7 Culture Expansion of Human Wharton's Jelly MSCs in Native ECM Scaffold

1. Transfer the native decellularized ECM scaffold dishes to the biosafety cabinet from 4 °C storage (*see* Note 20).
2. Remove the PBS containing antibiotics and wash twice with 1× PBS.
3. Add 10 ml PBS and pre-warm the dish in 37 °C incubator for 15 min.
4. Transfer the dish to the biosafety cabinet and remove the PBS.

5. Quick thaw the cryopreserved hWJ MSCs vial in a water bath and seed the cells in a native ECM dish containing 15 ml of MSC expansion media.
6. Incubate the dish in the humidified 37 °C tissue culture incubator (5% CO_2).
7. Next day, remove the media and add 15 ml of MSC expansion media.
8. Check the morphology and confluency of the cells.
9. Upon reaching 70–80% confluency, dissociate the cells using trypsin EDTA and process the cells for characterization by both flow cytometry and in vitro differentiation.
10. Cryopreserve the cells by mixing with MSC freezing media and transfer the cryovial to the –80 °C freezer in the freezing container followed by storage in the liquid nitrogen tank until further use.

4 Notes

1. Ethical approval from the respective institutional review board is necessary for using human tissue samples in research projects.
2. Patients consent is essential for using their tissue sample in a research project. The format of the consent form should be approved by the institutional review board, and consent should be obtained 1 day before the cesarean procedure.
3. Ethylene gas–sterilized plastic bag can also be used to transport a human umbilical cord tissue to the laboratory. Alternatively, PBS-containing centrifuge tube (50 ml) also can be employed for the umbilical cord transport.
4. Wash the tissue with an antibiotic and antifungal solution to avoid contamination in subsequent processing and culturing of tissues.
5. Intermittent mixing will allow equal enzymatic digestion of all the tissue pieces.
6. Adherence of enzymatically processed tissue to the surface of the dish is necessary for the successful isolation of MSCs.
7. Extended action of trypsin will change the surface marker expression profile of cells. Cells should be monitored under the microscope to avoid excess damage by trypsin enzyme. Conversion of spindle-shaped monolayer cells into rounded individual cells marks the successful dissociation of cells by trypsin.

8. Cryovials should be transferred to a liquid nitrogen tank for long-term storage. Extended stay at −80 °C freezer will gradually reduce the viability of cells.
9. Thawing of cryopreserved cells should be done quickly to avoid the damage to cells caused by the DMSO at room temperature.
10. MSCs adhere to the surface of the flask with in 3–4 h of seeding. Change the media after this point of time to remove the presence of DMSO in culture condition. Alternatively, during the thawing process, centrifuge the vial to pellet down the cells. Remove the supernatant and seed the cells into the flask.
11. Avoid exposing the tubes to direct light before the acquisition, as antibodies conjugated with fluorochrome are sensitive to the light. Wrap the tubes with aluminum foil throughout the process.
12. Prepare fresh differentiation media for every media change. Components in the differentiation cocktail are not stable for a long time in a 4 °C refrigerator.
13. Change the cap of the centrifuge tube with a vented cap.
14. Pre-coating the culture dish with 0.2% gelatin before seeding the cells for ECM induction is necessary to avoid the dislodging of the cell sheet during decellularization process.
15. Prepare fresh ECM induction media for every media change for a successful induction of ECM deposition.
16. ECM extraction buffer contains detergent which will degrade the ECM structure gradually upon exposure for an extended time. Standardization of decellularization time is necessary to remove the cytoplasmic and nuclear contents while preserving the architecture of the ECM.
17. Preparation of fresh ECM extraction buffer for each experiment is needed for an efficient decellularization process.
18. Decellularized ECM is very prone to dislodging from the surface. Wash carefully to maintain the structure.
19. Disintegrated nuclear fragments stick to the ECM surface. Removal of these fragments is essential before using this ECM scaffold for culture expansion of hWJ MSCs.
20. Check the ECM for any contaminants by observing under phase-contrast microscope before the culture expansion of MSCs.

Acknowledgments

We would like to thank the University Grants Commission (UGC) for granting Junior Research Fellowship (JRF)/Senior Research

Fellowship (SRF) to Balasubramanian Sundaram. We would like to thank the Department of Biotechnology (DBT), India, for granting Ramalingaswami Fellowship and research support grant to Sanjay Kumar. We would like to thank Merylin Nixon, Department of Community Health, Christian Medical College Vellore Hospital, Vellore, India, for assisting us in obtaining umbilical cord tissue samples. We would like to appreciate the technical assistance provided by core facilities of Centre for Stem Cell Research, A unit of inStem Bengaluru, Christian Medical College, Vellore, Bagayam, Tamil Nadu, 632002, India.

References

1. Squillaro T, Peluso G, Galderisi U (2016) Clinical trials with mesenchymal stem cells: an update. Cell Transplant 25:829–848
2. Sabapathy V, Sundaram B, VM S et al (2014) Human Wharton's jelly mesenchymal stem cells plasticity augments scar-free skin wound healing with hair growth. PLoS One 9:1–10
3. Monsel A, Zhu Y, Gudapati V et al (2016) Mesenchymal stem cell derived secretome and extracellular vesicles for acute lung injury and other inflammatory lung diseases. Expert Opin Biol Ther 2598:1–13
4. Watt FM, Huck WTS (2013) Role of the extracellular matrix in regulating stem cell fate. Nat Rev Mol Cell Biol 14:467–473
5. Hynes RO (2009) Extracellular matrix: not just pretty fibrils. Science 326:1216–1219
6. Gattazzo F, Urciuolo A, Bonaldo P (2014) Extracellular matrix: a dynamic microenvironment for stem cell niche. Biochim Biophys Acta Gen Subj 1840:2506–2519
7. Li B, Moshfegh C, Lin Z et al (2013) Mesenchymal stem cells exploit extracellular matrix as mechanotransducer. Sci Rep 3:2425
8. Kusuma GD, Carthew J, Lim R et al (2017) Effect of the microenvironment on mesenchymal stem cells paracrine signalling: opportunities to engineer the therapeutic effect. Stem Cells Dev 26(9):617–631. https://doi.org/10.1089/scd.2016.0349
9. McBeath R, Pirone DM, Nelson CM et al (2004) Cell shape, cytoskeletal tension, and RhoA regulate stem cell lineage commitment. Dev Cell 6:483–495
10. Kilian KA, Bugarija B, Lahn BT et al (2010) Geometric cues for directing the differentiation of mesenchymal stem cells. Proc Natl Acad Sci U S A 107:4872–4877
11. Nam J, Johnson J, Lannutti JJ et al (2011) Modulation of embryonic mesenchymal progenitor cell differentiation via control over pure mechanical modulus in electrospun nanofibers. Acta Biomater 7:1516–1524
12. Oh S, Brammer KS, Li YSJ et al (2009) Stem cell fate dictated solely by altered nanotube dimension. Proc Natl Acad Sci 106:2130–2135
13. Xing Q, Yates K, Tahtinen M et al (2015) Decellularization of fibroblast cell sheets for natural extracellular matrix scaffold preparation. Tissue Eng Part C Methods 21:77–87
14. Liu X, Zhou L, Chen X et al (2016) Culturing on decellularized extracellular matrix enhances antioxidant properties of human umbilical cord-derived mesenchymal stem cells. Mater Sci Eng C 61:437–448
15. Rakian R, Block TJ, Johnson SM et al (2015) Native extracellular matrix preserves mesenchymal stem cell "stemness" and differentiation potential under serum-free culture conditions. Stem Cell Res Ther 6:235

Methods in Molecular Biology (2018) 1577: 55–69
DOI 10.1007/7651_2017_30

Published online: 10 May 2017

Decellularized Iliotibial Band Recolonized with Allogenic Homotopic Fibroblasts or Bone Marrow-Derived Mesenchymal Stromal Cells

Clemens Gögele, Silke Schwarz, Benjamin Ondruschka, Niels Hammer, and Gundula Schulze-Tanzil

Abstract

Decellularized scaffolds present promising biomimetic approaches in various fields of tissue engineering. Different tissues have been selected for decellularization, among them extracellular matrix (ECM)-rich tissues such as tendons, ligaments and cartilage. The dense ECM of ligaments is particularly challenging to achieve a completely non-immunogenic ECM void of any cells. Here, the methods for decellularization adapted to ligamentous tissue of the iliotibial band (ITB) are presented along with cell isolation and several recolonization techniques using allogenic ITB-derived fibroblasts or mesenchymal stromal cells (MSCs).

Keywords: Acellularization, Decellularization, Explant culture, Iliotibial band, Ligamentocytes, Mesenchymal stromal cells, Recellularization

Abbreviations

AB	Alcian blue
ACL	Anterior cruciate ligament
dest	Aqua destillata
BSA	Bovine serum albumin
DAPI	4′,6-Diamidino-2-phenylindol
DMSO	Dimethylsulfoxide
DMEM	Dulbecco's Minimum Essential Medium
DP	Decellularization process
ECM	Extracellular matrix
EDTA	Ethylenediaminetetraacetic acid
FBS	Foetal bovine serum
FDA	Fluorescein diacetate
HE	Hematoxylin and eosin
ITB	Iliotibial band
MSC	Mesenchymal stromal cells

Clemens Gögele and Silke Schwarz contributed equally to this work.

MTS	3-(4,5-Dimethylthiazol-2-yl)-5-(3-carboxymethoxyphenyl)-2-(4-sulfophenyl)-2H-tetrazolium
PBS	Phosphate buffered saline
PET	Polyethylene terephthalate
PI	Propidium iodide
rpm	Rounds per minute
RPMI	Roswell Park Memorial Institute medium
RT	Room temperature
SDS	Sodium dodecyl sulphate
3D	Three-dimensional
TBS	Tris buffered saline
TRIS	Tris-(hydroxymethyl)-aminomethane

1 Introduction

The iliotibial band (ITB) may be a suitable allograft for surgical anterior cruciate ligament (ACL) reconstruction in cases where no other grafts are available [1]. The ITB is already in use for Achilles tendon reconstruction in children [2]. Utilizing allografts instead of autografts for transplantation reduces surgery time, limits the site of surgery to the actual site of injury and omits the risk of donor site morbidity since no graft needs to be harvested from the patient [1]. Furthermore, the ITB, consisting of dense connective tissue with interwoven type I collagen bundles is fairly large and provides a versatile and stable scaffold. It consists of both a thicker proximal part and distally of a large thinner flat part, which both can be tailored in a desired shape as required for transplantation (Fig. 1A1, A2). It contains ligamentocytes, which represent a specialized fibroblast population. Allografts bear the risk of immunogenic rejection.

The immunogenicity of allogenic grafts such as the ITB can be reduced by removing resident cells, as cells expose most of the donor-specific antigens. Using common protocols for decellularization, the ECM is highly conserved across individuals and between mammal species. By decellularization biomimetic scaffolds can be created for many tissue engineering based approaches. It is well reported in scientific literature that the cell-free ECMs of ECM-rich tissues maintain most of their specific properties as reviewed by [3, 4]. However, dense and ECM-rich tissues such as the ITB are challenging for decellularization. We described previously the decellularization and properties of ECM-rich tissues such as tendon [5], cartilage [6] and ITB [7]. Depending on the specific tissue, the decellularization protocol needs to be adapted carefully. Several agents required for successful cell removal from tissues bear cytotoxic properties. For medical applications, acellular ECM needs to be recolonized by host cells either in vitro or in vivo. A successful recellularization largely depends on the cytocompatibility of the

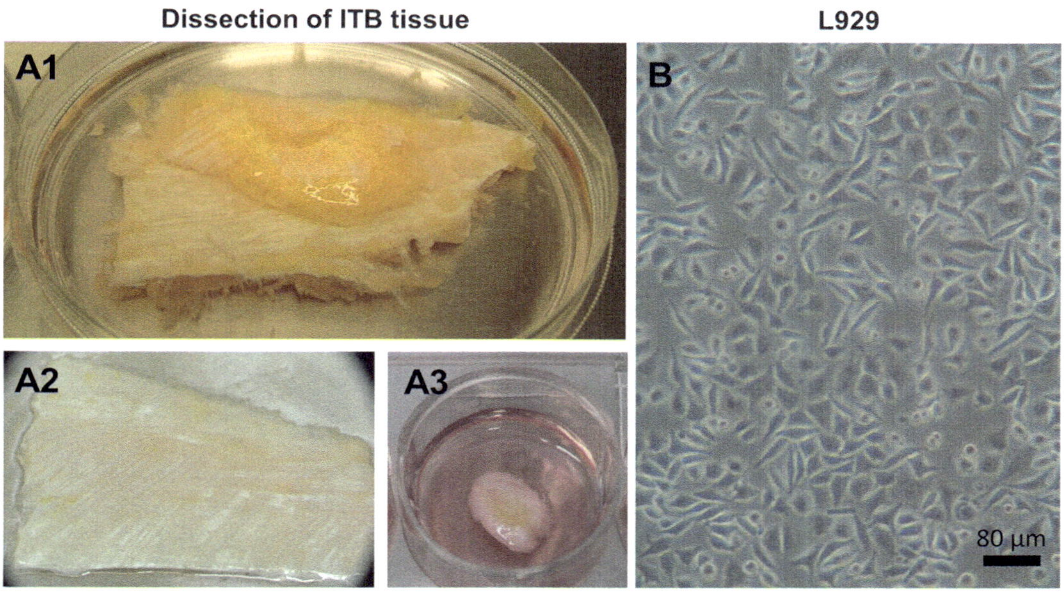

Fig. 1 Macroscopic images of ITB prepared for decellularization and microscopical view of L929 cells. (**A1**) Dissection of the explanted ITB. (**A2**) Dissected ITB. (**A3**) Macroscopic view on a decellularized tissue cylinder (6 mm diameter) prepared for cell seeding. (**B**) Expanded L929 fibroblasts for cytotoxicity testing. Scale bar: 80 μm. *ITB* iliotibial band

cell-free ECM and the maintenance of binding motives for cells, which might be cell-type specific.

Bone marrow-derived mesenchymal stromal cells (MSCs) are easier to harvest in contrast to homotopic fibroblasts, easy to propagate and capable for differentiating into fibroblasts [8]. Therefore, we propose here two different approaches for cell recolonization: using bone marrow-derived MSCs and allogenic ITB-derived cells.

In this chapter, we describe a method to prepare cell-free and cyto-compatible ECM derived from the ITB as a scaffold to support ligamentocyte growth and recolonization strategies with homotopic fibroblasts and MSCs, which could be promising for ligament reconstruction.

2 Materials

2.1 ECM Preparation

1. Dissect human ITB under sterile conditions and remove connective tissue (Fig. 1A1, A2).
2. *Dulbecco's Phosphate-Buffered Saline (PBS):* 1× sterile, without calcium and magnesium (Merck, Darmstadt, Germany, Cat.: L1825).
3. *Transport medium:* PBS supplemented with 1% (v/v) penicillin/streptomycin (Merck, Cat.: A2213) and 1% (v/v) gentamycin (Merck, Cat.: A2712).

4. Biopsy punch to prepare 6-mm diameter pieces for decellularization (Fig. 1A3).
5. Cell culture dishes: (Sarstedt, Nümbrecht, Germany; 92 × 17 mm, Cat.: 83.3902.300), sterile.
6. Centrifugation tubes: 50 mL Falcon tubes, sterile (Becton Dickinson Lab ware, Erembodegem, Belgium) for washing and digestion of tissue samples.
7. Sterile stainless steel tweezers.
8. Scalpel (B. Braun, Melsungen, Germany), sterile.
9. H_2O_{bidest}: Sterile, pyrogen free, hypotonic water (Carl Roth, Karlsruhe, Germany, Cat.: 3255.1).

2.2 Decellularization Buffers

1. *Decellularization buffer:* 1% sodium dodecyl sulphate (SDS, Carl Roth, Cat.: 0183.1), 0.2% sodium acid, 5 mM EDTA (pH 8.0), freshly added 0.4 mM phenylmethanesulfonyl fluoride (Sigma-Aldrich, Munich, Germany, Cat.: P7626) and one Complete Protease Inhibitor tablet (Roche, Mannheim, Germany, Cat.: 11836153001)/10 mL decellularization buffer in TBS buffer pH 8.
2. *Trypsin/EDTA solution*: 0.05% trypsin/0.02% EDTA (w/v) (Merck, Cat.: L2143).
3. *TBS Buffer:* Tris-(hydroxymethyl)-aminomethane, CELLPURE® ≥99.9% (Carl Roth, Cat.: 3170), dissolve 6.05 g of TRIS in 30 mL H_2O_{deion}, adjust pH to 8 with HCl, fill up with H_2O_{deion} to a final volume of 50 mL.
4. Bovine serum albumin (BSA, Sigma-Aldrich, Cat.: A4919-16).

2.3 Cell Culture

1. *Cultivation and expansion medium for ITB-derived fibroblasts:* DMEM/Ham's F12 medium (1:1) (Merck, Cat.: FG4815), supplemented with 10% foetal bovine serum (FBS, Merck, Cat.: S0615), 1% penicillin/streptomycin solution, 25 μg/mL ascorbic acid (*see* **Note 1**) (Sigma-Aldrich, Cat.: A4544), 2.5 μg/mL amphotericin B (Merck, Cat.: A2612), MEM amino acid solution (Sigma-Aldrich, Cat.: M7145).
2. *PBS:* 1× sterile, w/o Ca^{2+}/Mg^{2+}.
3. *Trypsin/EDTA solution:* 0.05% trypsin/0.02% EDTA (w/v).
4. Hemocytometer.
5. *Trypan blue exclusion dye:* Dilute 0.4% trypan blue solution (Sigma-Aldrich, Cat.: T8154) with 1× PBS (Merck, Cat.: L1825) to an end concentration of 0.04% trypan blue solution.
6. *Optional: Human Bone marrow:* Fresh unprocessed human bone marrow aspirates (Lonza, Basel, Switzerland, Cat.: 1M-125) or freshly isolated human bone marrow aspirates checked to be free of HIV and hepatitis viruses.

7. *Cultivation medium human bone marrow-derived stem cells:* DMEM (Merck, Cat. FG0445) supplemented with 1% L-glutamine, 1% penicillin/streptomycin, 10% FBS and additionally 1% sodium pyruvate (Merck, Cat. L 0473).
8. *L929:* Cell line murine fibrosarcoma-derived fibroblasts (DSMZ No.: ACC2), fibroblasts growing as monolayer.
9. *L929 Cultivation medium:* RPMI 1640 (Thermo Fisher Scientific, Germany, Cat.: 11875093) supplemented with 10% FBS, 100 U/mL penicillin/streptomycin solution.
10. Primary ITB-derived fibroblasts.
11. Biocoll separating solution (Merck, Cat.: L6113).
12. Cell culture flasks: 25 cm^2. T25 (Sarstedt, Cat.: 83.3910.302).
13. Cell culture flasks: 75 cm^2. T75 (Sarstedt, Cat.: 83.3911.302).
14. Cell culture flasks: 175 cm^2. T175 (Sarstedt, Cat.: 83.3912.302).
15. Cell culture dishes: 92 × 17 mm (Sarstedt, Cat.: 83.3902.300), sterile.
16. Collagenase solution: 0.075% collagenase NB5 derived from *Clostridium histolyticum* (Serva Electrophoresis GmbH, Heidelberg, Germany, Cat.: 17459).
17. Centrifugation tubes: 50 mL Falcon tubes sterile for washing and digestion of tissue samples.
18. Tweezers: A pair of sterile stainless steel tweezers.
19. Stainless steel sieve (mesh size 100 μm) for tissue dissociation.
20. Syringe (B. Braun), 10 mL, sterile.
21. Scalpel (B. Braun), sterile.
22. Rotatory device (Cat.: SU1400, Sunlab, Sustainable Lab Instruments, Aschaffenburg, Germany).
23. 50 mL bioreactor tubes (tube spine bioreactor50, TPP, Trasadingen, Switzerland, Cat.: 87050).

2.4 Cytotoxicity Testing According to International Standard ISO 10993-5:2009

1. *Extraction medium:* L929 Cultivation medium. RPMI 1640 (Thermo Fisher Scientific, Darmstadt, Germany, Cat.: 11875093) supplemented with 10% FBS, 1% penicillin/streptomycin solution.
2. *Positive control:* 10% DMSO solution (Sigma-Aldrich, Cat.: D2650) in L929 cultivation medium.
3. *Negative control*: ThinCert™ polyethylene terephthalate (PET) membranes (3 μm) (Greiner Bio-One GmbH, Frickenhausen, Germany, Cat.: 665630).
4. 96-Well cell culture plates (Sarstedt).

5. CellTiter 96® AQueous One Solution Cell Proliferation Assay (3-(4,5-dimethylthiazol-2-yl)-5-(3-carboxymethoxyphenyl)-2-(4-sulfophenyl)-2H-tetrazolium, inner salt, MTS), (Promega, Mannheim, Germany, Cat.: G3581).
6. Photometer.

2.5 Histochemical Stainings

1. Microtome to prepare paraffin sections.
2. Horizontal shaker.
3. Microwave.
4. Microscope.
5. Pipettes.
6. Beaker glass.
7. *Paraformaldehyde solution:* 4% in PBS, pH 7.0–7.6, (VWR, Langenfeld, Germany, USB Molecular Biology Reagents, USBM199431LT).
8. Isopropanol (Sigma-Aldrich, Cat.: 33539), 2.5 L.
9. Entellan (Merck, Cat.: 1.07960.0500).
10. Tap water.
11. Deionized water (H_2O_{deion}).
12. Xylol (Carl Roth, Cat.: 9713) >98%, absolute, for histology.
13. Isopropanol (2-Propanol, Sigma-Aldrich, Cat.: 278475), anhydrous, 99.5%, 100% and diluted with H_2O_{deion} to 90 and 70%.

2.5.1 Hematoxylin Eosin (HE) Staining

1. *Hematoxylin solution:* Harris Hematoxylin, Hemalum, ready for use (Carl Roth, Cat.: X903).
2. *Eosin solution*: Eosin G-solution 0.5%, ready for use (Carl Roth, Cat.: X883).

2.5.2 Alcian Blue (AB) Staining

1. Alcian blue 8 GS. C.l. 74240 (Carl Roth, Cat.: 3082).
2. 3% Acetic acid.
3. *Alcian blue staining solution:* Solve 0.5 g Alcian blue 8 GS in 50 mL 3% acetic acid solution. Adjust pH to 2.5.
4. *Nuclear Fast Red solution:* Counterstain, Nuclear Fast Red, ready for use (Carl Roth, Cat.: N069).

2.5.3 DAPI Staining

1. 4′,6-Diamidino-2-phenylindol (DAPI, Roche Diagnostics GmbH, Cat.: 10236276001).
2. TBS, tris buffered saline (tablets, Medicago, Uppsala, Sweden, Cat.: 097500-5).
3. 100% Triton X-100 (Sigma-Aldrich, Cat.: 106KD177).

2.6 Life/Dead Staining

1. *Fluorescein diacetate (FDA) stock solution*: FDA (Sigma-Aldrich, Cat.: 31545) 3 mg/mL dissolved in 1 mL acetone. Store at +4 °C in darkness.
2. *Propidium iodide (PI) stock solution*: PI solution 1%, ready for use (Carl Roth, Cat.: 2218), 1% = 10 mg/mL.
3. *FDA/PI staining solution*: Add 5 μL of the FDA stock solution and 1 μL of the PI stock solution to 1 mL PBS.

3 Methods

3.1 ECM Decellularization and Sterilization

1. Dissect the human ITB under aseptic conditions and remove adjacent connective tissue (Fig. 1A1, A2).
2. For decellularization of ITB, punch out cylinders with a predefined diameter, e.g. 6 mm (Fig. 1A3). Immerse tissue cylinders in sterile pyrogen-free H_2O_{bidest} and freeze them at −20 °C for 4 h. Thaw the tissue samples at 37 °C in a water bath for 30 min before freezing them again at −20 °C for 4 h.
3. Repeat the freeze/thaw cycle six times.
4. Incubate the tissue samples in decellularization buffer at room temperature (RT) for 24 h.
5. Rinse three times with sterile pyrogen-free H_2O_{bidest} (*see* **Note 2**).
6. Incubate in 0.05% trypsin/0.02 mM EDTA at RT for 24 h.
7. Rinse three times with sterile pyrogen-free H_2O_{bidest}.
8. Finally immerse in 3% Triton-X100 for 48 h at 37 °C.
9. Sterilize the decellularized ITB samples in 70% ethanol with continuous shaking on a rotatory device at RT (not necessary if everything was conducted under sterile conditions).
10. Rinse the decellularized extracellular matrices (ECM) three times in 1× PBS w/o Ca^{2+}/Mg^{2+} at RT on a rotatory device.
11. Freeze and store the decellularized tissue samples at −20 °C in sterile, pyrogen-free H_2O_{bidest} until further use.

3.2 Cytotoxicity Testing According to International Standard ISO 10993-5:2009 (E)

1. Use murine fibrosarcoma cell line L929 for biological evaluation of medical devices according to international standard ISO 10993-5:2009 (Fig. 1B).
2. Seed thawed L929 cells with an initial density of 1.0×10^4 cells/cm^2 in cell culture flasks and cultured in RPMI 1640 medium supplemented with FBS (10%), penicillin (100 U/mL) and streptomycin (100 U/mL) until 80–90% confluence at 37 °C and 5% CO_2.
3. Change medium three times a week.

3.2.1 Preparation of Liquid Extracts of Processed Cartilage Matrix

1. To determine potential cytotoxic effects (*see* **Note 3**) due to decellularization process (DP) liquid extracts of decellularized ITB scaffolds have to be prepared.
2. Use the L929 culture medium as extraction medium.
3. Incubate sterile ITB scaffolds in extraction medium at 37 °C and 5% CO_2 for 24 h under aseptic conditions using sterile, chemically inert cell culture plates.
4. Culture medium is used as extraction vehicle because of its potential to extract polar and non-polar substances.

3.2.2 Cytotoxicity Test Using L929

1. Detach amplified cells with 0.05% trypsin/0.02% EDTA solution and determine cell numbers by trypan blue exclusion dye.
2. Seed the L929 cells in 96-well cell culture plates with an initial density of 1×10^4 cells per well and incubate the cells for 24 h at 37 °C and 5% CO_2 to allow cell adherence.
3. Remove the medium and incubate the murine L929 cells with 100 μL extract or control solutions per each well for 24 h at 37 °C and 5% CO_2.
4. Use a 10% DMSO solution in L929 cultivation medium as positive control.
5. Apply the liquid extract of ThinCertTM PET membranes as negative control.
6. After 24 h of incubation with the respective extracts, discard the medium completely.
7. Add 20 μL MTS solution (CellTiter 96® AQueous One Solution Cell Proliferation Assay to each well.
8. Incubate the cells for additional 2 h under standard culture conditions and measure absorbance photometrically at a wavelength of 490 nm.
9. Check the decellularization success using histochemical staining (Fig. 2A1–C2, see **Note 4**).

3.3 Cell Isolation and ECM Recellularization

3.3.1 Isolation of ITB-Derived Ligamentocytes

1. Human primary ITB-derived cells, ligamentocytes, can be isolated from ITBs from healthy middle-aged donors. For harvesting human derived tissues, tissue removal has to be approved by the local Ethics Committee.
2. Remove the peri-ligamentous connective and fatty tissue of human ITB carefully (Fig. 1A2).
3. Cut the ITB in small pieces with a size of 4 × 4 mm and transfer the tissue explants to T25 cell culture flasks (Fig. 3A1, A2).
4. Culture the ligament explants in cultivation medium for ITB-derived fibroblasts for several days and change medium two times a week.

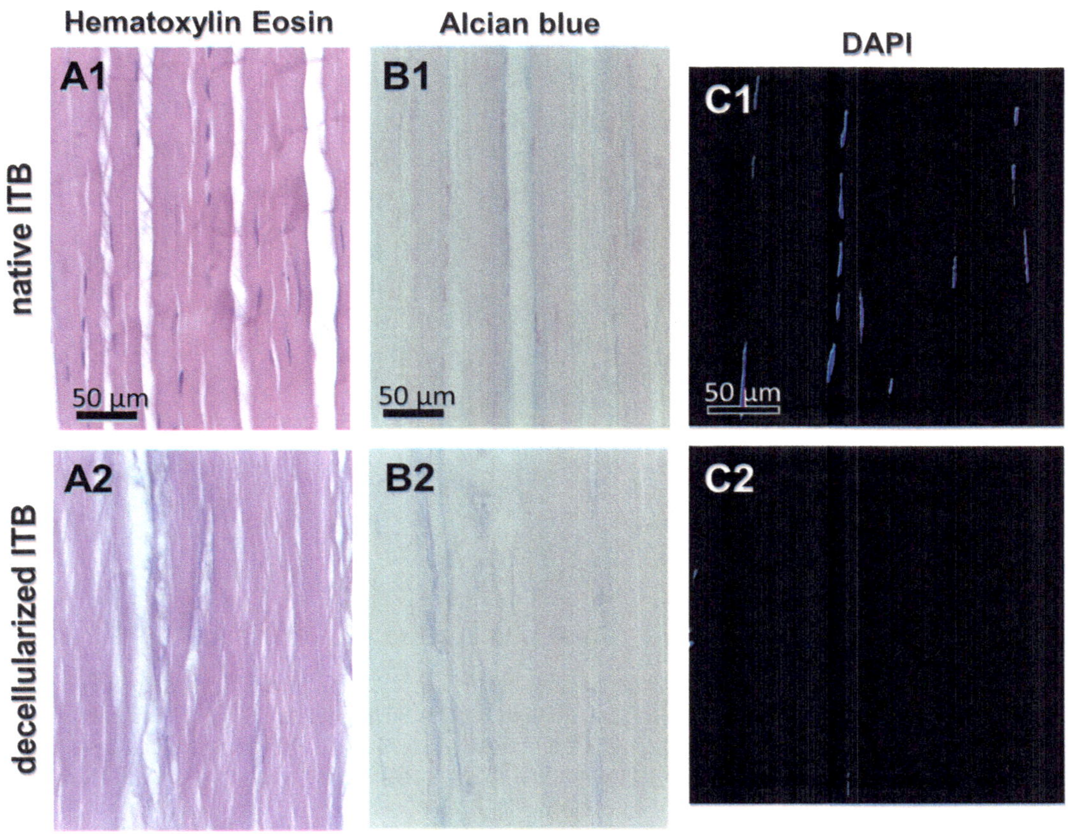

Fig. 2 Histological images of native and decellularized ITB tissues. Native (**A1–C1**) and decellularized (**A2–C2**) tissues were stained with Hematoxylin Eosin (**A1, A2**), Alcian blue (**B1, B2**) and DAPI (**C1, C2**). Scale bars: 50 μm. *ITB* iliotibial band

5. Seven to fourteen days after the initial ligamentous explant transfer, ligamentocytes start to continuously grow out from these explants and adhere to the cell culture surface (Fig. 3A2). Medium has to be changed twice a week and cells propagated until reaching 80–90% confluence.

3.3.2 Cell Expansion

1. When outgrowing ITB-derived ligamentocytes reach a confluence of 80–90% they have to be detached by trypsinization and cell number as well as cell viability has to be determined by trypan blue exclusion dye using a hemocytometer.
2. Isolated ligamentocytes can be expanded in cultivation medium (see **Note 5**). Once the ITB-derived ligamentocytes reach a confluence between 80 and 90%, cells can be detached either with a conventional Trypsin/EDTA solution (0.05%/ 0.02%) or with Accutase (Thermo, PAA) (see **Note 6**).

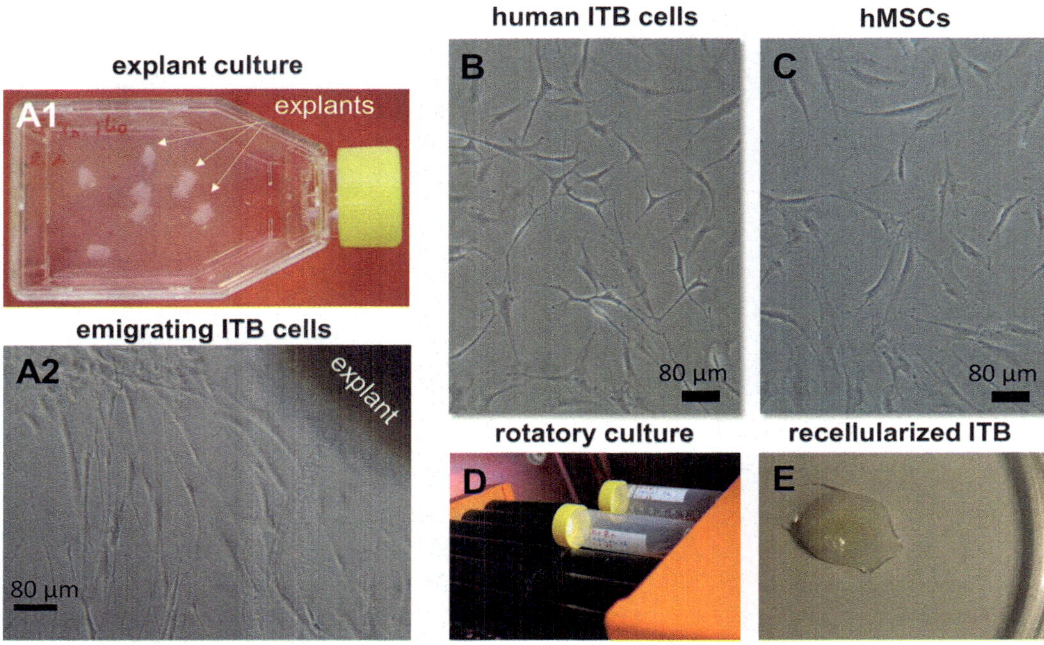

Fig. 3 ITB explant culture and dynamical cell seeding procedure of decellularized ITB tissue. (**A1**) Explant cultures prepared from ITB tissue. (**A2**) Cell outgrowth from the explants. Scale bar: 80 μm. ITB: iliotibial band. (**B, C**) Expanded ITB-derived fibroblasts and hMSCs in monolayer culture. (**D**) Dynamical seeding using 5 mL cell suspension in a bioreactor tube on a rotator device. (**E**) Reseeded tissue piece after 1 week of culture. Scale bars: 80 μm. *ITB* iliotibial band. *MSCs* mesenchymal stromal cells

3. Expand the cells first in a T75 cell culture flask and transfer them to T175 cell culture flasks. The ligamentocytes are amplified in monolayer culture until a sufficient cell number is reached for further cell seeding and 3D long-term culture. Incubate cultures at 37 °C in a humidified atmosphere with 5% CO_2 and change the medium two times a week.

3.3.3 Pre-incubation of ECM

1. Thaw the decellularized and sterilized ITB scaffolds at 4 °C.
2. Rinse the decellularized ECM three times in 1× PBS w/o Ca^{2+}/Mg^{2+} at RT on a rotatory device.
3. Directly pre-incubate ECM scaffolds in 100% inactivated FBS for 5 h at 37 °C and 5% CO_2 on a rotatory device (36 rpm for 24 h) to reduce cytotoxicity due to decellularization process.
4. Optionally: pre-incubate the decellularized ECM scaffolds in 5% bovine serum albumin (BSA) solubilized in sterile pyrogen-free H_2O_{bidest} at 37 °C and 5% CO_2 (22 rpm for 24 h) [6].

3.3.4 ECM Recellularization

1. Thaw ECM scaffolds at 4 °C and pre-incubate them with pure, inactivated FBS. Aspirate carefully the FBS from the scaffolds before cell seeding. Place them on a sterile pulp to soak excess fluid but do not dry them completely.

2. Upon 80% confluence, trypsinize ligamentocytes or MSCs (Fig. 3B, C) and calculate cell numbers and scaffold volume as described previously (*see* **Note 7**) [9].
3. *Statistical seeding procedure*:

   Place the dried ECM (*see* **Note 8**) on the non-adherent (e.g. agarose coated) well of a six well plate. For each ECM scaffold resuspend 0.5–1 × 10^6 cells in a volume of 60 μL of the respective cultivation medium (for MSCs or ligamentocytes) equal to the scaffold void volume. Let the ECM surface directly absorb 30 μL of the prepared cell suspension by pipetting this suspension volume on it. Incubate for 30 min to allow cell adherence to the ECM surface (*see* **Notes 8** and **9**). Then, turn the ECM and add 30 μL to the side, which was previously the bottom. Wait again 30 min before adding 0.5 mL of fresh cultivation medium to each well. After 2 h, add 2.5 mL cultivation medium to achieve the final culturing volumes in the wells. Incubate the cultures at 37 °C in 5% CO_2 and change two times a week.
4. If MSCs were seeded on the scaffold expose them to chondrogeneic differentiation for minimum 14 days to show multipotency (as detailed previously [9]). Culture undifferentiated controls for comparison.
5. Optionally: centrifuge the cells on the ECM construct at 300 × *g* for 5 min. Turn the construct and centrifuge once again.
6. *Dynamical seeding procedure*:
   Suspend 0.5–1 × 10^6 cells in 5 mL of the respective cultivation medium in a bioreactor tube together with the ECM scaffold and place the tube on a rotatory device, rotating at 36 rpm (angle 10°) (Fig. 3D).

3.4 Isolation of Human Bone Marrow-Derived Mesenchymal Stem Cells (hMSCs)

1. Human bone marrow-derived mesenchymal stem cells (hMSCs) can be harvested from human femoral heads.
2. Curette whole femoral head spongiosa from the interior of the femoral neck and head, transfer to a course mesh and press the tissue through the pores of the mesh using a stamp. Rinse the spongiosa fragments with iced 1× PBS w/o Ca^{2+} and Mg^{2+}. Using a syringe, press the liquid cell suspension through a 140 μm pore diameter sterilized filter membrane. Perform every step under sterile conditions.
3. To increase the cell yield, digest the remaining tissue fragments in the sieve for 25 min at 37 °C under continuous gentle shaking using 0.075% collagenase solution. Stop the digestion by adding 5% FBS. Filter the cell suspension through the

140 μm pore diameter filter and add it to the aforementioned cell suspension from **Step 1**.

4. Wash the isolated cell suspension with 1× PBS w/o Ca^{2+} and Mg^{2+} and centrifuge at 200 × *g* and 4 °C for 5 min. Resuspend the purified cell pellet and overlay it cautiously on the biocoll separating solution before centrifuging at 250 × *g* and 4 °C without brake. After 20 min, remove the supernatant and extract the interphase containing the MSCs carefully using a 2 mL plastic pipette. Add 15 mL of iced 1× PBS w/o Ca^{2+} and Mg^{2+} to the MSC layer and centrifuge for 5 min at 200 × *g* and 4 °C.
5. Subsequently, resuspend the MSCs in 20 mL of *cultivation medium for hMSCs* and seed them into T175 culture flasks.
6. hMSC monolayer expansion culture should be performed in *cultivation medium for hMSCs* at 37 °C, 90% air humidity and 5% CO_2.
7. After 3 days, discard non-adherent cells and culture the adherent hMSCs to 80% confluence, change the medium every 2–3 days.
8. The cells should be expanded to sufficient numbers for scaffold seeding by regular passaging using standard protocols. Harvest the hMSCs by applying a conventional Trypsin/EDTA solution (0.05% trypsin/0.02% EDTA) or with Accutase (Thermo, PAA) (*see* **Note 6**) and determine cell number as well as cell viability by trypan blue staining using a hemocytometer.

Optional:

9. Bone marrow aspirates from human donors can be obtained from Lonza (Basel, Switzerland) or local institutes of transfusion medicine.
10. Seed 5–6 mL of bone marrow aspirate in T175 cell culture flasks and cover with 30 mL cultivation medium for hMSCs.
11. To allow hMSCs to adhere within the first 48–96 h, the medium may remain unchanged.
12. After 4 days, remove and discard the cell culture supernatant containing the hematopoietic cell fraction of bone marrow. hMSCs are isolated by plastic adherence.
13. Subsequently, medium has to be changed twice a week and isolated MSCs are maintained and amplified in cultivation medium until they reach a confluence of 80–90%.
14. Detach hMSCs by applying a conventional Trypsin/EDTA solution (0.05% trypsin/0.02% EDTA) or with Accutase (Thermo, PAA) (*see* **Note 6**) and determine cell number as well as cell viability by trypan blue staining using a hemocytometer.

3.5 Cytocompatibility Testing: Life/Dead Staining

1. To rapidly estimate the cell viability in reseeded decellularized ECMs (Fig. 3E), FDA/PI staining can be performed (Fig. 4A1–A4). Incubate the scaffolds in FDA/PI solution for 2 min in the dark. Rinse the scaffolds gently with PBS (*see* **Note 9**).
2. Place the wet scaffolds on a slide (26 × 76, 0.16–0.19 thickness, suitable for confocal laser scanning microscopy, Langenbrinck, Emmendingen, Germany) and monitor the green (living cells, FDA) or red (dead cells, PI) fluorescence by fluorescence or by confocal laser scanning microscopy.
3. Perform HE staining and DAPI stainings (Figs. 2A, B and 4B) of paraffin sections after embedding scaffolds in paraffin and following standard protocols.

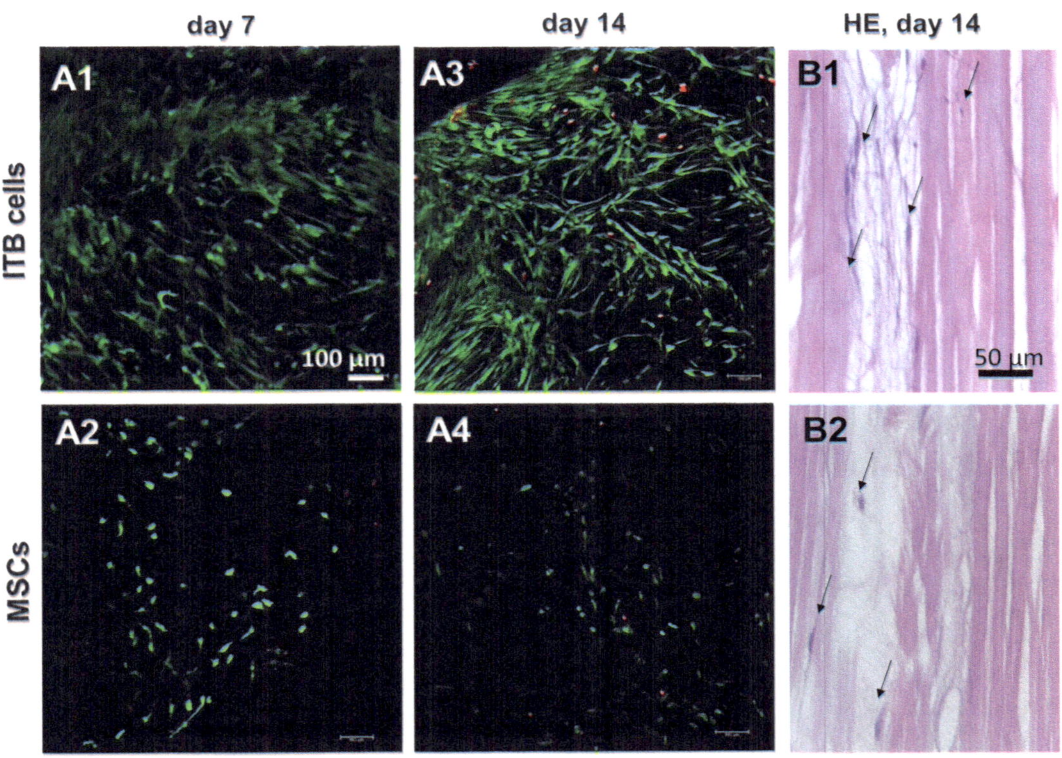

Fig. 4 Live/Dead and HE stainings of dynamically recellularized ITB tissue. Decellularized ITB was recellularized with allogenic ITB-derived fibroblasts (**A1, A3, B1**) or MSCs (**A2, A4, B2**). (**A1–A4**) Cell vitality was assessed at day 7 (**A1, A2**) and day 14 (**A3, A4**) using Live/Dead staining. Dead cells are *red* and vital cells *green*. (**B1, B2**) HE staining. Scale bars (**A1–A4**): 100 µm. Scale bars (**B1, B2**): 50 µm. *MSCs* mesenchymal stromal cells. *ITB* iliotibial band

3.6 Histochemical Staining

1. Samples were dehydrated in ascending alcohol series followed by xylol (Carl Roth, Cat.: 9713.3) and embedded in paraffin wax (Merck KGaA).
2. 5–8 μm thick paraffin sections were prepared using a microtome.

3.6.1 Hematoxylin Eosin Staining

1. Deparaffinized sections were incubated for 6 min in Harry's hematoxylin (Sigma-Aldrich), before rinsed in water and counterstained for 4 min in eosin (Carl Roth).

3.6.2 Alcian Blue Staining

1. Sections were incubated for 3 min in 1% acetic acid, incubated for 30 min in 1% AB (Carl Roth GmbH) and after rinsing in 3% acetic acid and 2 min washing in A. dest., cell nuclei were counterstained in nuclear fast red aluminium sulphate solution (Carl Roth GmbH) for 5 min.

3.6.3 DAPI

1. Incubate deparaffinized section with a 0.1% Triton X-100 containing TBS solution with DAPI (1 μg/mL stock solution, diluted 1:100 in TBS).

4 Notes

1. The addition of ascorbic acid seems to be advisable, because ascorbic acid is required as co-factor for optimal function of prolyl-hydroxylase, which plays a key role in collagen synthesis [10].
2. Rinsing steps, e.g. to remove decellularization buffer components, particularly SDS should be performed in sufficient volume and 50 mL Falcon tubes.
3. Cytotoxicity testing is required before recolonization since there is a high risk of cytotoxic SDS captured within the decellularized ECM.
4. The AB staining allows to compare sulphated glycosaminoglycan content before and after decellularization.
5. MSCs and ligamentocytes for ECM seeding should not be too extensively expanded (MSCs < passage 5; ligamentocytes < passage 7) before seeding on the scaffolds since MSCs lose their differentiation capability during culturing. Cells should not derive from old donors (<50 years).
6. Detachment with Accutase might be more gentle to the cells.
7. The total volume of the 6-mm diameter ECM scaffold is calculated as described previously [9].
8. To avoid cell migration from the ECM to the culture dishes, they should be coated with cell culture compatible agarose

(low gelling temperature, A9045-5G, Sigma), alginate (alginic acid sodium salt, Sigma, A2158-100G) or using low attachment plates (Costar, Ultra low attachment surface, Corning New York, USA).

9. ECM constructs seeded with cells should be gently handled to avoid cell loss.

Acknowledgement

We would like to thank Tobias Schneider for his support. The authors acknowledge the financial support from Manfred Roth foundation. We would like to thank the University of Salzburg and the Austrian Agency for International Cooperation in Education and Research (OeAD-GmbH) for the scholarship and financial support of Mr. Clemens Gögele.

References

1. Delcroix GJ, Kaimrajh DN, Baria D, Cooper S, Reiner T, Latta L et al (2013) Histologic, biomechanical, and biological evaluation of fan-folded iliotibial band allografts for anterior cruciate ligament reconstruction. Arthroscopy 29:756–765
2. Hu R, Ren YJ, Yan L, Xiao ZH, Ding F, Li F et al (2015) A free anterolateral thigh flap and iliotibial band for reconstruction of soft tissue defects at children's feet and ankles. Injury 46:50–67
3. Schulze-Tanzil G, Al-Sadi O, Ertel W, Lohan A (2012) Decellularized tendon extracellular matrix-a valuable approach for tendon reconstruction? Cell 1:1010–1028
4. Huang Z, Godkin O, Schulze-Tanzil G (2016) The challenge in using mesenchymal stromal cells for recellularization of decellularized cartilage. Stem Cell Rev 13:50–67
5. Lohan A, Stoll C, Albrecht M, Denner A, John T, Kruger K et al (2013) Human hamstring tenocytes survive when seeded into a decellularized porcine Achilles tendon extracellular matrix. Connect Tissue Res 54:305–312
6. Zhao Huang BK, Kokozidou M, Arens S, Schulze-Tanzil G (2016) Establishment of a cytocompatible cell-free intervertebral disc matrix for chondrogenesis with human bone marrow mesenchymal stromal cells. Cells Tissues Organs 201:354–365
7. Hammer N, Huster D, Boldt A, Hadrich C, Koch H, Mobius R et al (2016) A preliminary technical study on sodium dodecyl sulfate-induced changes of the nano-structural and macro-mechanical properties in human iliotibial tract specimens. J Mech Behav Biomed Mater 61:164–173
8. Gaspar D, Spanoudes K, Holladay C, Pandit A, Zeugolis D (2015) Progress in cell-based therapies for tendon repair. Adv Drug Deliv Rev 84:240–256
9. Nooeaid P, Schulze-Tanzil G, Boccaccini AR (2015) Stratified scaffolds for osteochondral tissue engineering. Methods Mol Biol 1340:191–200
10. Vitagliano L, Berisio R, Mazzarella L, Zagari A (2001) Structural bases of collagen stabilization induced by proline hydroxylation. Biopolymers 58:459–464

Methods in Molecular Biology (2018) 1577: 71–86
DOI 10.1007/7651_2017_62

Published online: 10 August 2017

Preparation of Cell-Derived Decellularized Matrices Mimicking Native ECM During the Osteogenesis and Adipogenesis of Mesenchymal Stem Cells

Takashi Hoshiba, Naoki Kawazoe, and Guoping Chen

Abstract

The extracellular matrix (ECM) is an important extracellular microenvironmental factor that regulates stem cell differentiation. The ECM is remodeled according to stem cell differentiation progression to precisely regulate the differentiation. Thus, it is expected that the matrices mimicking native ECM surrounding differentiating cells at each differentiation stage provide a favorable microenvironment to promote stem cell differentiation. However, it is difficult to prepare matrices mimicking native ECM using chemical methods because the ECM has a complicated composition. The decellularization technique is useful to prepare such matrices. In this chapter, we described the protocol to prepare matrices mimicking native ECM surrounding cells that are differentiating from mesenchymal stem cells to either osteoblasts or adipocytes via stem cell differentiation culture and a detergent- and nuclease-based decellularization technique.

Keywords Adipogenesis, Cell culture, Decellularization, Extracellular matrix, Mesenchymal stem cells, Osteogenesis

1 Introduction

The extracellular matrix (ECM) is a special architecture to support or surround the cells as a substrate for cell adhesion in vivo [1]. In addition, the ECM regulates many cellular functions, such as cell survival, proliferation, migration, morphogenesis, and differentiation [1, 2]. To precisely regulate cellular functions, the ECM consists of many ECM molecules, including collagens, proteoglycan, glycoproteins, and glycosaminoglycans [1, 3]. Moreover, the ECM compositions differ according to tissue types, pathological conditions, and developmental stages [3–7].

Stem cell differentiation is also regulated by the ECM [4, 5]. The ECM compositions are dynamically remodeled during the progression of stem cell differentiation [4, 5]. Thus, it is expected that matrices mimicking native ECM according to stem cell differentiation stages can provide a favorable microenvironment to promote stem cell differentiation.

However, it is difficult to prepare matrices mimicking native ECM surrounding differentiating stem cells using conventional chemical methods because the ECM consists of many ECM molecules that have not been completely identified. Thus, decellularized matrices have been utilized as native ECM models [8, 9]. Decellularized matrices are mainly prepared from the ECM in tissues/organs and the ECM produced by cultured cells. Both of these sources have advantages and disadvantages [9]. In particular, cultured cell-derived decellularized matrices are suitable for mimicking native ECM in limited regions, such as the stem cell niche and differentiating stem cells, because it is difficult to identify and isolate such ECM from tissues and organs [9, 10].

In this chapter, we described a protocol to prepare cell-derived decellularized matrices called as "tissue-development-mimicking matrices," which mimic native ECM surrounding differentiating stem cells at different developmental stages through their differentiation. In particular, the protocols are described to prepare the cell-derived decellularized matrices mimicking native ECM surrounding differentiating mesenchymal stem cells (MSCs) at different developmental stages via stepwise osteogenesis and adipogenesis [11–14]. We call these matrices "stepwise osteogenesis-mimicking matrices" and "stepwise adipogenesis-mimicking matrices," respectively. These matrices consist of "stem cell matrices," "osteogenic or adipogenic early stage matrices," and "osteogenic and adipogenic late stage matrices" (Table 1). Stem cell matrices are cell-

Table 1
The types and characteristics of stepwise osteogenesis- and adipogenesis-mimicking matrices

	Stem cell	Osteogenic		Adipogenic	
Differentiation stage	Undifferentiation stage	Early stage	Late stage	Early stage	Late stage
Matrix name	Stem cell matrix	Osteogenic early-stage matrix	Osteogenic late-stage matrix	Adipogenic early-stage matrix	Adipogenic late-stage matrix
Gene expression level	*ALP*: low	*ALP*: high	*ALP*: high	*ALP*: low	*ALP*: low
	IBSP: low	*IBSP*: low	*IBSP*: high	*IBSP*: low	*IBSP*: low
	LPL: low	*LPL*: low	*LPL*: low	*LPL*: high	*LPL*: high
Tissue-specific character	ALP staining: (−)	ALP staining: (+)	ALP staining: (+)	ALP staining: (−)	ALP staining: (−)
	Alizarin red S staining: (−)	Alizarin red S staining: (−)	Alizarin red S staining: (+)	Alizarin red S staining: (−)	Alizarin red S staining: (−)
	Nile red staining (−)	Nile red staining (−)	Nile red staining (−)	Nile red staining (−)	Nile red staining (+)

derived decellularized matrices mimicking native ECM surrounding undifferentiated MSCs. Osteogenic or adipogenic early-stage matrices are cell-derived decellularized matrices mimicking the native ECM surrounding the cells at the early stage of either osteogenesis or adipogenesis. Osteogenic or adipogenic late-stage matrices are cell-derived decellularized matrices mimicking native ECM surrounding the cells at the late stage of either osteogenesis or adipogenesis.

2 Materials

All solutions were prepared with deionized water. All reagents and equipment for cell culture and matrix preparation must be sterile. The sterilization method should be properly selected to avoid the loss of the reagents' activities. In addition, aseptic techniques should be used accordingly. The cells were cultured on tissue culture polystyrene (TCPS) plates, dishes, and flasks (Becton Dickinson) for all cell culture processes. The size of the TCPS cell culture containers should be properly selected according to the purpose of the study using cell-derived decellularized matrices. In addition to the reagents described below, phosphate buffered saline (PBS, Nacalai tesque) was used for the wash processes.

2.1 Cell Culture for Cell-Derived Decellularized Matrices Formation

2.1.1 MSC Passage Culture

1. MSCs: human MSCs derived from bone marrow (passage 2, Lonza).
2. MSC Growth Medium (Lonza).
3. Trypsin-EDTA: fivefold dilution of 0.25% trypsin-EDTA with PBS.
4. Phase Contrast Microscope (Olympus).

2.1.2 MSC Culture Without Differentiation Induction for the Preparation of Stem Cell Matrices

1. Undifferentiation culture medium: Low-glucose Dulbecco's modified Eagle medium (Sigma) supplemented with 10% fetal bovine serum (Equitech-Bio), 584 mg/mL L-glutamine (Gibco), 100 units/mL penicillin (Gibco), 100 mg/mL streptomycin (Gibco), 0.1 mM non-essential amino acids (Gibco), 1 mM sodium pyruvate (Gibco), 0.4 mM proline (Sigma), and 50 mg/mL ascorbic acid (Sigma) (see Note 1).

2.1.3 Osteogenic Differentiation Culture for the Preparation of Stepwise Osteogenesis-Mimicking Matrices

1. Osteogenic culture medium: Low-glucose Dulbecco's modified Eagle medium (Sigma) supplemented with 10% fetal bovine serum (Equitech-Bio), 584 mg/mL L-glutamine (Gibco), 100 units/mL penicillin (Gibco), 100 mg/mL streptomycin (Gibco), 0.1 mM non-essential amino acids (Gibco), 1 mM sodium pyruvate (Gibco), 0.4 mM proline (Sigma), 50 mg/mL ascorbic acid (Sigma), 10 nM dexamethasone (Sigma), and 10 mM β-glycerophosphate (Sigma) (see Note 2).

2.1.4 Adipogenic Differentiation Culture for the Preparation of Stepwise Adipogenesis-Mimicking Matrices

1. Expansion culture medium: High-glucose Dulbecco's modified Eagle medium (Sigma) supplemented with 10% fetal bovine serum (Equitech-Bio), 584 mg/mL L-glutamine (Gibco), 100 units/mL penicillin (Gibco), 100 mg/mL streptomycin (Gibco), 0.1 mM non-essential amino acids (Gibco), 1 mM sodium pyruvate (Gibco), 0.4 mM proline (Sigma), and 50 mg/mL ascorbic acid (Sigma).
2. Adipogenic culture medium: Expansion culture medium supplemented with 1 μM dexamethasone (Sigma), 0.5 mM methylisobutylxanthine (Sigma), 10 μg/mL insulin, and 100 mM indomethacin.

2.2 Confirmation of Osteogenic and Adipogenic Stages

2.2.1 Isolation of mRNA

1. ISOGEN (Nippon Gene).
2. Chloroform (Wako).
3. Isopropanol (Wako).
4. DEPC-treated water (Nacalai tesque).
5. 70% Ethanol: add approximately 15 mL of DEPC-treated water into 35 mL of ethanol (Wako).
6. Cooled centrifuge.
7. Spectrophotometer.

2.2.2 cDNA Synthesis by Reverse Transcription Reaction

1. M-MLV Reverse Transcriptase: One of the components of an M-MLV Reverse Transcriptase kit (Invitrogen).
2. DTT: 0.1 M dithiothreitol, one of the components of an M-MLV Reverse Transcriptase kit.
3. First-Strand Buffer: One of the components of an M-MLV Reverse Transcriptase kit.
4. dNTP: mixture of dATP, dTTP, dCTP, and dGTP at a concentration of 5 mM (each).
5. Random hexamer: 50 μM random primer.
6. RNase H solution: 2 units/mL RNase H.

7. DEPC-treated water.
8. Thermal cycler.

2.2.3 Real-Time Polymerase Chain Reaction

1. TaqMan Universal PCR Master Mix (Applied Biosystems).
2. Specific primers and probe set (Table 2) (see Note 3).
3. Real-Time PCR System.
4. 96-Well PCR plate.

2.2.4 Alkaline Phosphatase (ALP) Staining for the Detection of Cells at the Osteogenic Early Stage

1. Formaldehyde solution: 10% formaldehyde containing PBS.
2. ALP substrate solution: 0.1% naphthol AS-MX phosphate (Sigma) and 0.1% fast red violet LB salt (Sigma) in a 56 mM 2-amino-2-methyl-1,3-propanediol aqueous solution (Sigma) (see Note 4).
3. Optical microscope.

2.2.5 Alizarin Red S Staining for the Detection of Cells at the Osteogenic Late Stage

1. Formaldehyde solution (see above).
2. Alizarin red S solution: 0.5% alizarin red S (Acros Organics) dissolved in deionized water.
3. Optical microscope.

Table 2
Primers and probes for real-time PCR analysis

mRNA		Oligonucleotide
18S rRNA	Forward	5′-GCCGCTAGAGGTGAAATTCTTG-3′
	Reverse	5′-CATTCTTGGCAAATGCTTTCG-3′
	Probe	5′-CCGGCGCAAGACGGACCAGA-3′
GAPDH	Forward	5′-ATGGGGAAGGTGAAGGTCG-3′
	Reverse	5′-TAAAAGCAGCCCTGGTGACC-3′
	Probe	5′-CGCCCAATACGACCAAATCCGTTGAC-3′
ALP	Forward	5′-GACCCTTGACCCCCACAAT-3′
	Reverse	5′- GCTCGTACTGCATGTCCCCT-3′
	Probe	5′-TGGACTACCTATTGGGTCTCTTCGAGCCA-3′
IBSP	Forward	5′-TGCCTTGAGCCTGCTTCC-3′
	Reverse	5′-GCAAAATTAAAGCAGTCTTCATTTTG-3′
	Probe	5′-CTCCAGGACTGCCAGAGGAAGCAATCA-3′
LPL		TaqMan Gene Expression Assays (Hs00173425_m1)

2.2.6 Nile Red Staining for the Detection of Adipogenic Late Stage

1. Formaldehyde solution (See above).
2. Nile red solution: 100 ng/mL nile red (Sigma) in PBS. The stock solution of nile red is prepared in acetone at a concentration of 1 mg/mL. The stock solution is diluted in PBS to obtain the final concentration. After dilution, the working nile red solution should be used immediately. The stock solution should be stored at 4 °C in a dark box.
3. Hoechst 33258 solution: 10 μg/mL Hoechst 33258 (Dojindo) in PBS.
4. Fluorescent microscope (see Note 5).

2.3 Decellularization

1. Decellularization solution: 0.5% Triton X-100 (Sigma) and 20 mM NH_4OH (Wako) containing PBS. Used immediately after preparation. The stock aqueous solution of NH_4OH is prepared and stored at room temperature.
2. Nuclease solution: PBS containing 100 μg/mL DNase I (Roche Diagnostic), 100 μg/mL RNase A (Nacalai tesque), and 10 mM $MgCl_2$. Used immediately after preparation. The stock solutions of DNase I and RNase A are separately prepared in PBS at concentrations of 10 mg/mL. The stock aqueous solution of $MgCl_2$ is prepared at a concentration of 50 mM. The stock solutions of DNase I and RNase A are stored at −20 °C. The stock solution of $MgCl_2$ is stored at room temperature.
3. Glutaraldehyde solution: 0.1% glutaraldehyde containing PBS. Dilute 25% glutaraldehyde (Wako) with PBS.
4. Glycine solution: 0.1 M glycine (Wako) containing PBS.
5. Refrigerator.
6. −80 °C freezer.

2.4 Confirmation of Decellularization

1. Glutaraldehyde solution (see above).
2. Glycine solution (see above).
3. Permeabilization solution: 1% Triton X-100 containing PBS.
4. Phalloidin solution: 0.2 unit/mL Alexa Fluor 488 phalloidin containing PBS.
5. Hoechst 33258 solution (see above).
6. Nile red solution (see above).
7. Fluorescent microscope.

3 Methods

All cell culture processes and matrix preparation must be performed under sterile conditions using aseptic techniques. All cell culture incubations were performed in a 37 °C, 5% CO_2 incubator. To identify the stages of osteogenesis and adipogenesis, the expression of several differentiation markers and tissue-specific characters was confirmed, as shown in Table 1.

3.1 Cell Culture for Cell-Derived Decellularized Matrices Formation

3.1.1 MSC Passage Culture

1. Seed MSCs (passage 2) at a density of 5000 cells/cm^2.
2. Culture the MSCs in MSC growth medium.
3. After reaching 70–80% confluence, wash the cells with PBS once and then detach them from the substrata by incubation with 5 mL of trypsin/EDTA for 5 min in a CO_2 incubator.
4. After the confirmation of cell detachment using a phase contrast microscope, add 5 mL of fresh MSC growth medium to the cell suspension to terminate trypsin activity.
5. Centrifuge at 1200 rpm at 4 °C for 5 min.
6. Aspirate the supernatant, and re-suspend the cell precipitation with fresh MSC growth medium.
7. Seed MSCs (passage 3) in MSC growth medium at a density of 5000 cells/cm^2 for further passage culture (see Note 6).

3.1.2 MSC Culture Without Differentiation Induction for the Preparation of Stem Cell Matrices

1. Seed MSCs at passage 4 in desirable sized-TCPS cell culture containers at a density of 5000 cells/cm^2.
2. Culture the cells in undifferentiation culture medium for 1 week (see Note 7).
3. Change the medium to fresh undifferentiation culture medium twice a week.

3.1.3 Osteogenic Differentiation Culture for the Preparation of Stepwise Osteogenesis-Mimicking Matrices

1. Seed MSCs at passage 4 in desirable sized-TCPS cell culture containers at a density of 5000 cells/cm^2.
2. Culture the cells in osteogenic culture medium for either 1 or 3 weeks (see Notes 8 and 9).
3. Change the medium to fresh osteogenic culture medium twice a week.

3.1.4 Adipogenic Differentiation Culture for the Preparation of Stepwise Adipogenic-Mimicking Matrices

1. Seed MSCs at passage 4 in desirable sized-TCPS cell culture containers at a density of 20,000 cells/cm^2.
2. Culture the cells in expansion culture medium for 2 days.
3. Change the medium to adipogenic culture medium after 2 days of expansion culture.

4. Culture the cells in adipogenic culture medium for an additional 3 and 10 days (see Note 10).
5. Change the medium to fresh adipogenic culture medium twice a week.

3.2 Confirmation of Osteogenic and Adipogenic Stages

3.2.1 Isolation of mRNA

1. Add 1 mL of ISOGEN into a culture and scrape the cells to solubilize.
2. Collect the cell lysate in ISOGEN in a fresh 1.5-mL tube.
3. Add 200 μL of chloroform, and mix vigorously.
4. Centrifuge the samples at 12,000 rpm for 15 min at 4 °C.
5. Collect the water phase carefully, and transfer it into a fresh 1.5-mL tube.
6. Add 500 μL of isopropanol, and mix vigorously.
7. Incubate the samples at 4 °C for 20 min.
8. Centrifuge the samples at 12,000 rpm at 4 °C for 10 min.
9. Discard the supernatant, and then add 1 mL of 70% ethanol.
10. Centrifuge the samples at 12,000 rpm at 4 °C for 5 min.
11. Discard the supernatant.
12. Dissolve the precipitation with DEPC-treated water (see Note 11).
13. Measure the absorbance at a wavelength of 260 nm to calculate the concentration of total RNA.

3.2.2 cDNA Synthesis by Reverse Transcription Reaction

1. Add 0.66 μg of isolated total RNA to 1 μL of random hexamer.
2. Add DEPC-treated water to the samples up to 11 μL.
3. Incubate the samples for 10 min at 70 °C.
4. Cool the samples on ice for 5 min.
5. Add 4 μL of First-Strand Buffer, 2 μL of DTT solution, and 2 μL of dNTP solution to the samples.
6. Incubate the samples for 5 min at 42 °C.
7. Add 1 mL of M-MLV Reverse Transcriptase to the samples.
8. Incubate the samples for 50 min at 42 °C.
9. Incubate the samples for 15 min at 70 °C to stop the reaction.
10. Add 1 μL of RNase H solution to the samples.
11. Incubate the samples for 20 min at 37 °C.

3.2.3 Real-Time Polymerase Chain Reaction

1. Dilute the cDNA samples five times with autoclaved deionized water.
2. Add 1 μL of diluted cDNA samples to 96-well PCR plate.
3. Add autoclaved deionized water to specific primers and probe set, and add water up to 11.5 μL.
4. Add 12.5 μL of TaqMan Universal PCR Master Mix (see Note 12).
5. Mix by pipetting gently and seal.
6. Spin-down.
7. Start the reaction (see Note 13).
8. Analyze the date using the comparative Ct method (Figs. 1a, b and 2a. See Note 14).

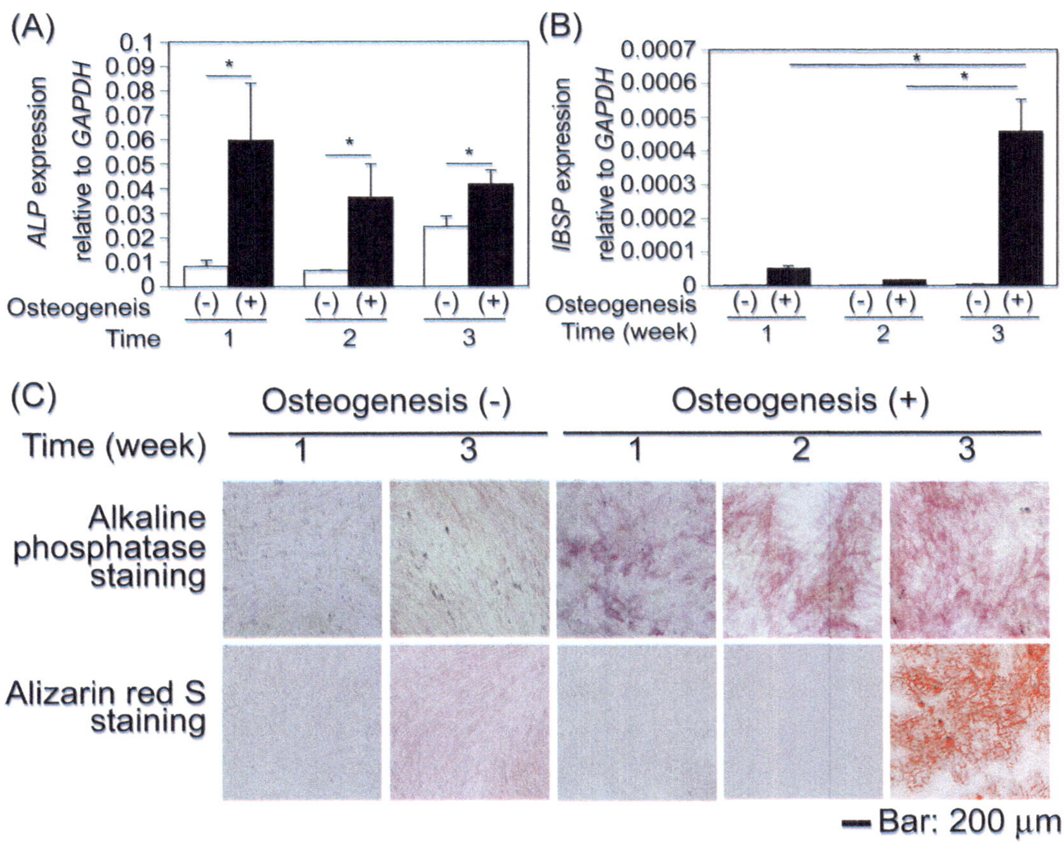

Fig. 1 Confirmation of stepwise osteogenesis of MSCs in osteogenic differentiation culture for the preparation of stepwise osteogenesis-mimicking matrices. (**a**) *ALP* expression level. (**b**) *IBSP* expression level. Data represent the mean ± SD ($n = 3$), *: $P < 0.05$. (**c**) ALP staining. Bar indicates 200 μm. This figure is reproduced from ref. [11] with permission from the American Society for Biochemistry and Molecular Biology and is rearranged for this chapter

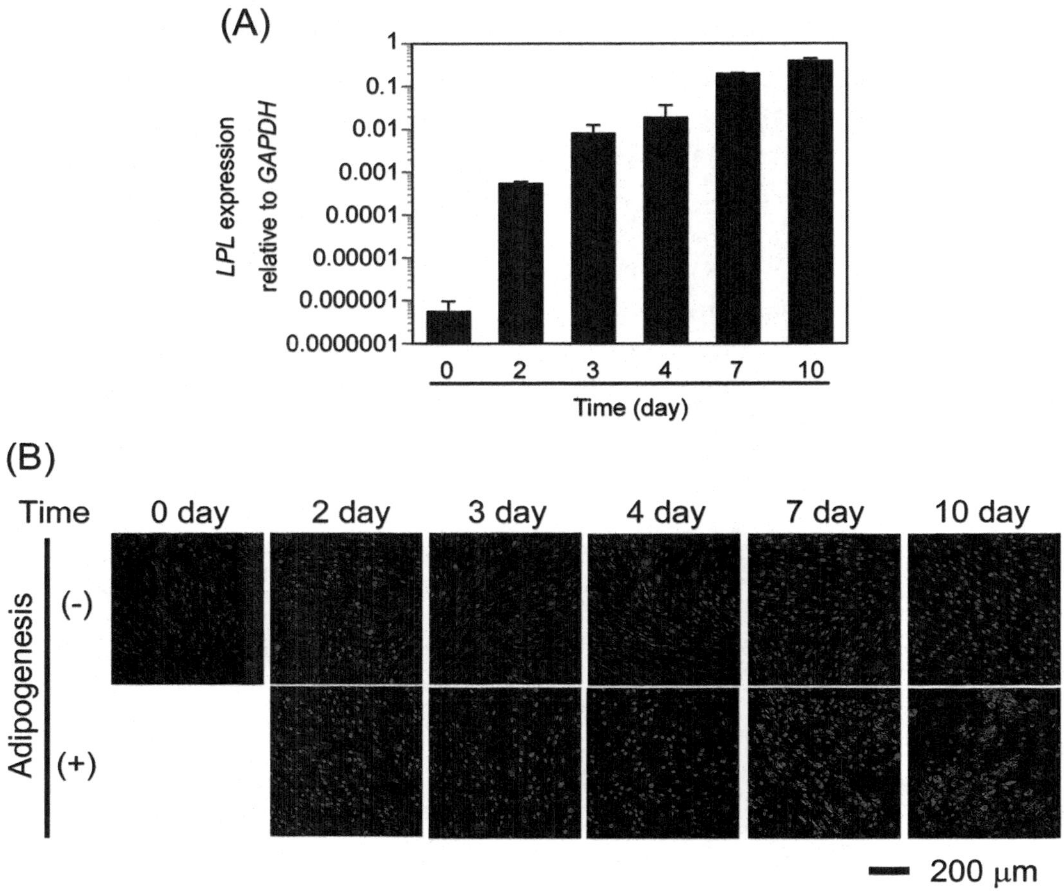

Fig. 2 Confirmation of stepwise adipogenesis of MSCs in adipogenic differentiation culture for the preparation of stepwise adipogenesis-mimicking matrices. (**a**) *LPL* expression level. Data represent the mean ± SD ($n = 3$). (**b**) Nile red staining. Bar indicates 200 μm. This figure is reproduced from ref. [12] with permission from John Wiley and Sons and is rearranged for this chapter

3.2.4 ALP Staining for Detection of the Cell at the Osteogenic Early Stage

1. Aspirate the culture medium, and wash the cells with PBS once.
2. Add formaldehyde solution and incubate the samples at room temperature for 30 min (see Note 15).
3. Aspirate formaldehyde solution, and wash the samples with PBS twice.
4. Add the ALP substrate solution, and incubate the samples for 10 min at room temperature.
5. Wash the samples with deionized water twice, and air-dry them to store (see Note 16).
6. Observe the samples using an optical microscope (Fig. 1c).

3.2.5 Alizarin Red S Staining for the Detection of Cells at the Osteogenic Late Stage

1. Aspirate the culture medium, and wash the cells with PBS once.
2. Add formaldehyde solution, and incubate the samples at room temperature for 30 min.
3. Aspirate formaldehyde solution, and wash the samples with PBS twice.
4. Add alizarin red S solution, and incubate the samples for 30 min at room temperature.
5. Aspirate alizarin red S solution, wash the samples with deionized water twice, and air-dry them to store (see Note 17).
6. Observe the samples with an optical microscope (Fig. 1c).

3.2.6 Nile Red Staining for the Detection of Adipogenic Late Stage

1. Aspirate the culture medium, and wash the cells with PBS once.
2. Add formaldehyde solution, and incubate the samples at 4 °C overnight.
3. Aspirate formaldehyde solution, and wash the samples with PBS twice.
4. Add nile red solution, and incubate the samples for 5 min at room temperature under dark conditions.
5. Aspirate the nile red solution.
6. Add Hoechst 33258 solution, and incubate the samples for 5 min at room temperature under dark conditions.
7. Aspirate the Hoechst 33258 solution, and wash the samples with PBS twice.
8. Observe the samples with a fluorescence microscope (Fig. 2b).

3.3 Decellularization

1. Aspirate the culture medium after the culture for matrices formation (see above).
2. Wash the samples with PBS twice (see Note 18).
3. Add decellularization solution to the samples, and incubate the samples for 5 min at 37 °C.
4. Aspirate the decellularization solution, and wash with PBS twice (see Note 19).
5. Add nuclease solution to the samples, and incubate for 1 h at 37 °C.
6. Aspirate the nuclease solution, and wash with PBS twice.
7. Repeat the freeze-thaw cycle (4 to −80 °C) three times (see Notes 20 and 21).
8. Add glutaraldehyde solution to the samples, and incubate for 6 h at 4 °C.
9. Aspirate the glutaraldehyde solution, and wash with PBS once.
10. Add glycine solution to the samples to quench unreacted aldehyde groups.

11. Incubate the samples at 4 °C for 3 h or longer.
12. Aspirate the glycine solution, and wash with PBS once.
13. Add PBS to the samples to store.
14. Freeze the samples at −80 °C, and store them at −80 °C until use (Note 22).

3.4 Confirmation of Decellularization

1. Culture the cells for the preparation of the samples with and without the treatment with decellularization and nuclease solutions (see Note 23).
2. For the samples with decellularization, perform the decellularization as described above.
3. For the samples without decellularization, wash the cells with PBS twice, and incubate them in glutaraldehyde solution for 6 h at 4 °C (Note 24). After the incubation in glutaraldehyde solution, incubate the samples in glycine solution for at least 3 h at 4 °C.
4. Add permeabilization solution for 2 min at room temperature.
5. Aspirate permeabilization solution, and wash with PBS twice.
6. Add phalloidin solution to the samples, and incubate for 1 h at room temperature under dark conditions.
7. Aspirate the phalloidin solution, and wash with PBS twice.
8. Add Hoechst 33258 solution, and incubate for 5 min at room temperature under dark conditions.
9. Aspirate the Hoechst 33258 solution, and wash with PBS twice.
10. Add nile red solution to the samples, and incubate for 5 min at room temperature under dark conditions (see Notes 25 and 26).
11. Replace the nile red solution with fresh PBS after washing with PBS twice.
12. Observe the samples using a fluorescence microscope (Figs. 3 and 4).

4 Notes

1. Instead of low-glucose DMEM, high-glucose DMEM can be used.
2. For the osteogenesis of human MSCs, medium containing 100 μg/mL is often used. However, the cells tended to shrink and detach from the substrate by the exposure of dexamethasone at a concentration of 100 μg/mL. Thus, 10 μg/mL

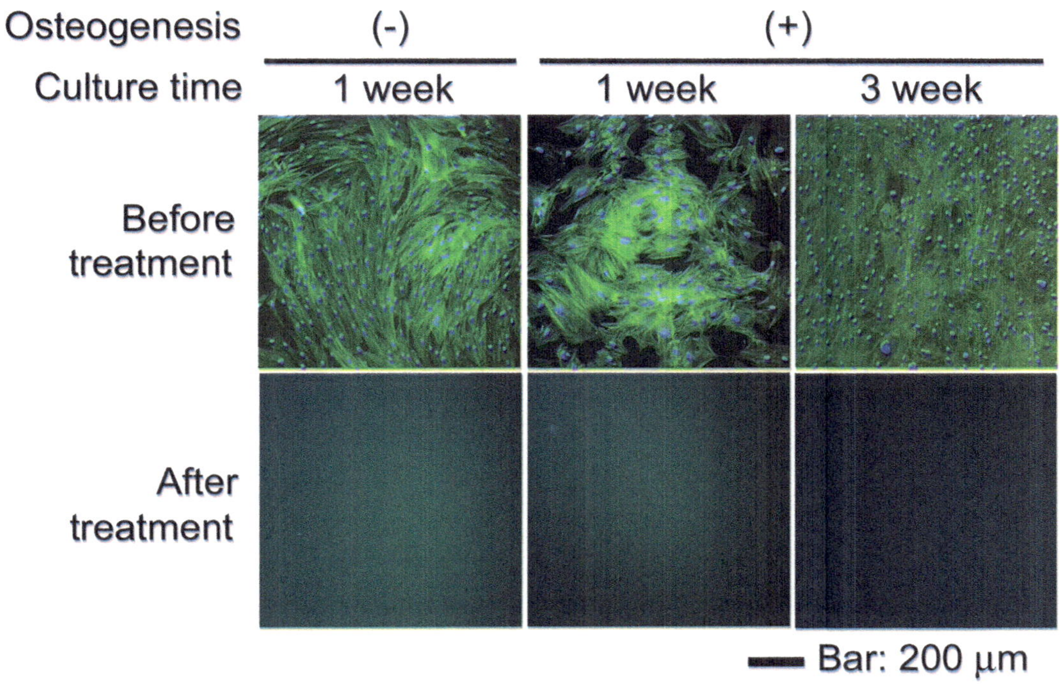

Fig. 3 Confirmation of decellularization of stepwise osteogenesis-mimicking matrices. *Blue* and *green* indicate cell nuclei and actin fibers, respectively. Bar indicates 200 μm. This figure is reproduced from ref. [11] with permission from the American Society for Biochemistry and Molecular Biology

dexamethasone-containing medium is used to prevent cell shrinkage and spontaneous detachment.

3. Other osteogenic and adipogenic differentiation markers may be available. However, the markers should be able to indicate the differentiation stages.
4. Instead of fast red violet LB salt, fast blue BB salt can be used.
5. Instead of nile red staining, oil red O staining can be used to detect lipid droplet accumulation.
6. The cells at passage 3 can also be frozen to store. However, all experiments were performed using the MSCs at passage 4.
7. If MSCs were cultured in high-glucose DMEM-based undifferentiation culture medium (see also Note 1), seed the cells at a density of 20,000 cells/cm^2 and culture for 12 days. The medium should be changed to fresh high-glucose DMEM-based undifferentiation culture medium twice a week.
8. Osteogenic early- and late-stage matrices can be obtained from cells cultured after 1 and 3 weeks, respectively.
9. If the cells experienced spontaneous shrinkage and became detached from the substrate according to the progression of

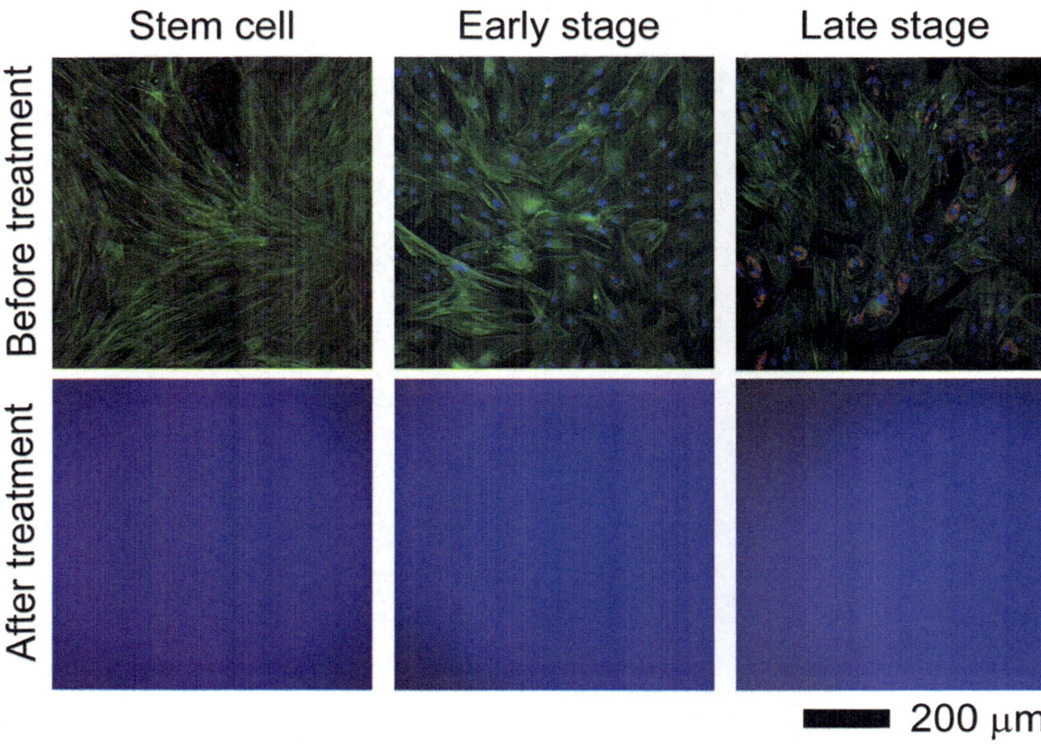

Fig. 4 Confirmation of decellularization of stepwise adipogenesis-mimicking matrices. *Blue*, *green*, and *red* indicate cell nuclei, actin fibers, and lipid droplets, respectively. Bar indicates 200 μm. This figure is reproduced from ref. [12] with permission from John Wiley and Sons

the osteogenesis of MSCs, then the culture can be terminated after a shorter time period (e.g., 2.5 weeks). However, it is necessary to confirm whether the cells are positively stained using alizarin red S staining. Alizarin red S can indicate calcium deposition, an indicator of osteogenic late stage.

10. Adipogenic early- and late-stage matrices can be obtained from cells cultured after 3 and 10 days in adipogenic culture medium, respectively.
11. The volume of DEPC-treated water should be changed according to the estimated amount of total RNA.
12. Final concentrations were as follows: primers: 300 nM and probe: 150 nM. The TaqMan Gene Expression Assay was performed according to the manufacturers' recommendations.
13. Reaction conditions were as follows: Step 1: 42 °C for 10 min, and Step 2: 95 °C for 30 s and 60 °C for 30 s (repeat 40 times).
14. 18S rRNA expression is used as an endogenous control. Gene expression levels are expressed as the values relative to *GAPDH* expression.

15. Longer incubation with formaldehyde solution causes a decrease in ALP activity and results in weak staining.
16. After incubation with ALP substrate solution, a red precipitate will develop in the samples. Thus, the samples can be directly observed in PBS or deionized water after washing with either PBS or deionized water. In this case, the air-dry step can be skipped.
17. After the incubation with alizarin red S solution, a red color will develop in the samples. Thus, the samples can be directly observed in PBS or deionized water after washing with either PBS or deionized water. In this case, the air-dry step can be skipped.
18. PBS, decellularization solution, and nuclease solution should be warmed at 37 °C before starting the decellularization.
19. If Triton X-100 remained in the samples, then it will inhibit the activities of the nucleases. Thus, Triton X-100 should be carefully washed from the samples.
20. Lipid droplets cannot be removed from the samples completely by treatment with the decellularization and nuclease solutions. Freeze-thaw cycles are required to remove lipid droplets in the cells at the adipogenic late stage. For the preparation of stepwise osteogenesis-mimicking matrices, stem cell matrices, and adipogenesis early-stage matrices, this step can be skipped.
21. The decellularized matrices are sometimes detached from the substrates. In this case, such samples cannot be used for further experiments.
22. It is not clear how long the samples can be stored at −80 °C. In our case, the samples can be used for further experiments after storage for 6 months or a longer period.
23. The samples without decellularization are used as experimental controls. These samples should be prepared by the cells cultured under same conditions for the preparation of corresponding matrices.
24. PBS for washing should be warmed at 37 °C prior to the treatment.
25. This step can be skipped for the confirmation of decellularization in stepwise osteogenesis-mimicking matrices.
26. This step can be performed at the same time as the incubation with Hoechst 33258 solution.

Acknowledgments

This work was supported by a Grant-in-Aid for Young Scientists (A) (26702016) and a Grant-in-Aid for Challenging Exploratory Research (16K15209), funded by MEXT, Japan.

References

1. Hynes RO (2009) The extracellular matrix: not just pretty fibrils. Science 326:1216–1219
2. Giancotti FG, Ruoslahti E (1999) Integrin signaling. Science 285:1028–1032
3. Manabe R, Tsutsui K, Yamada T, Kimura M, Nakano I, Shimono C, Sanzen N, Furutani Y, Fukuda T, Oguri Y, Shimamoto K, Kiyozumi D, Sato Y, Sado Y, Senoo H, Yamashina S, Fukuda S, Kawai J, Sugiura N, Kimata K, Hayashizaki Y, Sekiguchi K (2008) Transcriptome-based systematic identification of extracellular matrix proteins. Proc Natl Acad Sci U S A 105:12849–12854
4. Daley WP, Peters SB, Larsen M (2008) Extracellular matrix dynamics in development and regenerative medicine. J Cell Sci 121:255–264
5. Page-McCaw A, Ewald AJ, Werb Z (2007) Matrix metalloproteinases and the regulation of tissue remodelling. Nat Rev Mol Cell Biol 8:221–233
6. Ioachim E, Charchanti A, Briasoulis E, Karavasilis V, Tsanou H, Arvanitis DL, Agrantis NJ, Pavlidis N (2002) Immunohistochemical expression of extracellular matrix components tenascin, fibronectin, collagen type IV and laminin in breast cancer: their prognostic value and role in tumour invasion and progression. Eur J Cancer 38:2362–2370
7. Kwon S-Y, Chae SW, Wilczynski SP, Arain A, Carpenter PM (2012) Laminin 322 expression in breast carcinoma. Appl Immunohistochem Mol Morphol 20:159–164
8. Hoshiba T, Lu H, Kawazoe N, Chen G (2010) Decellularized matrices for tissue engineering. Expert Opin Biol Ther 10:1717–1728
9. Hoshiba T, Chen G, Endo C, Maruyama H, Wakui M, Nemoto E, Kawazoe N, Tanaka M (2016) Decellularized extracellular matrix (ECM) as an in vitro model to study the comprehensive roles of the ECM in stem cell differentiation. Stem Cells Int. Article ID 6397820 doi: 10.1155/2016/6397820
10. Kerever A, Schnack J, Vellinga D, Ichikawa N, Moon C, Arikawa-Hirawasa E, Efird JT, Mercier F (2007) Novel extracellular matrix structures in the neural stem cell niche capture the neurogenic factor fibroblast growth factor from the extracellular milieu. Stem Cells 25:2146–2157
11. Hoshiba T, Kawazoe N, Tateishi T, Chen G (2009) Development of stepwise osteogenesis-mimicking matrices for the regulation of mesenchymal stem cell functions. J Biol Chem 284:31164–31173
12. Hoshiba T, Kawazoe N, Tateishi T, Chen G (2010) Development of extracellular matrices mimicking stepwise adipogenesis of mesenchymal stem cells. Adv Mater 22:3042–3047
13. Hoshiba T, Kawazoe N, Chen G (2011) Mechanism of regulation of PPARG expression of mesenchymal stem cells by osteogenesis-mimicking extracellular matrices. Biosci Biotechnol Biochem 75:2099–2104
14. Hoshiba T, Kawazoe N, Chen G (2012) The balance of osteogenic and adipogenic differentiation in human mesenchymal stem cells by matrices that mimic stepwise tissue development. Biomaterials 33:2025–2031

Methods in Molecular Biology (2018) 1577: 87–93
DOI 10.1007/7651_2017_28

Published online: 28 April 2017

Mouse Skeletal Muscle Decellularization

Martina Piccoli, Caterina Trevisan, Edoardo Maghin, Chiara Franzin, and Michela Pozzobon

Abstract

Natural acellular matrices obtained from decellularization procedures are biocompatible and non-immunogenic materials considered promising tools for regenerative medicine purposes. Before in vivo implantation, these matrices must be efficiently decellularized, removing all the cellular components to avoid any immunogenic reaction. At the same time, it is important to maintain the original three-dimensional structure of the specific tissue. Here we describe a method: (1) to decellularize mouse quadriceps using a detergent-enzymatic treatment (DET) and (2) to assess decellularization efficiency and scaffold properties.

Keywords: Decellularization, Extracellular matrix, Mouse skeletal muscle, Volumetric muscle loss, Tissue engineering

1 Introduction

The extracellular matrix (ECM) is a very complex structure composed of functional and structural proteins, glycosaminoglycans (GAG), glycoproteins, and other small molecules secreted by tissue-resident cells [1]. ECM is fundamental to provide structural support to the tissues and it is also a cell behavior key regulator [2]. It can be obtained through decellularization of tissues or organs, a procedure that aims at removing all the donor cellular components, maintaining the original three-dimensional structure and composition [3]. There are several methods for decellularizing tissues, like physical, chemical, and enzymatic, and they can be combined, depending on the tissue type, in order to obtain the highest decellularization efficiency [4]. In a previous work, we obtained and characterized a decellularized mouse diaphragmatic muscle using a detergent-enzymatic treatment (DET) [5]. Other decellularization methods have been applied for skeletal muscles such as *tibialis anterior* [6, 7] and *cricoarytenoid dorsalis* [8]. The naturally derived acellular matrices are biocompatible materials that are considered suitable tools for tissue engineering applications [1]. The ECM, alone or repopulated with cells, has been applied for the regeneration of several tissues such as trachea [9, 10] and skin [11]. In addition, ECM derived from different sources, like porcine small intestinal submucosa or porcine urinary bladder

[12, 13], has been used specifically for skeletal muscle regeneration after traumatic- or surgery-related volumetric muscle loss (VML) in which there is a depletion of skeletal muscle tissue with resultant functional impairment. We have considered the skeletal muscle-derived ECM as a tissue specific scaffold suitable for stimulating muscle regeneration after VML. Here we describe a protocol to decellularize mouse quadriceps and to assess the efficiency of the procedure by: (1) quantifying the DNA amount and remaining nuclei after decellularization, (2) analyzing the ECM structure through histological stain, immunofluorescence, scanning electron microscopy (SEM), and (3) quantifying the ECM components before and after decellularization.

2 Materials

2.1 Equipment and Supplies

1. Biologic safety cabinet
2. Serological pipettes (5, 10, and 25 mL)
3. 20, 200, and 1,000 μL micropipettes
4. 20, 200, and 1,000 μL micropipette tips
5. 1.5 mL microcentrifuge tubes
6. Polystyrene conical tubes (15 and 50 mL)
7. Nanodrop 2000 spectrophotometer (Thermo Scientific) or comparable for DNA quantification
8. Electrophoresis chamber and power supply (Bio-Rad) for agarose gel electrophoresis
9. Microscope slides (frosted end)
10. Cryostat (Leica CM1520) or comparable for tissue slide preparation
11. Stove for immunostaining
12. Inverted light microscope (Olympus IX71) or comparable for histological analyses
13. Inverted fluorescent microscope (Leica DMI6000B) or comparable for immunofluorescent analyses
14. Scanning electron microscope (JEOL JSM-6490) or comparable for ultrastructure analyses
15. Spectrophotometer (SpectraMax Plus) or comparable for protein quantification

2.2 Decellularization Reagents

1. 1× sterile phosphate-buffered saline (PBS)
2. Penicillin/streptomycin (pen/strep) (GIBCO, Thermo Fisher Scientific)

3. 4 % sodium deoxy cholate (SDC) (Sigma-Aldrich)
4. 2,000 kU Deoxyribonuclease 1 (DNase 1) (Sigma-Aldrich) in 1 M NaCl solution

2.3 DNA Extraction and Quantification

1. DNeasy Blood & Tissue kit (Qiagen) for DNA extraction
2. 1 % agarose gel (in 1× TBE) for electrophoresis. Running buffer is 1× TBE: 0.5 M Tris base, 0.5 M boric acid, 0.01 M EDTA (disodium salt)
3. SYBR® Safe DNA gel stain (Thermo Fisher Scientific) for visualization of DNA in agarose gel

2.4 Samples Fixing and Sectioning

1. 4 % paraformaldehyde (PFA) for fixation
2. 10, 15, and 30 % sucrose solutions for dehydration
3. Optimal cutting temperature (O.C.T., Kaltek) for embedding
4. Sectioning samples in 6–8 μm slices using a cryostat

2.5 Immunofluorescence Analyses

1. 0.5 % Triton X-100 in 1× PBS for permeabilization
2. 10 % horse serum in 1× PBS for saturation
3. Primary antibody: polyclonal rabbit anti-mouse laminin (Sigma-Aldrich), dilution 1:200. Use 1 % bovine serum albumin (BSA) in 1× PBS for dilution
4. Secondary antibody: Alexa Fluor chicken anti-rabbit 488 (Life Technologies), dilution 1:200 in 1 % BSA.
5. Fluorescent mounting medium plus 100 ng/ml 4′,6-diamidino-2-phenylindole (DAPI, Sigma) for nuclei counterstaining

2.6 Histological Analyses

1. Hematoxylin & Eosin (H&E) kit for rapid frozen section (Bio-Optica).

2.7 Scanning Electron Microscopy

1. 4 % PFA for 24 h at 4 °C for fixation
2. Ethanol gradient (30–50–70–90–96–100 %) for dehydration
3. A gold layer for sample coating using an ion beam coater

2.8 Extracellular Matrix Component Quantification

1. Sircol™ Soluble Collagen Assay (Biocolor) for collagen quantification
2. Blyscan GAG Assay Kit (Biocolor) for sulphated glycosaminoglycan (sGAG) quantification.

3 Methods

3.1 Decellularization Procedure

1. Collect quadriceps muscles and carry them in sterile 1× PBS to the laboratory processing room (*see* **Note 1**).
2. Quickly rinse the muscles in 1× PBS with 3 % pen/strep.

3. Transfer the samples on a petri dish using sterile disposable plastic pincers and cut each into two halves following fibers direction (*see* **Note 2**).
4. Put the samples in 40 mL of sterile and deionized water with 3 % pen/strep at 4 °C, for 24 h (*see* **Note 3**).
5. Put samples in 30 mL of 4 % SDC for 4 h at room temperature (RT) and in gentle agitation (*see* **Note 4**). SDC is used to disrupt cell membranes.
6. Wash samples from three to five times with sterile and deionized water, to remove the detergent.
7. Put samples in 40 mL of 2,000 kU DNase 1 in 1 M NaCl for 3 h at RT and in gentle agitation in order to breakdown DNA.
8. Repeat the steps from point 5 to point 8 other two times.
9. Wash samples for at least 3 days in sterile 1× PBS and 3 % pen/strep (*see* **Note 5**).
10. Freeze in liquid nitrogen samples dedicated for DNA and ECM components quantification or fix in 4 % PFA, dehydrate in increasing sucrose gradient, embed in O.C.T., and then section for histological, immunofluorescence analyses or alternatively dehydrate with ethanol gradient and process samples for SEM.

3.2 Gross Appearance, DNA, and Nuclei Quantification

1. Morphological aspect and nuclei quantification (Fig. 1a, c): after tissue (fresh quadriceps, FQ, and decellularized quadriceps, DQ) sectioning, proceed with the immunostaining protocol using the antibodies described in the items 3, 4, and 5 of Sect. 2.5 (incubate laminin and secondary antibody for 1 h at 37 °C). For nuclei counting, perform counts on five random pictures of up to three independent experiments of both samples.
2. DNA quantification (Fig. 1b): proceed with DNA isolation protocol following manufacturer's instruction of item 1 of Sect. 2.3 dissolving about 25 mg of tissue. After decellularization, remaining genetic material should be less than 50 ng DNA/mg tissue and if DNA fragments are present, they should have a maximum length of 200 bp [14]. Use items 7 and 8 of Sect. 2.1 and items 2 and 3 of Sect. 2.3 to quantify DNA amount and fragments length.

3.3 Extracellular Matrix Three-Dimensional Structure and Components Quantification

1. Cross sectional area (CSA) and number of fibers evaluation (Fig. 1d, e): calculate the mean fiber CSA of both FQ and DQ samples measuring fibers area on five random pictures of up to three independent experiments. On the same slide, count the number of fibers per field.
2. ECM three-dimensional structure analysis (Fig. 2a, b): through item 1 of Sect. 2.6, determine FQ and DQ morphological aspect and confirm nuclei depletion after decellularization. With SEM analysis, evaluate the ultrastructure maintenance of decellularized samples.

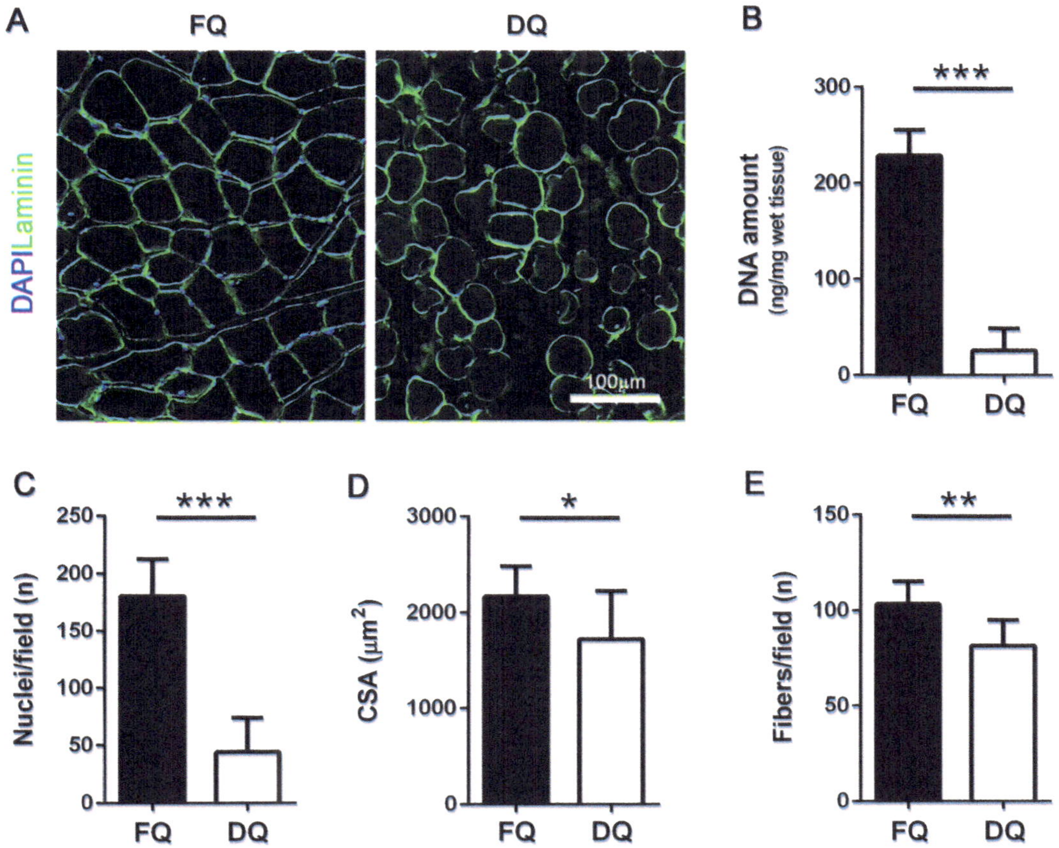

Fig. 1 Decellularization efficiency. (**a**) Representative images of fresh and decellularized samples. (**b**) DNA amount quantification. (**c**) Quantification of the number of nuclei per field. (**d**) Evaluation of cross sectional area (CSA) of muscle fibers. (**e**) Quantification of the number of fibers per field. * $p < .05$; ** $p < .01$; *** $p < .001$

3. ECM components quantification (Fig. 2c, d): using items 1 and 2 of Sect. 2.8 to quantify collagen and sGAG in FQ and DQ samples, in order to verify the ECM integrity not only from the morphological point of view but also from the quality of retained proteins.

4 Notes

1. If it is not possible to decellularize the sample immediately, it can be stored in sterile deionized water with 3 % pen/strep at 4 °C not longer than 72 h.
2. Note that this protocol has been optimized using a piece of muscle weighting around 250 mg (generally half quadriceps).
3. Sterility could represent a very important issue when ECM matrices are repopulated with cells or used in vivo. In order to use

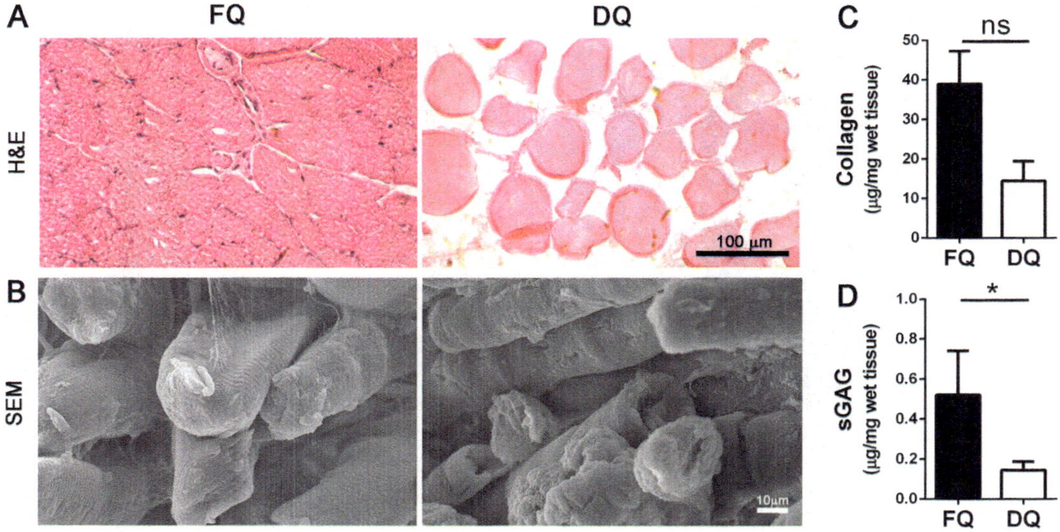

Fig. 2 Extracellular matrix (ECM) preservation. (**a**) Representative images of Hematoxylin & Eosin (H&E) stain on fresh and decellularized samples. (**b**) Scanning electron microscopy (SEM) images of fresh and decellularized quadriceps. (**c**) Quantification of collagen on muscle samples. (**d**) Quantification of sulphated glycosaminoglycan (sGAG) of fresh and decellularized muscles. * $p < .05$

ECM as scaffold to seed cells, it is mandatory to work in aseptic conditions with sterile solutions and material. To this end, the procedure should be performed using clean supplies from the specimen collection (i.e. remove the hairs and the skin from the mouse hindlimbs) to the decellularization process (i.e., sterile reagents and environment).

4. In order to obtain an efficient decellularization, do not use more than six halves quadriceps in a volume of 30 mL of SDC or 40 mL of DNase 1.

5. Washing step is mandatory to obtain a suitable sample both for in vitro and in vivo tests. If necessary, extend the washing period by changing the solution twice a day.

Acknowledgments

Piccoli M and Franzin C are supported by Istituto di Ricerca Pediatrica Città della Speranza, Grant Number 16/02; Trevisan C and Maghin E are supported by Istituto di Ricerca Pediatrica Città della Speranza. Pozzobon M is supported by University of Padova, Grant number GRIC15AIPF, Assegno di ricerca Senior. Franzin C, Piccoli M, and Pozzobon M are coinventors of Italian Patent N. 0001422436 entitled "Matrice acellulare per ricostruzione in vivo di muscolo scheletrico."

References

1. Badylak SF (2004) Xenogeneic extracellular matrix as a scaffold for tissue reconstruction. Transpl Immunol 12:367–377. doi:10.1016/j.trim.2003.12.016
2. Daley WP, Peters SB, Larsen M (2008) Extracellular matrix dynamics in development and regenerative medicine. J Cell Sci 121:255–264. doi:10.1242/jcs.006064
3. Badylak SF, Taylor D, Uygun K (2011) Whole-organ tissue engineering: decellularization and recellularization of three-dimensional matrix scaffolds. Annu Rev Biomed Eng 13:27–53. doi:10.1146/annurev-bioeng-071910-124743
4. Gilbert TW, Sellaro TL, Badylak SF (2006) Decellularization of tissues and organs. Biomaterials 27:3675–3683. doi:10.1016/j.biomaterials.2006.02.014
5. Piccoli M, Urbani L, Alvarez-Fallas ME, Franzin C, Dedja A, Bertin E, Zuccolotto G, Rosato A, Pavan P, Elvassore N, De Coppi P, Pozzobon M (2016) Improvement of diaphragmatic performance through orthotopic application of decellularized extracellular matrix patch. Biomaterials 74:245–255. doi:10.1016/j.biomaterials.2015.10.005
6. Gillies AR, Smith LR, Lieber RL, Varghese S (2011) Method for decellularizing skeletal muscle without detergents or proteolytic enzymes. Tissue Eng Part C Methods 17:383–389. doi:10.1089/ten.TEC.2010.0438
7. Kasukonis BM, Kim JT, Washington TA, Wolchok JC (2016) Development of an infusion bioreactor for the accelerated preparation of decellularized skeletal muscle scaffolds. Biotechnol Prog 32:745–755. doi:10.1002/btpr.2257
8. Fishman JM, Lowdell MW, Urbani L, Ansari T, Burns AJ, Turmaine M, North J, Sibbons P, Seifalian AM, Wood KJ, Birchall MA, De Coppi P (2013) Immunomodulatory effect of a decellularized skeletal muscle scaffold in a discordant xenotransplantation model. Proc Natl Acad Sci U S A 110:14360–14365. doi:10.1073/pnas.1213228110
9. Elliott MJ, De Coppi P, Speggiorin S, Roebuck D, Butler CR, Samuel E, Crowley C, McLaren C, Fierens A, Vondrys D, Cochrane L, Jephson C, Janes S, Beaumont NJ, Cogan T, Bader A, Seifalian AM, Hsuan JJ, Lowdell MW, Birchall MA (2012) Stem-cell-based, tissue engineered tracheal replacement in a child: a 2-year follow-up study. Lancet 380:994–1000. doi:10.1016/S0140-6736(12)60737-5
10. Macchiarini P, Jungebluth P, Go T, Asnaghi MA, Rees LE, Cogan TA, Dodson A, Martorell J, Bellini S, Parnigotto PP, Dickinson SC, Hollander AP, Mantero S, Conconi MT, Birchall MA (2008) Clinical transplantation of a tissue-engineered airway. Lancet 372:2023–2030. doi:10.1016/S0140-6736(08)61598-6
11. Wainwright DJ (1995) Use of an acellular allograft dermal matrix (AlloDerm) in the management of full-thickness burns. Burns 21:243–248
12. Sicari BM, Agrawal V, Siu BF, Medberry CJ, Dearth CL, Turner NJ, Badylak SF (2012) A murine model of volumetric muscle loss and a regenerative medicine approach for tissue replacement. Tissue Eng Part A 18:1941–1948. doi:10.1089/ten.TEA.2012.0475
13. Sicari BM, Rubin JP, Dearth CL, Wolf MT, Ambrosio F, Boninger M, Turner NJ, Weber DJ, Simpson TW, Wyse A, Brown EH, Dziki JL, Fisher LE, Brown S, Badylak SF (2014) An acellular biologic scaffold promotes skeletal muscle formation in mice and humans with volumetric muscle loss. Sci Transl Med 6:234ra258. doi:10.1126/scitranslmed.3008085
14. Crapo PM, Gilbert TW, Badylak SF (2011) An overview of tissue and whole organ decellularization processes. Biomaterials 32:3233–3243. doi:10.1016/j.biomaterials.2011.01.057

Methods in Molecular Biology (2018) 1577: 95–102
DOI 10.1007/7651_2017_31

Published online: 30 April 2017

Extracellular Matrix from Whole Porcine Heart Decellularization for Cardiac Tissue Engineering

Matthew J. Hodgson, Christopher C. Knutson, Nima Momtahan, and Alonzo D. Cook

Abstract

Decellularization of whole porcine hearts followed by recellularization with fully differentiated cells made from patient-specific human induced pluripotent stem cells (iPSCs) may provide the ultimate solution for patients with heart failure. Decellularization is the process of completely disrupting all cells and removing the cellular components (e.g., antigenic proteins, lipids, DNA) from organic tissue, leaving only the extracellular matrix (ECM). The decellularization of porcine hearts can be accomplished in 24 h and results in 98% DNA removal with only 6 h of detergent exposure. Automatically controlling the pressure during decellularization reduces the detergent exposure time while still completely removing immunogenic cell debris.

Keywords Acellular biological matrices, Automation, Cardiomyocytes, Decellularization, Differentiation, Extracellular matrix, Heart, Induced pluripotent stem cells

1 Introduction

The cardiac extracellular matrix (ECM) consists primarily of proteins such as collagen and elastin that provide a mechanical scaffold for homing and 3D functioning of the beating heart muscle cells (cardiomyocytes). The ECM also contains glycosaminoglycans (GAGs) and growth factors that assist in signaling and communication between the cells and promote cell proliferation and differentiation. The mechanical properties of the ECM can be altered by injury or diminished by environmental factors and age [1]. When this occurs it is necessary to surgically repair or replace the damaged heart. Several strategies have been proposed to find alternatives to donated human heart transplants, including decellularization of porcine hearts followed by recellularization with patient-specific human cardiomyocytes made from induced pluripotent stem cells (iPSCs) [2].

The heart is uniquely suited for decellularization because retrograde perfusion through the aorta can facilitate vascular perfusion [3]. The decellularization process removes cellular components and factors that may induce a chronic immune response, such as DNA or GAL epitopes, while the non-immunogenic matricellular proteins and growth factors of the ECM are preserved [4]. A key challenge in decellularization is to avoid damage to the ECM from overexposure of common cell removal reagents such as the detergent sodium dodecyl sulfate (SDS). Although SDS is considered one of the most effective detergents for removing cells and immunogenic components from the ECM, excess exposure creates a risk of removing the desirable GAGs and growth factors as well as denaturing ligands and proteins such as collagen, elastin, and fibronectin that are necessary for mechanical support, cell attachment, and tissue regeneration [5, 6].

Decellularization is accomplished in multiple stages. The initial stage of decellularization consists of perfusion of phosphate-buffered saline (PBS) alternating with distilled water, resulting in a hypotonic shock which prepares the cell membranes for lysis. After this initial stage, the heart is perfused with alternating cycles of SDS and distilled water in the decellularization apparatus. The SDS is the main decellularizing agent and water serves to wash out residual SDS and cellular debris. Following the cycles of SDS, the heart is perfused with Triton X-100, another lysing agent, and water.

Automated systems assist in standardizing the decellularization process [7] and a pressure controlled system can assure effective perfusion through the coronary vessels during decellularization [8, 9]. Following decellularization, assays can be performed to measure residual DNA, collagen, and glycosaminoglycans [9]. Decellularized porcine cardiac tissue scaffolds should preserve the vasculature, mechanical integrity, and biocompatibility while also reducing the magnitude of an immunogenic response after transplantation when seeded with terminally differentiated autologous human induced pluripotent stem cells (iPSCs) [10, 11]. The following protocol has been successful in producing decellularized whole porcine hearts.

2 Materials

Unless otherwise stated, prepare all solutions using type 1 distilled water and analytical grade reagents. Use an analytical scale to weigh amounts of reagents needed for solutions.

1. 1 L 5.0% wt/vol SDS (see Note 1): Fill a 2 L beaker with 500 mL of water, add a stir bar, and place on stir plate. Weigh 50.0 g SDS (Fisher Scientific, UN2926 Fair Lawn, NJ), or 60.0 g if SDS is 85% pure. Add the measured SDS to the water while stirring. Fill the beaker to 1 L and continue stirring. 3 L will be needed to complete the decellularization process. Store at room temperature.

2. 4 L of 10× PBS: Fill a 4 L beaker with 3.5 L of deionized water, add a stir bar, and place on stir plate. Measure the following salts on the analytical balance using weigh boats: 320.0 g NaCl, 8.0 g KCl, 108.8 g $Na_2HPO_4\,(H_2O)_7$, and 9.6 g KH_2PO_4. Add all of the salts to the water while stirring. Add 5 mL of 50 wt/wt% NaOH solution. Add water to a total volume of 4.0 L. Only 1 L will be needed for each decellularization process. The remainder can be used at the slaughterhouse to rinse out the porcine hearts. Store at room temperature.
3. Triton X-100: 10% Solution in 1 L of water. Measure 100 mL Triton X-100 (Fisher Scientific, BP151-500) in a 2 L beaker with 500 mL of water, add a stir bar, and place on a hot plate at 50 °C. After solution is well mixed, fill the beaker to 1 L and continue stirring. Store at room temperature.
4. Whole Porcine Heart: Use porcine hearts harvested from slaughter-weight, 6-months old, mature female Hampshire, Yorkshire or Spot crossbreed swine at a local slaughterhouse according to approved protocols for safety and animal care, or other types of porcine hearts may also be used. The following materials are used in slaughterhouse preparation; 1 L 10× PBS (see PBS, step 2 above), 3 glass vials of 50,000 U heparin in 10 mL, 1× Antibiotic solution (100 U/mL penicillin, 100 μg/mL streptomycin, 25 μg/mL amphotericin B, Anti-Anti, Gibco®, Billings, MT), a 1-gal pressure sprayer, sleeve gloves, examination gloves, scissors, and scalpels.
5. A Langendorff system based on Radnoti® (Monrovia, CA) porcine working heart system can be modified to be used as the decellularization apparatus (see Fig. 1). The modified apparatus should consist of a 5 L reservoir for the perfusate solution, a bubble trap, and the main 5 L decellularization chamber to submerge the heart in the perfusate solution. A peristaltic pump (Masterflex® HV-07557-00, Vernon Hills, IL) with Tygon L/S 36 tubing (Masterflex® E-3603) is used to produce a pressurized pulsatile flow [9].
6. A negative feedback control system is implemented to maintain critical parameters during the decellularization process. The system consists of a computer running National Instruments® (Austin, TX) LabView (2013) software, and containing hardware to collect pressure data while sending control signals to the pump. The pressure controller is designed to control the pressure in the bubble trap. The pressure gauge is placed in line of the flow system at the bubble trap, right before the flow enters the heart through the aorta [9]. A proportional–integral–derivative (PID) control algorithm monitors the pressure signal every tenth of a second and adjusts the pump signal to maintain the desired pressure. The program also includes functions for running timed experiments, automatically adjusting pressure

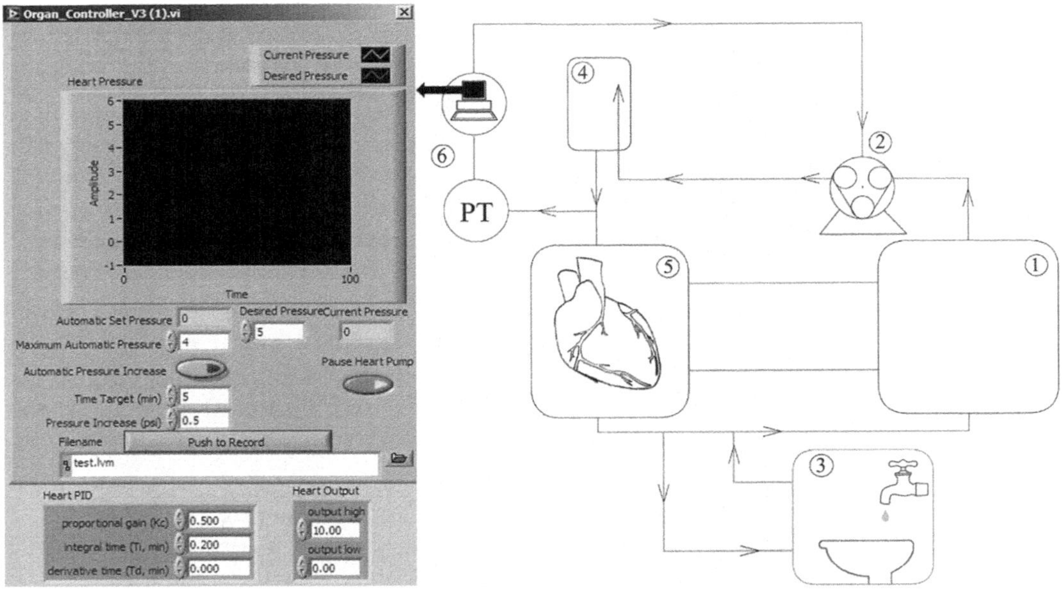

Fig. 1 Automated controller interface (*left*) and schematic view of the automated pressure controlled bioreactor for porcine heart decellularization (*right*). (1) Solution reservoir; (2) peristaltic pump; (3) faucet and sink; (4) bubble trap; (5) decellularization chamber; (6) control loop

levels, and switching fluid streams via electrically controlled solenoid valves, as well as a manual control mode.

3 Methods

3.1 Acquiring a Heart

1. Obtain porcine hearts from a local slaughterhouse, or buy from an online supplier. To harvest a heart from swine after exsanguination, open the chest cavity using a 12″ Finochietto retractor and dissect the heart from the lung while preserving as much of the aorta, pulmonary artery, vein, and caudal and cranial vena cava as possible.
2. Immediately after removing the heart, use a pressurized hand action pump filled with a solution of 1× PBS, heparin (10 U/mL) and 1× antibiotic solution to rinse all the chambers and vessels of the heart via antegrade perfusion through the vena cava and pulmonary vein and retrograde perfusion through the aorta.
3. Manually massage the hearts while they are being washed to remove any static blood in the coronary vessels to prevent further thrombosis.
4. Transport hearts in cold heparinized (10 U/mL) type 1 distilled water and store frozen at −20 °C.

3.2 Preparing to Decellularize

1. Thaw the heart at 4 °C overnight and remove from the 10 U/mL heparinized water (see Notes 2 and 3).
2. Trim the excess fat and separate the aorta from the pulmonary artery at the brachiocephalic trunk.
3. Partially fill the decellularization chamber and solution reservoir with type 1 distilled water. Add 500 mL of 10× PBS to the reservoir and 500 mL to the chamber of the decellularization apparatus. Fill the remaining volume with type 1 distilled water while still leaving sufficient volume for the heart to be placed in the decellularization chamber.
4. Cannulate the aorta with a Radnoti® (Monrovia, CA) 1/2″ male tube and connect the heart to the apparatus (see Note 4).
5. Submerge the heart in the 1× PBS solution to avoid axial stress on the aorta due to the heart's weight.
6. Prepare the software program by adjusting its settings so that the desired pressure is 0 PSI, with a maximum pressure of 5 PSI, a pressure increase of 0.3 PSI and interval to 5 min.

3.3 Decellularization Process

1. After attaching the heart and adding PBS solution, run the pump in the reverse direction to remove any air bubbles from within the heart. This should be done briefly for no longer than 20 s.
2. Begin by perfusing 1× PBS through the coronary vessels to remove the blood and thromboemboli for 1 h. During the 1 h of PBS perfusion, the pressure is gradually increased to 4 psi using the automated system. The system must be on a closed loop (see Note 5).
3. Replace the PBS solution with non-recycled type 1 distilled water by placing the system on an open loop (see Notes 6 and 7).
4. Perfuse the heart with distilled water at 5 psi for 1 h (see Note 8).
5. To prepare for the following step, remove or turn off the source of non-recycled type 1 distilled water. While still allowing water to exit through the decellularization chamber's tubing, kink the tubing so as to stop the flow and place the tubing in the solution reservoir. Once submerged, you may unkink the tubing. This prevents air bubbles from being trapped in the tubing and permits unobstructed flow.
6. Add 500 mL of 5.0% SDS in the reservoir container and 500 mL to the main chamber of the decellularization apparatus.
7. Perfuse the heart with SDS for 1 h at 5 psi. This must be done on a closed loop.

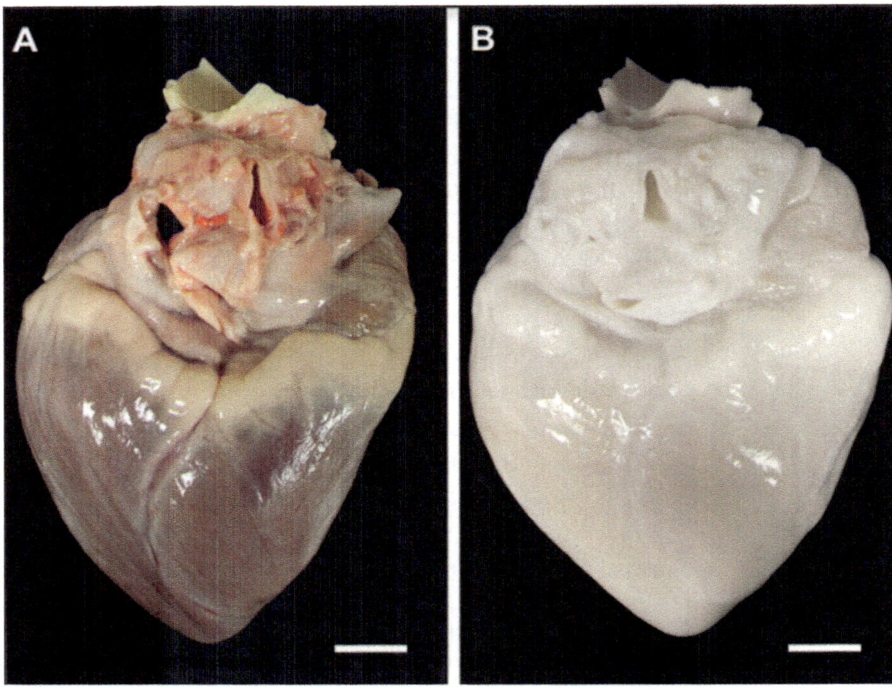

Fig. 2 Fresh porcine heart (**a**). Decellularized porcine heart (**b**)

8. Replace the SDS solution with non-recycled type 1 distilled water, similar to step 3, placing the system on an open loop.
9. Repeat steps 4–6.
10. Perfuse the heart with SDS for 2 h at 5 psi. This must be done on a closed loop.
11. Replace the SDS solution with non-recycled type 1 distilled water, similar to step 3, placing the system on an open loop.
12. Repeat again steps 4–6.
13. Perfuse the heart with SDS for 3 h at 5 psi. This must be done on a closed loop.
14. After the sixth total hour of SDS exposure, perfuse the heart again with distilled water for 4 h at 5 psi on an open loop.
15. After placing the system on a closed loop, replace the water with 10 L 1% (v/v) Triton X-100. This is done by adding 500 mL of 10% solution to both the main chamber and reservoir. Perfuse the heart while automatically increasing pressure to 5 psi for 2 h.
16. Finish the decellularization with an extensive wash of the heart with distilled water at 5 psi for at least 4 h (see Notes 9 and 10). The decellularized porcine heart should appear white, as shown in Fig. 2.

4 Notes

1. The SDS is created at a concentration ten times higher than needed. However, when the solution is added to the combined 10 L volume of the reservoir and the decellularization chamber, it is reduced to the proper concentration of 0.5%. SDS is a known respiratory, skin, and eye irritant. The fine powder can spread easily through the air. Wear a safety mask, gloves, and eye protection when weighing and mixing. The solution may require approximately 1 h to become homogenous.
2. When selecting a heart to decellularize, look down the aorta at the aortic valve to check that the valve is intact. The valve prevents backflow during decellularization and diverts flow into the coronary arteries for effective perfusion into the walls of the ventricles and atria of the heart.
3. This decellularization protocol has shown to be more effective with smaller porcine hearts than larger ones. Small porcine hearts have thinner tissue which facilitates more complete perfusion of decellularizing agents like SDS and Triton X-100.
4. Connecting the heart properly is a critical part of the decellularization process. While connecting the heart, be careful to attach the tubing such that it is not physically contacting the aortic valve to ensure that the coronary orifices are not covered, and to preserve structural integrity of the valve. Two to three zip-ties are used to form an airtight seal between the heart and the tubing that exits the bubble trap. The zip ties are tightened using an adjustable cable tie gun. Once submerged in the solution, we recommend gently shaking the heart to release any trapped air bubbles.
5. Closed loop refers to maintaining a constant solution concentration by emptying the decellularization chamber into the solution reservoir to then be pumped back into the decellularization chamber.
6. Open loop refers to a continuous flow of non-recycled water through the system. The open loop system empties the water into a waste container or sink at the same rate it is supplied from a fresh source. When changing solutions, maintain the liquid level above the heart to prevent it from being exposed to air. If air bubbles are trapped in the heart vasculature, they may prevent the solution from lysing cells and removing cellular debris. Source of distilled water can come from a tap, or be pumped from another external source.
7. The process of switching to an open loop can be hastened by siphoning large portions of the solution out of the solution reservoir and then filling it with type 1 distilled water. We've

found an efficient way of siphoning involves adding water to the reservoir through tubing that is connected to the tap, and then disconnecting the tubing from the tap. The change in pressure forces the flow in the opposite direction, which will begin draining the reservoir.

8. When cycling distilled water through the decellularization apparatus on an open loop, monitor the water level closely and make any necessary adjustments to make sure the system empties the water into a waste container or sink at the same rate it is supplied to prevent flooding.
9. Hearts should be stored appropriately depending on the subsequent procedures that will be performed using the decellularized porcine heart tissue. Seventy percent ethanol solution can be used to prevent contamination of the samples. We recommend storage in 5% ethanol solution for cryostat.
10. The following assays can be performed to evaluate the tissue samples: DNA analysis, collagen and GAGs measurements, macrophage stimulation assay, histology and scanning electron microscopy, mechanical properties assay, cytotoxicity assay, and thrombosis assay.

References

1. Hanson KP, Jung JP, Tran QA, Hsu SP, Lida R, Ajeti V, Campagnola PJ, Eliceiri KW, Squirrell JM, Lyons GE, Ogle BM (2013) Spatial and temporal analysis of extracellular matrix proteins in the developing murine heart: a blueprint for regeneration. Tissue Eng Part A 19:1132
2. Momtahan N, Sukavaneshvar S, Roeder BL, Cook AD (2015) Strategies and processes to decellularize and recellularize hearts to generate functional organs and reduce the risk of thrombosis. Tissue Eng Part B Rev 21:115
3. Guyette JP, Gilpin SE, Charest JM, Tapias LF, Ren X, Ott HC (2014) Perfusion decellularization of whole organs. Nat Protoc 9:1451
4. Ott HC, Matthiesen TS, Goh S, Black LD, Kren SM, Netoff TI, Taylor DA (2008) Perfusion-decellularized matrix: using nature's platform to engineer a bioartificial heart. Nat Med 14:213
5. Faulk DM, Carruthers CA, Warner HJ, Kramer CR, Reing JE, Zhang L, D'Amore A, Badylak SF (2014) The effect of detergents on the basement membrane complex of a biologic scaffold material. Acta Biomater 10:183
6. Gilbert TW, Sellaro TL, Badylak SF (2006) Decellularization of tissues and organs. Biomaterials 27:3675
7. Price AP, Godin LM, Domek A, Cotter T, D'Cunha J, Taylor DA, Panoskaltsis-Mortari A (2015) Automated decellularization of intact, human-sized lungs for tissue engineering. Tissue Eng Part C Methods 21:94
8. Huelsmann J, Aubin H, Kranz A, Godehardt E, Munakata H, Kamiya H, Barth M, Lichtenberg A, Akhyari P (2013) A novel customizable modular bioreactor system for whole-heart cultivation under controlled 3D biomechanical stimulation. J Artif Organs 16:294
9. Momtahan N, Poornejad N, Struk JA, Castleton AA, Herrod BJ, Vance BR, Eatough JP, Roeder BL, Reynolds PR, Cook AD (2015) Automation of pressure control improves whole porcine heart decellularization. Tissue Eng Part C Methods 21(11):1148–1161
10. Araki R, Uda M, Hoki Y, Sunayama M, Nakamura M, Shunsuke A, Sugiura M, Ideno H, Shimada A, Nifuji A, Abe M (2013) Negligible immunogenicity of terminally differentiated cells derived from induced pluripotent or embryonic stem cells. Nature 494:100
11. Lin B, Lu T-Y, Yang L (2014) Hear the beat: decellularized mouse heart regenerated with human induced pluripotent stem cells. Expert Rev Cardiovasc Ther 12:135

Methods in Molecular Biology (2018) 1577: 103–110
DOI 10.1007/7651_2017_96

Published online: 21 November 2017

In-Lab Manufacturing of Decellularized Rat Renal Scaffold for Kidney Bioengineering

Andrea Peloso, Antonio Citro, Valeria Corradetti, Szandra Brambilla, Graziano Oldani, Francesco Calabrese, Tommaso Dominioni, Marcello Maestri, and Lorenzo Cobianchi

Abstract

Whole-organ decellularization is recently gaining interest in the transplantation field as strategy to obtain acellular scaffold only composed by extracellular matrix. These structures, that still remain organ-specific in terms of biological cues and tridimensional morphology could be then recellularized with patient's autologous cells. The final result should be a nwe transplantable autologous organ that should by-pass, at the same time, the problem of organ shortage and secondly the consequences related to the immunosuppression need. Herein we describe the protocol to manufacture a whole-organ transplantable rat kidney scaffold by a dual-detergent (Triton X-100 and SDS) arterial peristaltic perfusion. Final results show whole-renal acellular scaffold with contextual preservation of tridimensional architecture and biological properties deriving from the extracellular matrix composition.

Keywords: Extracellular matrix, Kidney scaffold, Kidney transplantation, Organ bioengineering, Rat model, Regenerative medicine

1 Introduction

Organ transplantation is currently recognized as the gold standard treatment for end-stage kidney diseases (ESKD) [1] as regards survival, quality of life, and economics management. Unfortunately the actual clinical panorama is harshly afflicted by the limited supply of kidneys suitable for transplantation due to an unbridgeable gap between donors and recipients. Moreover shortage of organs grew worse in the last decades leading to a steady increase of deaths on the waiting list [2] alongside a future estimation of 5.4 billion of expected people needing dialysis or transplantation in 2030 [3]. This context makes essential to find alternative strategies able to obtain functional kidneys suitable for transplantation.

Regenerative medicine and organ bioengineering are two new fields of research explored as potential solutions, with their shared final aim to create in-lab transplantable organs.

Decellularization technology is one of the most followed approaches providing the removal of the entire cellular components leaving intact the surrounding extracellular matrix (ECM) [4]. ECM represents miscellanea of different components produced by cells in order to create a function-specific microenvironment. This environment keeps the potential to support organ development, function, and repairing [5]. Moreover, in terms of component composition and geometrical architecture, each scaffold is organ-specific. For these reasons ECM-based scaffold can represent a theoretically perfect scaffold for organ bioengineering, offering at the same time both organ-specific structural and biological cues. Several treatments have been proposed to decellularize tissues or organs such as chemical [6–8] and biological agents [9, 10]. Many techniques to apply decellularization agents have been investigated too ranging from immersion and mechanical agitation [11], pressure gradient force [12], to whole-organ anterograde or retrograde perfusion [13]. In particular, in order to create whole-organ scaffold, detergent perfusion via vascular innate system guarantees numerous advantages such as, firstly, a most homogenous diffusion of the detergent and, secondly, an outflow system for perfused detergents and cellular debris.

This technology, properly modified, has been already applied to different organs such as the liver [14], heart [15, 16], kidney [17], intestine [18], and pancreas [19], both on small and large animal and human-discarded organs [20–22], demonstrating feasibility of its application.

Herein we describe our protocol to produce renal acellular scaffold in rodent model. This dual-detergent protocol permits to eliminate vascular and parenchymal renal cells although preserving the organ-specific tridimensional extracellular matrix template.

2 Materials

2.1 Kidney Harvesting

1. Stereomicroscopy for microsurgical procedure (surgical loop in alternative).
2. Isofluorane.
3. Microsurgical equipment.
4. Surgical suture material.
5. 24 gauge needle.
6. Low-weight heparin.
7. 0.9% NaCl solution.

2.2 Whole-Kidney Decellularization Solutions

1. Milli-Q purified water (or equivalent).
2. Phosphate-buffered saline (PBS).
3. Antibiotic/antimycotic solution (Sigma-Aldrich, St. Louis, MO, USA).

4. Triton X-100 (TNX) 1% (Sigma-Aldrich, St. Louis, MO, USA).
5. Sodium dodecyl sulfate (SDS) 1% (Sigma-Aldrich, St. Louis, MO, USA).
6. Peristaltic pump (Minipuls 2—Gilson, Inc., Middleton, WI, USA).

2.3 Morphometric Analysis Scanning Electron Microscopy (SEM) Resin Cast Analysis

1. NaCl 0.9% solution.
2. Specific resin cast solution (Batson's #17 Monomer Base Solution).
3. Specific resin cast promoter (Batson's #17 Anatomical Corrosion Kit Promoter).
4. Specific resin cast catalyst (Batson's #17 Anatomical Corrosion kit Catalyst).
5. Concentrated sodium hydroxide solutions (10% and 5%, respectively).

3 Methods

3.1 Kidney Harvesting and Bench Procedure

1. Adult male Lewis rats (weights ranging from 250 to 350 g, aged 3–4 months) have been used to obtain whole-kidney acellular scaffold.
2. Anesthesia is provided by isofluorane delivered via cone masks (3% for induction, 2% during the operation, 1 l/min air flow, FiO_2 70–75%).
3. Perform a midline xyphopubic incision from sternum to pubis and through the displacing to the right of loops of the small and large intestines, expose the left kidney retroperitoneal area. [**Note 1**]—Why use left kidney instead of right kidney?]
4. Isolate, tie, and section lumbar arteries using 7-0 silk sutures.
5. Gently mobilize and dissect suprarenal, infrarenal, and subrenal segments of the aorta from the vena cava.
6. Isolate the renal artery down to the renal hilum.
7. Isolate and section the ureter close to the bladder [**Note 2**]—seeding via ureter].
8. Dissect and remove all the perinephric fat tissue.
9. Inject 10UI of low-weight heparin into the animal vascular system.
10. Ligate the cranial stump of the aorta, and insert a 24 G intravenous cannula into the aortic distal segment, secured by 7-0 silk sutures [**Note 3**—The use of aortic patch for scaffold in-vivo transplantation].

11. Remove the graft en bloc.
12. Transfer the removed graft to a Petri dish and store it with 0.9% NaCl solution at 4 °C.
13. Flush the left kidney with 30 ml of 4 °C low-weight heparized 0.9% saline solution in order to eliminate remaining intraluminal blood and clots eventually formed.
14. Use a 14 G plastic tube to create a cuff for the renal vein [**Note 3**]—A circumferential incision was made around the outer surface of the cuff to improve its adherence with the vein.

3.2 Rat Kidney Decellularization

3.2.1 Detergent Solutions and Intermediate Solutions Preparation

1. Prepare 100 ml of Triton X-100 in Milli-Q purified water (or equivalent) with a concentration of 1%.
2. Prepare 100 ml of sodium dodecyl sulfate (SDS).
3. Prepare 100 ml of PBS in two 50 ml beakers.
4. Prepare 100 ml of 0.9% NaCl solution adding antibiotic/antimycotic solution for the final perfusion.

3.2.2 Whole-Kidney Decellularization

1. Prior starting decellularization protocol, the entire system must be filled with the detergents [**Note 4**]—avoid the presence of air bubble in the circuit: it could wreck the entire experiment.
2. Connect the 24 G aortic cannula to the in-flow peristaltic system [**Note 5**]—put a three-way stopcock just before the 24 G cannula in order to avoid air perfusion and subsequently air embolism.
3. Perfuse the kidney with 100 of ionic TNX 1% solution at room temperature setting the peristaltic pump to 70 ml/h (total duration 90 min).
4. Stop the peristaltic pump [**Note 6**]—avoid the entrance of air during solutions switching, and change the solution with the first intermediate solution PBS for a total of 50 ml at the same flow rate (total duration 30 min).
5. Start the anionic detergent cycle with 100 ml of SDS 1% at room temperature by flow rate of 70 ml/h (total duration 90 min) [**Note 7**]—at this point a correct procedure must coincide with the macroscopic change of the color of the kidney (from natural reddish color to whitish translucent aspect).
6. Perform the second intermediate perfusion with residual 50 ml of PBS (room temperature; 70 ml/h) (total duration 30 min).
7. Finally wash the scaffold with 60 ml of 0.9 NaCl solution (flow rate 60 ml/h) (total duration 60 min), added with antibiotic/antimycotic, and store the scaffold at 4 °C. (See Figs. 1 and 2.)

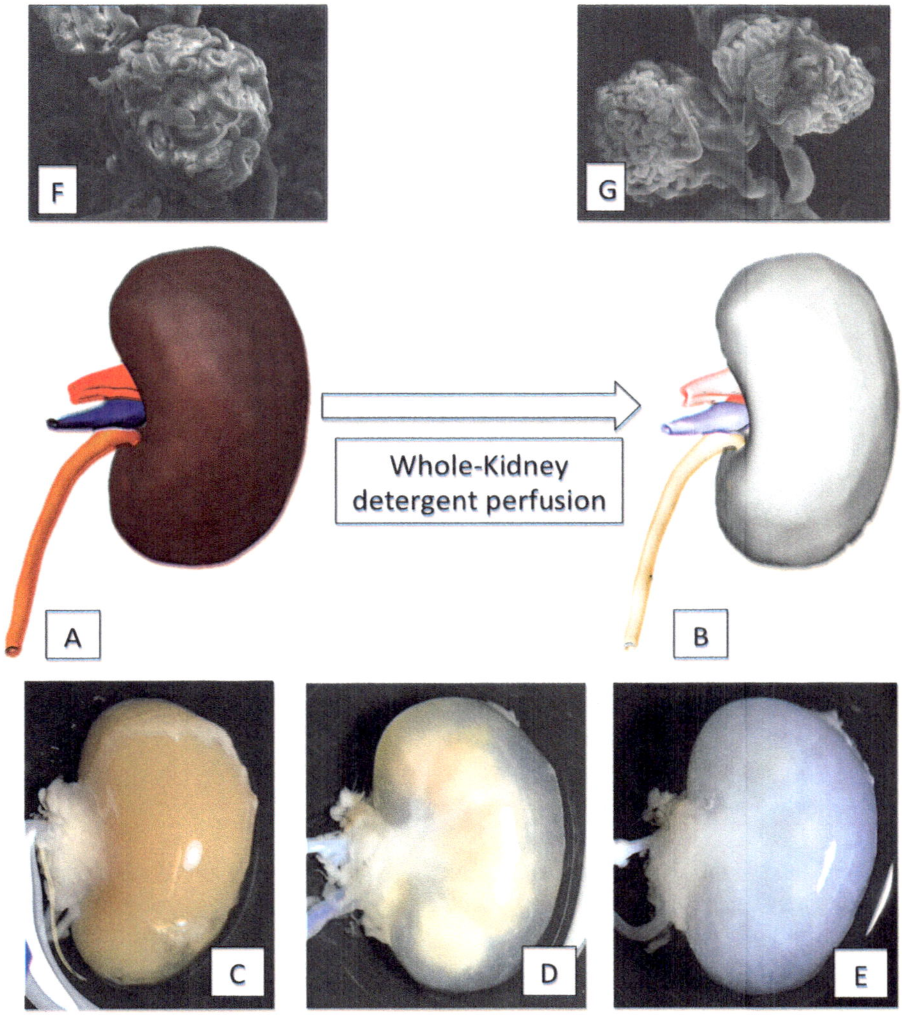

Fig. 1 (**a**, **b**) Graphical images of the native kidney (**a**) and final decellularized scaffold (**b**). No macroscopic changes are observed in terms of organ-specific shape and characteristic, whereas the color changes turning from natural reddish (**c**) to almost-transparent whitish one (**e**). During perfusion decellularization, the cortical renal area is the first to be decellularized (**d**) uncovering the medullar area that will be then decellularized. The analysis of corrosion resin cast of the native organ (**f**) and the renal scaffold (**g**) permits to investigate the morphometric preservation of the glomerular scaffold comparing it with the native

3.3 Morphometric Analysis: Scanning Electron Microscopy (SEM) Resin Cast Analysis

1. Prepare 10 ml of solution composed of specific resin cast solutions (main solution, promoter and catalyst solution).
2. For each kidney, slowly inject 4–7 ml of above-described solution via arterial system until an adequate venous outflow is obtained [**Note 8**].
3. Place the sample in deionized room temperature water overnight waiting for injected resin cast polymerization.

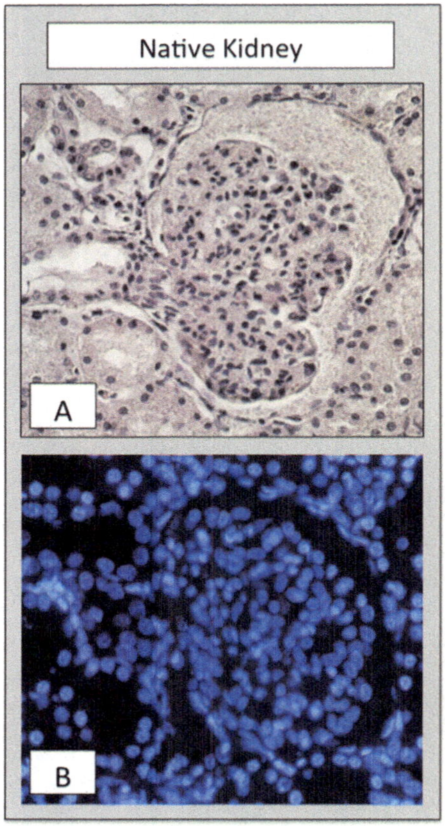

Fig. 2 H&E (**a**, **c**) and DAPI (**b**, **d**) stainings are essential histologic analyses to investigate decellularization effectiveness to remove cells and nuclear material and get a first impression of ECM quality [**Note 10**]

4. Rinse with concentrated sodium hydroxide solution (10%) for 24 h.
5. Rinse with concentrated sodium hydroxide solution (5%) for 24 h [**Note 9**].
6. Finally rinse in deionized water for 6 h.
7. Dry-air the sample.
8. Analyze the sample by scanning electron microscopy (SEM).

4 Notes

1. For anatomical reasons the use of the left kidney is preferred to obtain a transplantable renal scaffold. For each animal, the right kidney can be used as a control.
2. Even if in this procedure the ureter is not used, its anatomical maintenance is desired firstly in order to have a "complete" organoid with an excretory system and secondly to have a second route for cellular seeding during scaffold repopulation.

3. In our model venous anastomose is performed with a cuff-anastomose technique. For this reason we suggest to perform a shallow circumferential incision all around the external surface of the cuff. The renal vein is then inserted through the cuff to the renal hilum, and the protruding tip of the renal vein is then folded back to the cuff and secured with 7-0 silk suture.
4. Avoid the presence of air bubble in the circuit: it could wreck the entire experiment.
5. Put a three-way stopcock just before the 24 G cannula in order to avoid air perfusion and subsequently air embolism. This simple trick allows to spill out eventual air bubbles just before the organ.
6. Avoid the entrance of air during solution switching.
7. At this point a correct procedure must coincide with the macroscopic change of the color of the kidney (from natural reddish color to whitish translucent aspect).
8. In order to obtain a homogenous renal resin cast, it is preferable to gently flush the entire organ or scaffold with 10–20 ml of 0.9% NaCl solution. This shrewdness will remove residual blood cells or cellular debris from the vascular system.
9. Sodium hydroxide solution rinses are essential to detach all perivascular surrounding tissue, leaving the cast of all glomerular structure. These passages must be calibrated, in terms of duration, on the sample characteristics and final results. In case of unsatisfying final results, it is possible to repeat twice the rinse with 5% sodium hydroxide solution.
10. Often you can grossly regulate the harshness of the decellularization protocol simply by preliminary H&E staining observation. This can save time and money.

Acknowledgments

AP was supported by an Investigator Fellowship grant from Collegio Ghislieri, Pavia, Italy.

References

1. World Health Organization (2013) World Health Statistic. World Health Organization, Geneve, Switzerland
2. Beuscart JB, Pagniez D, Boulanger E, Duhamel A (2015) Registration on the renal transplantation waiting list and mortality on dialysis: an analysis of the French REIN registry using a multi-state model. J Epidemiol 25:133–141
3. Liyanage T, Ninomiya T, Jha V, Neal B, Patrice HM, Okpechi I, Zhao MH, Lv J, Garg AX, Knight J, Rodgers A, Gallagher M, Kotwal S, Cass A, Perkovic V (2015) Worldwide access to treatment for end-stage kidney disease: a systematic review. Lancet 385:1975–1982
4. Crapo PM, Gilbert TW, Badylak SF (2011) An overview of tissue and whole organ decellularization processes. Biomaterials 32:3233–3243

5. Page-McCaw A, Ewald AJ, Werb Z (2007) Matrix metalloproteinases and the regulation of tissue remodelling. Nat Rev Mol Cell Biol 8:221–233
6. Gilbert TW, Wognum S, Joyce EM, Freytes DO, Sacks MS, Badylak SF (2008) Collagen fiber alignment and biaxial mechanical behavior of porcine urinary bladder derived extracellular matrix. Biomaterials 29:4775–4782
7. Xu CC, Chan RW, Tirunagari N (2007) A biodegradable, acellular xenogeneic scaffold for regeneration of the vocal fold lamina propria. Tissue Eng 13:551–566
8. Dong X, Wei X, Yi W, Gu C, Kang X, Liu Y, Li Q, Yi D (2009) RGD-modified acellular bovine pericardium in a bioprosthetic scaffold for tissue engineering. J Mater Sci Mater Med 20:2327–2336
9. Petersen TH, Calle EA, Zhao L, Lee EJ, Gui L, Raredon MB, Gavrilov K, Yi T, Zhuang ZW, Breuer C, Herzog E, Niklason LE (2010) Tissue-engineered lungs for in vivo implantation. Science 329:538–541
10. Lehr EJ, Rayat GR, Chiu B, Churchill T, McGann LE, Coe JY, Ross DB (2010) Decellularization reduces immunogenicity of sheep pulmonary artery vascular patches. J Thorac Cardiovasc Surg 141:1056–1062
11. Dahl SL, Koh J, Prabhakar V, Niklason LE (2003) Decellularized native and engineered arterial scaffolds for transplantation. Cell Transplant 12:659–666
12. Bolland F, Korossis S, Wilshaw SP, Ingham E, Fisher J, Kearney JN, Southgate J (2007) Development and characterization of a full-thickness acellular porcine bladder matrix for tissue engineering. Biomaterials 28:1061–1070
13. Henderson PW, Nagineni VV, Harper A, Bavinck N, Sohn AM, Krijgh DD, Jimenez N, Weinstein AL, Spector JA (2010) Development of an acellular bioengineered matrix with a dominant vascular pedicle. J Surg Res 164:1–5
14. Shupe T, Williams M, Brown A, Willenberg B, Petersen BE (2010) Method for the decellularization of intact rat liver. Organogenesis 6:134–136
15. Akhyari P, Aubin H, Gwanmesia P, Barth M, Hoffmann S, Huelsmann J, Preuss K, Lichtenberg A (2011) The quest for an optimized protocol for whole-heart decellularization: a comparison of three popular and a novel decellularization technique and their diverse effects on crucial extracellular matrix qualities. Tissue Eng Part C Methods 17:915–926
16. Ott HC, Matthiesen TS, Goh SK, Black LD, Kren SM, Netoff TI, Taylor DA (2008) Perfusion-decellularized matrix: using nature's platform to engineer a bioartificial heart. Nat Med 14:213–221
17. Peloso A, Ferrario J, Maiga B, Benzoni I, Bianco C, Citro A, Currao M, Malara A, Gaspari A, Balduini A, Abelli M, Piemonti L, Dionigi P, Orlando G, Maestri M (2015) Creation and implantation of acellular rat renal ECM based scaffolds. Organogenesis 11:58–74
18. Chen MK, Badylak SF (2001) Small bowel tissue engineering using small intestinal submucosa as a scaffold. J Surg Res 99:352–358
19. Goh SK, Bertera S, Olsen P, Candiello JE, Halfter W, Uechi G, Balasubramani M, Johnson SA, Sicari BM, Kollar E, Badylak SF, Banerjee I (2013) Perfusion-decellularized pancreas as a natural 3D scaffold for pancreatic tissue and whole organ engineering. Biomaterials 34:6760–6772
20. Peloso A, Urbani L, Cravedi P, Katari R, Maghsoudlou P, Fallas ME, Sordi V, Citro A, Purroy C, Niu G, McQuilling JP, Sittadjody S, Farney AC, Iskandar SS, Zambon JP, Rogers J, Stratta RJ, Opara EC, Piemonti L, Furdui CM, Soker S, De Coppi P, Orlando G (2016) The human pancreas as a source of protolerogenic extracellular matrix scaffold for a new-generation bioartificial endocrine pancreas. Ann Surg 264:169–179
21. Peloso A, Petrosyan A, Da Sacco S, Booth C, Zambon JP, O'Brien T, Aardema C, Robertson J, De Filippo RE, Soker S, Stratta RJ, Perin L, Orlando G (2015) Renal extracellular matrix scaffolds from discarded kidneys maintain glomerular morphometry and vascular resilience and retains critical growth factors. Transplantation 99:1807–1816
22. Mazza G, Rombouts K, Rennie Hall A, Urbani L, Vinh Luong T, Al-Akkad W, Longato L, Brown D, Maghsoudlou P, Dhillon AP, Fuller B, Davidson B, Moore K, Dhar D, De Coppi P, Malago M, Pinzani M (2015) Decellularized human liver as a natural 3D-scaffold for liver bioengineering and transplantation. Sci Rep 5:13079

Methods in Molecular Biology (2018) 1577: 111–119
DOI 10.1007/7651_2017_72

Published online: 06 April 2018

Glutaraldehyde Cross-linking Modification of Decellularized Rat Kidney Scaffolds

Yaling Yu, Li Liu, Jianse Zhang, Zairong Wei, and Jin Mei

Abstract

The decellularized (DC) scaffolds retain three-dimensional structures for the stimulation of cell growth, with components of the extracellular matrix (ECM) relatively conserved between species. The strategy based on decellularized scaffolds provides a new way for organ regeneration, with a number of prominent advances having been reported in the past few years. While their lack of biomechanical strength and excessive degradation limit the clinical applications, therefore it is urgent to modify the DC scaffolds to improve the performance.

In this article we describe a simple and robust modification protocol for DC rat kidney scaffolds. To modify, we perfuse DC rat kidneys with glutaraldehyde through the perfusion circulation of the decellularization. After cross-linking, kidney scaffolds are harvested for evaluation of histology, structural stability, and biocompatibility, involving water absorption testing, biomechanical testing, scanning electron microscopy, and several different histological and immunofluorescent analyses.

Keywords Cross-linking, Decellularized scaffolds, Degradation, Kidney, Tissue/organ engineering

1 Introduction

Chronic kidney disease and the end-stage renal disease (ESRD) have been a major threat to the life of human beings, affecting between 8 and 16% of the global adult population [1]. Kidney transplantation remains the only available curative treatment. However, the short supply of donor kidneys is the bottleneck of the further development of organ transplantation; thus, a new methodology to alleviate the shortage of organ donors is badly in need. With the development of tissue engineering, the strategy of cells seeding into bio-scaffolds provides a new thinking for organ transplantation. Thereinto, the development and modification of ideal bio-scaffolds are of the first importance. Till now, decellularized (DC) scaffolds are considered to be one of the most potential biological materials. Researchers including our research group have fabricated DC kidney scaffolds [2–5] by using detergents

Yaling Yu and Li Liu contributed equally to this work.

perfusion, and the DC kidney scaffolds retain relatively intact extracellular matrix (ECM) structure and the vasculature, with low residual DNA, low immunogenicity, and good cytocompatibility. While their lack of biomechanical strength and excessive degradation limit the clinical applications, therefore it is urgent to modify the DC scaffolds to improve the performance. Chemical cross-linking has long been widely applied in the modification of DC scaffolds [6], especially glutaraldehyde (GA) [7, 8].

There are no reports about the cross-linking modification of DC kidney scaffolds, and the traditional shaking technique is not suitable for the cross-linking of the DC kidney scaffolds with complicated three-dimensional structure. In this article, we perfuse GA to cross-link the DC kidney scaffold to not only preserve its original unique characteristics but also improve its mechanical property. And thus the modified DC kidney scaffolds can meet the need for clinical applications.

2 Materials

2.1 Preparation of Decellularized (DC) Kidney Scaffolds

1. Deionized water (dH_2O).
2. 0.01 M phosphate-buffered saline, pH 7.4 (PBS).
3. 5% Chloral hydrate.
4. 24-Gauge cannula.
5. Peristaltic pump.
6. Antibiotics: 100 U/mL penicillin and 100 ng/mL streptomycin.

2.2 Cross-Linking Modification of DC Kidney Scaffolds

1. 25% Glutaraldehyde (GA).
2. 0.01 M phosphate-buffered saline, pH 7.4 (PBS).
3. Deionized water (dH_2O).
4. Antibiotics: 100 U/mL penicillin and 100 ng/mL streptomycin.
5. Peristaltic pump.
6. 0.9% Sodium chloride.

2.3 Histology Analysis

1. Dimethylbenzene.
2. Ethanol: 100, 90, 80, 70, and 50% v/v solutions of ethanol in dH_2O.
3. Harris hematoxylin.
4. 1% Hydrochloric acid in ethanol (v/v).
5. 1% Ammonia in dH_2O (v/v).
6. 1% Eosin in dH_2O (w/v).
7. Neutral balsam.
8. Bright-field microscope.

2.4 Scanning Electron Microscope Observation

1. 2.5% Glutaraldehyde in 0.1 M phosphate buffer: Combine 10 mL of 25% stock glutaraldehyde and 50 mL 0.2 M phosphate buffer, pH 7.4. Make up to 100 mL with dH_2O.
2. 1% Osmium tetroxide in dH_2O (w/v).
3. 3% Uranyl acetate in dH_2O (w/v).
4. Ethanol.
5. Isoamyl acetate.
6. Desiccator.
7. Aluminum tray.
8. Ion sputtering coating machine.
9. Scanning electron microscope.

2.5 Degradation In Vitro

1. Filter paper.
2. Electronic balance.
3. Centrifuge tube.
4. Trypsin containing ethylenediaminetetraacetic acid (EDTA).
5. Water bath constant temperature oscillator.

2.6 Water Absorption Assessment

1. Freeze dryer.
2. Electronic balance.
3. Centrifuge tube.
4. 0.01 M phosphate-buffered saline, pH 7.4 (PBS).
5. Shaker.

2.7 Biomechanical Test

1. Ophthalmic scissors.
2. Vernier caliper.
3. Force-measuring machine.

2.8 Histocompatibility Analysis

1. Autoclave.
2. Sprague-Dawley (SD) rats.
3. Chloral hydrate.
4. Surgical instruments.
5. Surgical suture.

2.9 Cytocompatibility Analysis

1. Freezing microtome.
2. Fibroblasts.
3. Culture medium.
4. Climbing glasses.
5. Inverted microscope.

3 Methods

3.1 Preparation of DC Kidney Scaffolds

1. Anesthetize rats with 5% chloral hydrate (0.6 mL/100 g body weight) via intraperitoneal injection.
2. Open abdominal cavity by a ventral midline incision, extending from the pubis to the xiphoid process.
3. Ligate suprarenal abdominal aorta, suprarenal inferior vena cava, lumbar arteries, testicular arteries of male rats, and ovarian arteries of female rats.
4. Insert a 24G cannula into the infrarenal abdominal aorta.
5. Remove bilateral kidneys with blood vessels connected.
6. Connect the cannula with the peristaltic pump to allow continuous rinsing with various detergents (*see* **Note 1**). Perfuse at approximately 8 mL/min in the following order:
   (a) 50 U/mL heparin in 0.01 M PBS for 30 min.
   (b) 0.1% Triton X-100 for 3 h.
   (c) dH_2O for 30 min.
   (d) 0.8% SDS in dH_2O (v/v) for 3 h.
   (e) dH_2O containing 100 U/mL penicillin and 100 mg/mL streptomycin for 24 h.

3.2 Cross-Linking Modification of DC Kidney Scaffolds

1. Perfuse 0.625% GA (*see* **Note 2**) through the decellularization circulator at the rate of 1 mL/min (*see* **Note 3**), 12 h.
2. Perfuse dH_2O containing penicillin-streptomycin at the rate of 1 mL/min, 12 h.
3. Once perfusion is complete, DC kidney scaffolds can be stored in 50 mL of dH_2O containing penicillin and streptomycin at 4 °C until use (*see* **Note 4**).

3.3 Histology Analysis by Hematoxylin-Eosin Staining

1. Dry kidney sections at 60 °C, 8 h.
2. Dewax with dimethylbenzene for 30 min twice.
3. Rehydrate with 100, 100, 90, 80, 70, and 50% alcohol, respectively, for 5 min.
4. Wash in distilled water, 5 min.
5. Immerse in hematoxylin, 2 min.
6. Rinse under running water, 5–10 s.
7. Color separate with 1% v/v hydrochloric acid in ethanol for 3 s.
8. Rinse under running water, 1–2 s.
9. Promote blue with 1% ammonia, 5–10 s.
10. Rinse under running water, 1–2 s.
11. Immerse in 1% v/v eosin in deionized water, 10 s.

12. Rinse under running water, 1–2 s.
13. Dehydrate with 80, 90, and 100% alcohol, respectively, for 2 min.
14. Immerse in dimethylbenzene for 5 min twice.
15. Mount with neutral balsam.
16. Examine and image slides with a brightfield microscope.

3.4 Scanning Electron Microscopy Observation

Scanning electron microscope was used to characterize the microstructure of the DC kidney scaffolds with very high spatial resolution.

1. Samples were fixed with 2.5% (v/v) glutaraldehyde in 0.1 M sodium cacodylate buffer overnight at 4 °C.
2. Wash in PBS, 3 × 15 min.
3. Post-fix with 1% osmium tetroxide for 1 h at 37 °C.
4. Wash in PBS, 2 × 15 min.
5. Stain with 2% uranyl acetate, 1 h.
6. Dehydrate with 70, 80, 90, 100, and 100% ethanol in 0.2 M PBS for 15 min each.
7. Displace with ethanol-isoamyl acetate (1:1) fluid, 15 min.
8. Displace with 100% isoamyl acetate, 15 min.
9. Dry in a desiccator, 8 h.
10. Stick samples to the aluminum tray.
11. Spray-gold with ion sputtering coating machine to coat the surface of the samples with a thickness of 1500 nm metal film.
12. Examine sections using a scanning electron microscope, and image by high resolution CCD digital camera.

3.5 Degradation In Vitro

1. Blot the surface moisture of samples.
2. Weigh and record (W_0).
3. Digest samples with EDTA-trypsin, 37 °C (*see* **Note 5**).
4. Take out samples and blot the surface moisture 0.5, 1, 2, 4, 8, 16, and 24 h post digestion.
5. Weigh and record (W_t).
6. Calculate the degradation rate according the formula

$$W\% = (W_0 - W_t)/W_0 \times 100\%.$$

3.6 Water Absorption Testing

1. Dry samples in freeze dryer (*see* **Note 6**).
2. Weigh and record (W_0).
3. Shake samples on the shaker at 12,000 bpm for 48 h.

4. Take out samples and blot the surface moisture.
5. Weigh and record (W_1).
6. Calculate the water absorption according the formula

$$W\% = (W_1 - W_0)/W1 \times 100\%.$$

3.7 Biomechanical Test

1. Shave samples into dumbbell shape.
2. Measure the length (a) and width (b) of the middle part of the samples (*see* **Note 7**).
3. Clamp the two sides of shaved samples.
4. Make sure the starting distance of the two sides (I_0) is 3 mm.
5. Pull samples with 1 N load tension and 1 mm/min stretching speed until broken the samples.
6. The computer records and calculates the load-displacement curve.
7. Draw the stress-strain diagram.
8. Calculate the ultimate tensile strength and modulus of elasticity.

3.8 Histocompatibility Test

1. Anesthetize rats with 5% chloral hydrate (0.6 mL/100 g) via intraperitoneal injection.
2. Remove the hair of the back.
3. Make two skin incisions along the posterior midline.
4. Implant the GA-treated scaffolds and non-GA scaffolds into the subcutaneous deep fascia.
5. Suture the skin wound, and disinfect the wound.
6. Take out the implanted scaffolds 4 days, 1 week, 2 weeks, and 4 weeks post implantation.
7. Perform H&E staining.

3.9 Cytocompatibility Test

1. Cut GA-treated scaffolds and non-GA scaffolds into 10 μm sections (*see* **Note 8**), and stick onto climbing glasses.
2. Put climbing glasses into 12-well plate.
3. Incubate with culture medium for 12 h (*see* **Note 9**).
4. Seed fibroblasts into scaffolds with the cell density of 2.5×10^5/well.
5. Add culture medium 3 h post cell seeding.
6. Observe under inverted microscope 4 and 24 h post seeding.

3.10 Analysis and Interpretation

DC kidney scaffolds have a somewhat transparent appearance (Fig. 1Ai). Modified with continuous GA perfusion, the cross-linked DC scaffolds retain the original appearance, with minor

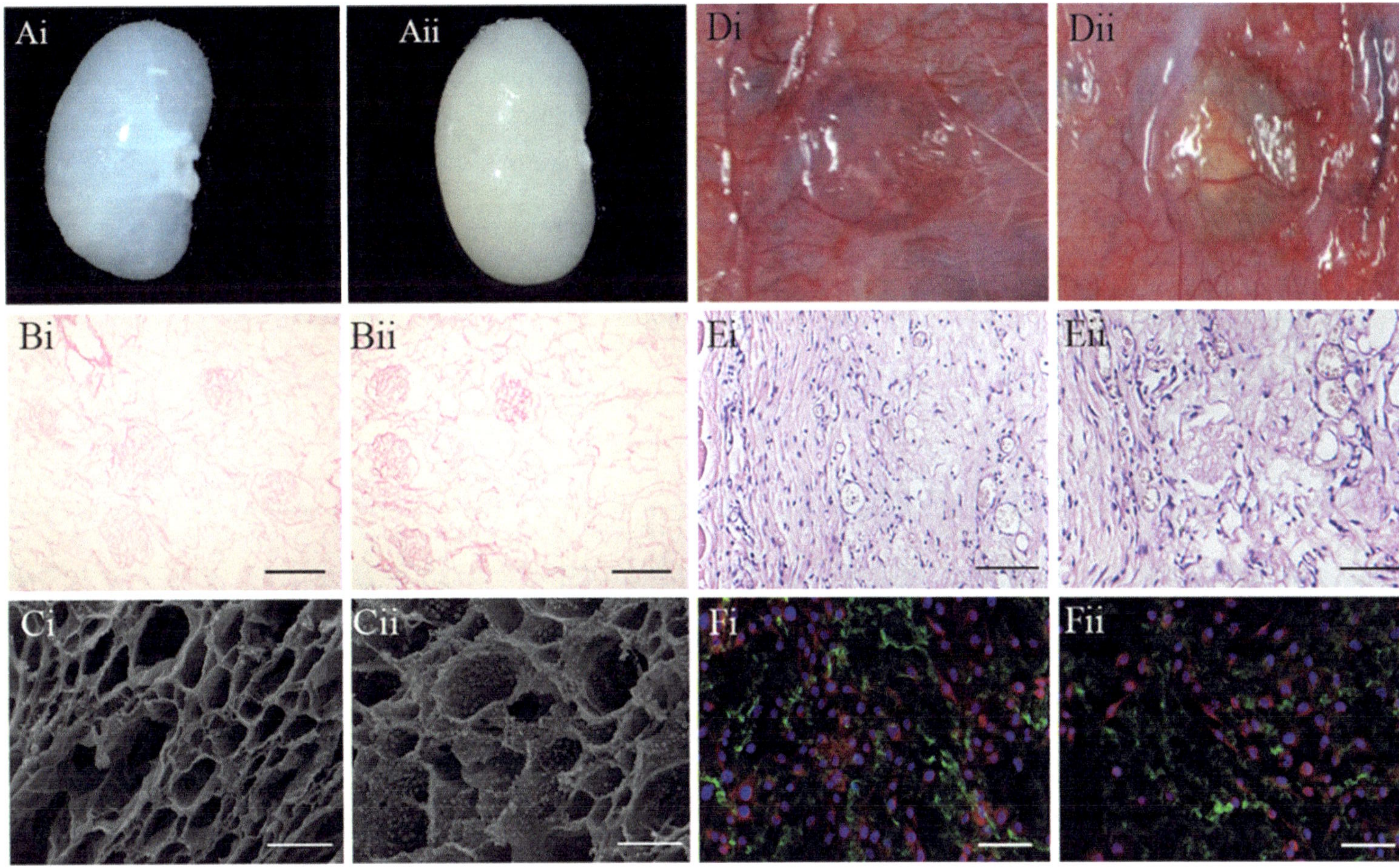

Fig. 1 Characterization of the GA-treated and non-GA DC kidney scaffolds. (**A**) Gross appearance of DC kidney scaffolds (**Ai**) and GA-treated DC kidney scaffolds (**Aii**). (**B**) H&E staining shows intact extracellular matrix in DC kidney scaffolds (**Bi**) and GA-treated DC kidney scaffolds (**Bii**). (**C**) Scanning electron microscope observation shows no interrupted microstructure in DC kidney scaffolds (**Ci**) and GA-treated DC kidney scaffolds (**Cii**). (**D**) DC kidney scaffolds (**Di**) and GA-treated DC kidney scaffolds (**Dii**) are implanted in vivo. (**E**) H&E staining shows DC kidney scaffold (**Ei**) is almost completely degraded, while GA-treated DC kidney scaffold retains almost original structure (**Eii**). (**F**) Immunofluorescent staining shows the *green* collagen IV in DC kidney scaffold (**Fi**) and GA-treated DC kidney scaffold (**Fii**) and the *red* α-SMA in fibroblasts. Scale bars = 100 μm (**B**, **E**, **F**) and 5 μm (**C**)

yellowish (Fig. 1Aii). The ECM structure observed by H&E staining (Fig. 1B) and SEM (Fig. 1C) in the GA-treated DC kidney scaffolds is well maintained compared with the DC kidney scaffolds. The implantations of both DC scaffolds in vivo verify the good histocompatibility (Fig. 1D, E). The fibroblasts show good attachment and proliferation post seeded onto the two DC scaffolds by immunofluorescence analysis (Fig. 1E).

4 Notes

1. All of the fluid ready for decellularization should be controlled at 25–30 °C.
2. The 0.625% GA is prepared by PBS solution, not dH_2O.
3. The velocity of cross-link is 1 mL/min, lower than that of decellularization.
4. The cross-linked DC scaffolds can be preserved for 1 week at 4 °C while 2 months at −80 °C.
5. The enzyme digestion is performed in thermostatic oscillator.
6. Dry samples in freeze dryer for 36 h.
7. During the biomechanical test, keep samples moist.
8. The scaffolds cutting into sections should be performed with freezing microtome.
9. The scaffolds both cross-linked and non-cross-linked are needed to be preincubated with culture medium.

Acknowledgments

This research was supported by National Natural Science Foundation of China (81570608).

References

1. Jha V, Garcia-Garcia G, Iseki K, Li Z, Naicker S, Plattner B et al (2013) Chronic kidney disease: global dimension and perspectives. Lancet 382:260–272
2. Katari R, Peloso A, Zambon JP, Soker S, Stratta RJ, Atala A et al (2014) Renal bioengineering with scaffolds generated from human kidneys. Nephron Exp Nephrol 126:119
3. Serban MA, Knight TA, Payne RG (2013) Preparation and evaluation of natural scaffold materials for kidney regenerative applications. Methods Mol Biol 1001:133–143
4. Song JJ, Guyette JP, Gilpin SE, Gonzalez G, Vacanti JP, Ott HC (2013) Regeneration and experimental orthotopic transplantation of a bioengineered kidney. Nat Med 19:646–651
5. YL Y, Shao YK, Ding YQ, Lin KZ, Chen B, Zhang HZ et al (2014) Decellularized kidney scaffold-mediated renal regeneration. Biomaterials 35:6822–6828
6. Zhai W, Chang J, Lin K, Wang J, Zhao Q, Sun X (2006) Crosslinking of decellularized porcine heart valve matrix by procyanidins. Biomaterials 27:3684–3690

7. Cheung DT, Perelman N, Ko EC, Nimni ME (1985) Mechanism of crosslinking of proteins by glutaraldehyde III. Reaction with collagen in tissues. Connect Tissue Res 13:109–115

8. Billiar K, Murray J, Laude D, Abraham G, Bachrach N (2001) Effects of carbodiimide crosslinking conditions on the physical properties of laminated intestinal submucosa. J Biomed Mater Res 56:101–108

Methods in Molecular Biology (2018) 1577: 121–127
DOI 10.1007/7651_2017_29

Published online: 30 April 2017

Preparation of Scaffolds from Decellularized Testicular Matrix

Yoni Baert and Ellen Goossens

Abstract

Biological scaffolds composed of extracellular matrix (ECM) are typically derived by processes that involve decellularization of tissues or organs. Here we describe a simple and robust methodology for the preparation of decellularized testicular matrix (DTM) scaffolds with minimal damage to the native three-dimensional structure and tissue-specific ECM components. Such DTM scaffolds can help to gain a better insight into the molecular composition and function of testicular ECM and to develop new tissue engineering approaches to treat various types of male fertility disorders.

Keywords: Testis, Extracellular matrix, Decellularization, Scaffold, In vitro spermatogenesis

1 Introduction

Spermatogenesis is a tightly regulated continuous process by which spermatozoa are produced from spermatogonial stem cells (SSCs). SSCs form thus the basis of spermatogenesis and are located at the basement membrane of seminiferous tubules where they need nursing by Sertoli cells and factors from the interstitial peritubular myoid cells and Leydig cells in order to differentiate [1]. Attempts on replicating this phenomenon in a culture dish, which is called in vitro spermatogenesis (IVS), are ongoing for nearly a century with limited success [2]. Going through the meiosis step of germ cell differentiation in vitro proved difficult and hindered the progress in this particular research area. These early studies—which were mainly using 2D culture systems—revealed that germ cells are only able to go beyond meiosis and traverse spermiogenesis in vitro if culture conditions are applied allowing the original testicular 3D cell arrangements [3].

In 2009, the first IVS system was reported using a rodent model. For this, testicular cell suspensions from prepubertal animals were combined with a 3D agarose matrix [3]. Morphologically normal mouse spermatozoa could be generated, but with low efficiency. Moreover, in a culture system consisting of immature rat testicular cells embedded in a 3D agarose matrix, only spermatogonia that maintained in close contact with Sertoli cells were developing [4]. Therefore, the specific cell orientation, which is normally provided by the basement membrane, might be crucial for successful IVS.

The problem of organization might be solved by using a scaffold that mimics the testicular architecture. Indeed, for tissue engineering purposes, it is already common to use scaffolds composed of biological extracellular matrix (ECM) [5]. In this context, we described for the first time the preparation of human decellularized testicular matrix (DTM) and its potential applications [6]. Firstly, DTM enables researcher to focus on the specific composition of the testicular matrix and further elucidate its role in spermatogenesis. Secondly, DTM provides the natural architectural arrangement that could allow the reassembling of testicular cells as well as the nutritive and endocrine support, which has been proven to be essential in murine IVS [3]. Therefore, scaffolds of DTM could be harnessed for the development of the first human IVS culture system, which is a missing link in male fertility preservation and could be a possible treatment for nonobstructive azoospermia [7].

The protocol to obtain DTM proved efficient elimination of endogenous cells, biocompatibility and maintenance of the 3D structure and the major ECM components, such as laminin and collagen I and IV (Fig. 1). The accessibility of the acellular interstitial space and seminiferous tubules of DTM fragments was improved by preparing thin scaffold discs [8]. The methodology to obtain biocompatible thin scaffold discs of DTM fragments is summarized here.

2 Materials

2.1 Decellularization of Testicular Tissue

1. Deionized water (dH_2O)
2. Sterile 0.01 M phosphate-buffered saline (PBS), pH 7.4
3. Sterile 1 % sodium dodecyl sulfate (SDS) in dH_2O (w/v) (*see* **Note 1**)
4. 4 % ethanol in dH_2O (v/v) with 0.1 % peracetic acid (v/v)
5. Sterile container for organ collection
6. Petri dishes with a diameter of 100 mm
7. 50-ml polypropylene tubes
8. Orbital shaker
9. Sterile surgical tweezer and scissor or scalpel
10. Cadaveric human testis (*see* **Note 2**)

2.2 Preparation of Thin Scaffold Discs

1. Sterile 0.7 % agarose in PBS (m/v)
2. Microwave
3. Sterile 6-mm biopsy punch

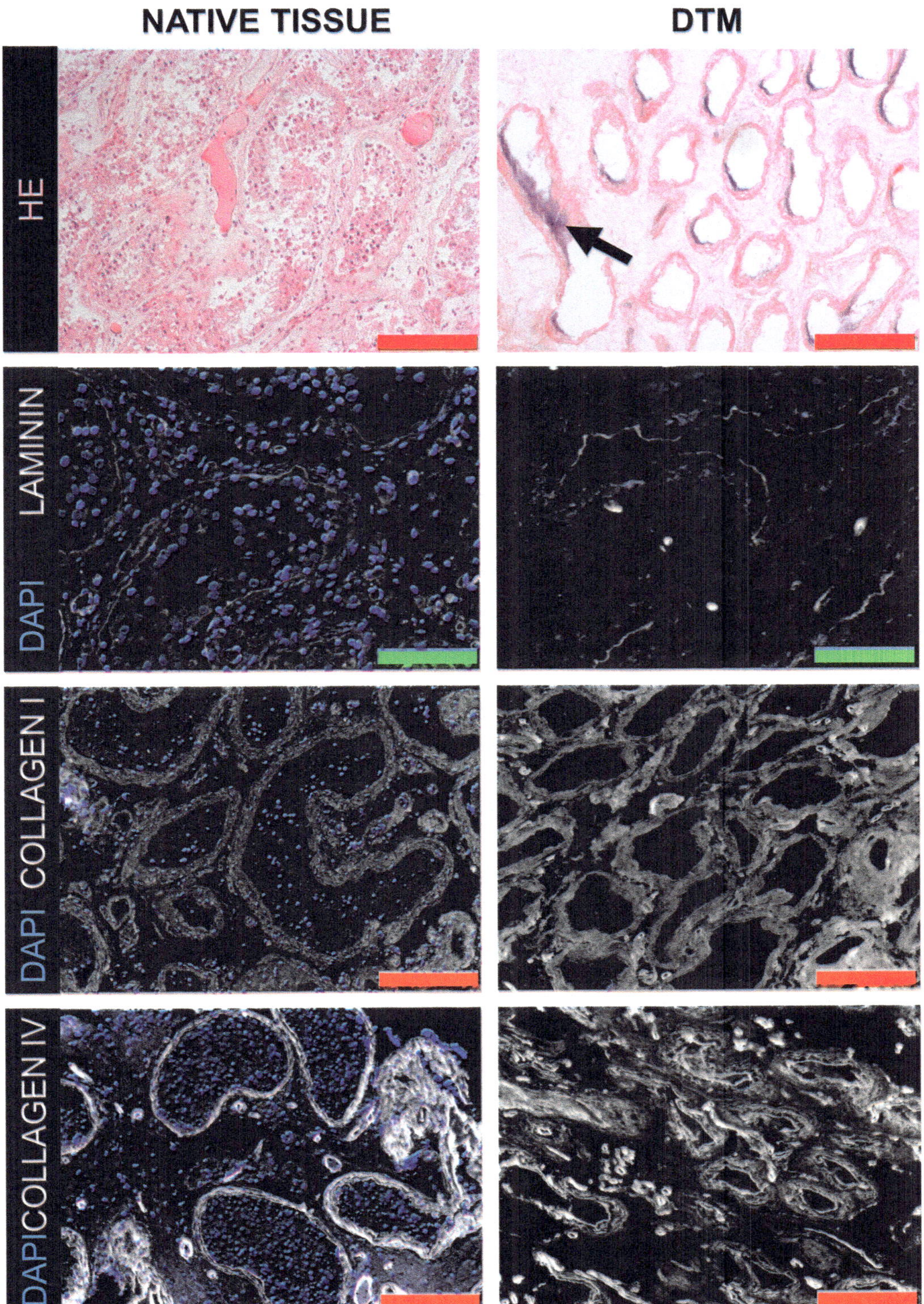

Fig. 1 Histology and immunofluorescent stainings for laminin, collagen I, and collagen IV before and after decellularization of testicular tissue. Hematoxylin–eosin (HE) staining (*first row*) of native testicular tissue (*left panel*)

4. 24-well plate
5. Hanging culture inserts for 24-well plate
6. Optimal cutting temperature (OCT) embedding compound
7. Small cryobrush
8. Cryostat-microtome

3 Methods

3.1 Decellularization of Testicular Tissue

1. Collect the testis in a sterile container containing PBS for transport.
2. Put the testes in a petri dish and remove the surrounding connective tissue including tunica albuginea to isolate the testicular tissue. Remove also the rete testis as this has a different histology compared to the seminiferous tubules.
3. Fragment the testicular tissue into pieces measuring 0.5–1 cm^3 (*see* **Note 3**).
4. Wash the testicular tissue fragments extensively with PBS to remove residual blood.
5. Transfer the fragments into 50-ml tubes and determine the weight, and add 1 ml of 1 % SDS in dH2O (m/v) per 10 mg of tissue (*see* **Note 4**).
6. Close the tubes tightly and seal with parafilm to prevent leakage.
7. Place and fix (e.g., with tape) the tubes vertically on the orbital shaker and shake for 24 h at room temperature to decellularize the testicular fragments.
8. Place the DTM fragments in new tubes and wash the fragments extensively with PBS for 4 h by mechanical agitation as described in **step** 7. Refresh the PBS every 30 min.
9. Disinfect the DTM fragments by exposure to 4 % ethanol in dH2O (v/v) with 0.1 % peracetic acid (v/v) for 2 h at room temperature.
10. Rinse DTM fragments with PBS three times for 5 min each.

Fig. 1 (Continued) and decellularized testicular matrix (DTM, *right panel*) shows a complete removal of cells from the testicular tissue after a 24-h mechanical agitation in 1 % sodium dodecyl sulfate (SDS), while maintaining the testicular architecture. Though, cellular debris can often be observed sticking to the DTM (*arrow*). Laminin (*second row*), collagen I (*third row*), and collagen IV (*fourth row*) are present in DTM with similar staining patterns as in control tissue, which indicates retention of important ECM proteins during decellularization. *Red scale bars*: 200 μm; *green scale bars*: 100 μm

3.2 Preparation of Thin Scaffold Discs

1. Heat the 0.7 % agarose in PBS (w/v) in the microwave for 1 min at 500 W until it is completely melted.
2. Dilute the 0.7 % agarose in PBS (w/v) with culture medium (1:2 dilution) (*see* **Note 5**).
3. Pipet 50 μl of the 0.35 % agarose solution in each hanging culture insert and let it jellify at room temperature. This will later serve as the support for the DTM scaffold discs (*see* **Note 6**).
4. Switch on the cryostat-microtome.
5. Clean and sterilize cryobrush, specimen holder, blade, and surfaces in the cryostat-microtome with 70 % ethanol.
6. Place the well plate with hanging culture inserts containing the agarose supports in the cryostat-microtome and take the lid off.
7. Wait until all materials acclimatize to the cryostat's subzero temperature.
8. Punch out 6-mm cylinders from the DTM fragments.
9. Add OCT embedding compound to the specimen holder and let it cool down in the cryostat-microtome. To glue the DTM cylinder to the specimen holder, bring the base of the DTM cylinder in contact with the OCT embedding compound just before it completely freezes. While the base is frozen in OCT embedding compound, the lid side of the DTM cylinder is accessible for cutting (Fig. 2a).
10. Make 90-μm thick sections of the DTM cylinder from the lid side.
11. Place the DTM scaffold discs carefully in the hanging culture inserts using the cryobrush (Fig. 2b).
12. Put the lid of the well plate back and let it adjust to room temperature. The DTM scaffold discs will melt and stick to the agarose support.
13. The DTM scaffold discs are now ready for experimentation (*see* **Note 7**).

4 Notes

1. SDS dissolves easier in dH_2O by heating the solution to 50 °C.
2. Human cadaveric testes are plentiful and therefore the material of choice for the derivation of DTM. Preferably, testes are used from donors who had normal spermatogenesis. Otherwise, the molecular analyses of DTM and the behavior of cells seeded on

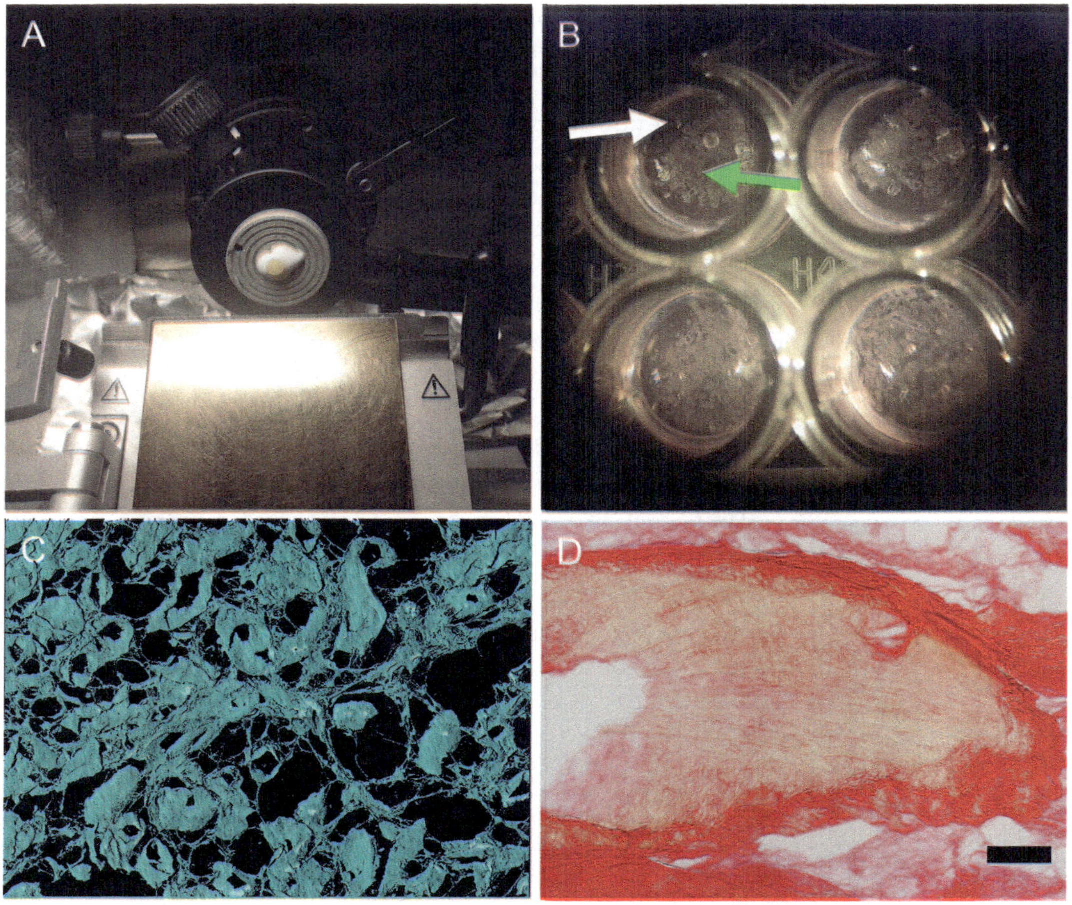

Fig. 2 Recommended cryostat-microtome setup for the sectioning of DTM cylinders to make scaffold discs. (**a**) Punched-out DTM cylinder glued to the specimen holder at the base using OCT embedding compound. (**b**) The 90-μm thick scaffolds (*green arrow*) are cut from the lid side and placed on an agarose support (*white arrow*). (**c**) Overview of a DTM scaffold stained with *picrosirius red* and imaged with fluorescence microscopy. (**d**) High magnification picture of the inside of a tubular fragment in a DTM scaffold stained with *picrosirius red* and imaged with bright field microscopy. Scale bar: 50 μm

DTM scaffolds could be influenced. That is because abnormal spermatogenesis is associated with changes in the ECM environment [9].

3. Decellularizing smaller pieces of testicular tissue by mechanical agitation can more easily lead to uncoiling of seminiferous tubules and loss of the 3D organization, while the elimination of cellular debris from the tubules can be less efficient when working with bigger pieces.
4. It is not recommended to completely fill the tubes. Leave the solution with fragments some room for shaking.

5. The agarose solution is diluted 1:2 in culture medium just before pipetting it into the hanging culture inserts to make a support for the DTM scaffolds. As such, nutrients are more rapidly available for cells used to recellularize a DTM scaffold.
6. The height of agarose support is supposed to be approximately 2 mm.
7. The diameter of the DTM scaffold can be changed by choosing for a biopsy punch with larger diameter. Importantly, the type of hanging culture inserts, amount of agarose solution, and size of well plates will need to be adapted accordingly.

Acknowledgements

This study was supported by research grants from the Flemish League Against Cancer-Public Utility Foundation (VLK), the Scientific Research Foundation Flanders (FWO), the Vrije Universiteit Brussel, the Swedish Research Council/Finnish Academy of Science, Emil och Wera Cornells Stiftelse, the Swedish Childhood Cancer Foundation as well as through the regional agreement on medical training and clinical research (ALF) between Stockholm County Council and Karolinska Institutet. Y.B. is a postdoctoral fellow of the FWO.

References

1. Schlatt S, Ehmcke J (2014) Regulation of spermatogenesis: an evolutionary biologist's perspective. Semin Cell Dev Biol 29:2–16
2. Martinovitch PN (1938) The development in vitro of the mammalian gonad. Ovary and ovogenesis. Proc R Soc Lond B Biol Sci 125(839):232–249
3. Stukenborg J-B et al (2009) New horizons for in vitro spermatogenesis? An update on novel three-dimensional culture systems as tools for meiotic and post-meiotic differentiation of testicular germ cells. Mol Hum Reprod 15 (9):521–529
4. Reda A et al (2014) In vitro spermatogenesis—optimal culture conditions for testicular cell survival, germ cell differentiation, and steroidogenesis in rats. Front Endocrinol (Lausanne) 5 (21):1–11
5. Brown BN, Badylak SF (2014) Extracellular matrix as an inductive scaffold for functional tissue reconstruction. Transl Res 163(4):268–285
6. Baert Y et al (2015) Derivation and characterization of a cytocompatible scaffold from human testis. Hum Reprod 30(2):256–267
7. Gassei K, Orwig KE (2016) Experimental methods to preserve male fertility and treat male factor infertility. Fertil Steril 105(2):256–266
8. Baert Y et al (2017) Primary human testicular cells self-organize into organoids with testicular properties. Stem Cell Reports 8(1):30–38
9. Volkmann J et al (2011) Disturbed spermatogenesis associated with thickened lamina propria of seminiferous tubules is not caused by dedifferentiation of myofibroblasts. Hum Reprod 26 (6):1450–1461

Methods in Molecular Biology (2018) 1577: 129–138
DOI 10.1007/7651_2017_33

Published online: 28 April 2017

Decellularization of Bovine Small Intestinal Submucosa

Mahmut Parmaksiz, Ayşe Eser Elçin, and Yaşar Murat Elçin

Abstract

Decellularization technology promises to overcome some of the significant limitations in the regenerative medicine field by providing functional biocompatible grafts. The technique involves removal of the cells from the biological tissues or organs for further use in tissue engineering and clinical interventions. There are significant differences between decellularization protocols due to the intrinsic properties of different tissue types and purpose of use. This multistep, chemical-solution-based protocol is optimized for the preparation of decellularized bovine small intestinal submucosa (SIS).

Keywords: Decellularization, Bovine, Small intestinal submucosa (SIS), Xenogenic biomaterials, Extracellular matrix (ECM), Regenerative medicine, Bioactive materials

1 Introduction

With the advent of decellularization techniques, last 10 years have witnessed a new impetus in the field of biomaterials, namely for the production of extracellular matrix (ECM) based scaffolds. These studies initiated a new stage that will enable the use of biological (xeno-, allo-, and auto-) ECM for clinical tissue engineering/regenerative medicine applications. The basic principle of the decellularization technique is to remove the cells from a tissue or organ from various sources to obtain a biocompatible natural three-dimensional structure [1, 2]. A body of work shows that upon decellularization, ECM-based biomaterials retain collagen types, glycosaminoglycans, and growth factors such as transforming growth factor-β (TGF-β), basic fibroblast growth factor (b-FGF), and vascular endothelial growth factor (VEGF) that play active roles in regeneration by contributing to the cell behavior [3–7]. Various tissue types including small intestinal submucosa (SIS), heart valve, skin, tendon, bladder, blood vessel, skeletal muscle, and others have been used in preclinical tissue engineering studies; thus some of them have been translated into clinical use [7, 8]. Among all, the SIS layer has attracted considerable attention, thanks to its rich ECM content and ease of procurement. Decellularized SIS is an ECM-based biomaterial, which is commonly isolated from porcine small intestine. It has been approved by the FDA for treating venous ulcers, diabetic ulcers, skin injuries, also for repairing bladder wall, hernia, vascular defects, and lastly cartilage regeneration [7, 9–13]. Chemical agents employed for decellularization of biological materials show variation, depending on the purpose of use and tissue source. Therefore, the decellularization protocol must be assessed by considering the structure and cellular content of the intended tissue or organ sample. For example, the effectiveness of the chemicals would possibly vary between tendon and liver decellularization [14–16]. In addition to cellular composition, the thickness, the compactness, the lipid composition, etc. of a tissue or organ will define the decellularization technique and choice of solutions [15]. Most decellularization protocols are multistep protocols involving physical, chemical, and additionally enzymatic digestion steps [17, 18]. Generally, physical methods such as agitation, sonication, electroporation, high pressure, or freezing/thawing comprise the initial step of the decellularization protocols. This stage aims to disrupt the cell membrane to release cellular components. Washing process removes the majority of the cellular components. However, physical methods are often not sufficient; hence, the efficiency of decellularization protocol is improved by the use of chemical or enzymatic agents [7, 19–21]. In this chapter, we provide an effective multistep protocol that utilizes physical and chemical agents for the decellularization of bovine SIS layers.

2 Materials

All solutions should be prepared with analytical grade reagents in ultrapure water (by purification of deionized water to obtain 18 MΩ-cm sensitivity at +25 °C). Unless otherwise indicated, prepare all reagents at room temperature and store at +4 °C until use. All the steps of isolation and decellularization should be carried out by shaking at room temperature. Waste disposal regulations should be carefully followed at all times.

2.1 Equipment

1. Shaking water bath
2. Magnetic stirrer
3. UV-Vis spectrophotometer
4. Micropipettes
5. Ultralow temperature freezer (−86 °C)
6. Liquid nitrogen tank
7. Surgical scissors, clamps, and forceps
8. Glass conical flasks and beakers in various volumes

2.2 Bovine Small Intestinal Submucosa Isolation and Decellularization Agents

1. Isotonic sodium chloride (NaCl) solution: Add 9 g NaCl to 1 L of distilled water inside an Erlenmeyer flask and solubilize it with a magnetic stirrer (*see* **Note 1**).
2. Sodium hydroxide (NaOH) solution: Dissolve 4 g NaOH pellets in 1 L distilled water, final concentration 0.1 M (*see* **Note 2**).
3. 0.1 M Ascorbic acid: 0.15 % Peracetic acid (PAA) (Sigma-Aldrich, cat. no. 77240) (3:1) (v:v) solution: First prepare 0.15 % PAA solution in 1 L distilled water. For this purpose, add 385 μL of PAA stock solution in 1 L distilled water and mix on a magnetic stirrer. Then, add 17.6 g ascorbic acid to the 0.15 % PAA solution and mix with a magnetic stirrer (*see* **Note 3**).
4. 0.15 % PAA solution in 70 % ethanol: Dilute 300 mL of 100 % ethanol stock solution in 700 mL distilled water. Later, add 385 μL PAA solution as described in the previous step (*see* **Note 4**).
5. 0.15 % PAA solution prepared in 10–15 % hydrogen peroxide (Sigma-Aldrich, cat. no. 216763): First, mix 500 mL H_2O_2 stock solution with 500 mL distilled water to 10–15 %. Subsequently, add 385 μL PAA in 10–15 % H_2O_2 solution and mix with a magnetic stirrer (*see* **Note 5**).

2.3 Spectrophotometric Determination of DNA Content

1. Digestion solution: 10 mM Tris–HCl, 50 mM potassium chloride (KCl), 1.5 mM magnesium chloride ($MgCl_2$), 0.5 % Tween-20, and 20 mg/mL proteinase K: add 0.1576 g Tris–HCl, 0.3727 g KCl, 0.0142 g $MgCl_2$, 500 μL Tween-20 to 100 mL distilled water and dissolve on magnetic stirrer. Add 2 g proteinase K to complete the preparation of the stock digestion solution (*see* **Note 6**). Prepare right before use.
2. Phenol:chloroform:isoamyl alcohol: add 2.5 mL phenol, 2 mL chloroform, and 0.5 mL isoamyl alcohol in a total volume of 5 mL (volumetric ratios are 25:24:1) (*see* **Note** 7).
3. 3 M Sodium acetate solution: Dissolve 1.2304 g sodium acetate crystals in 5 mL distilled water.

3 Methods

All procedures should be carried out at room temperature unless otherwise specified.

3.1 Isolation of the Bovine Small Intestinal Submucosa Layer

1. Wash bovine small intestine segments repeatedly with 0.9 % NaCl solution (*see* **Note 8**). Carefully remove the adipose tissue around the intestines without tearing the intestine and wash with 0.9 % NaCl solution (*see* **Note 9**).

2. Cut bovine small intestine segments into 10-cm-long pieces or according to the planned size of the final product; then cut vertically to bring into membrane form (Fig. 1a, b).
3. Gently remove the outermost layer (*Tunica serosa*) manually from the small intestine membranes (Fig. 1c).

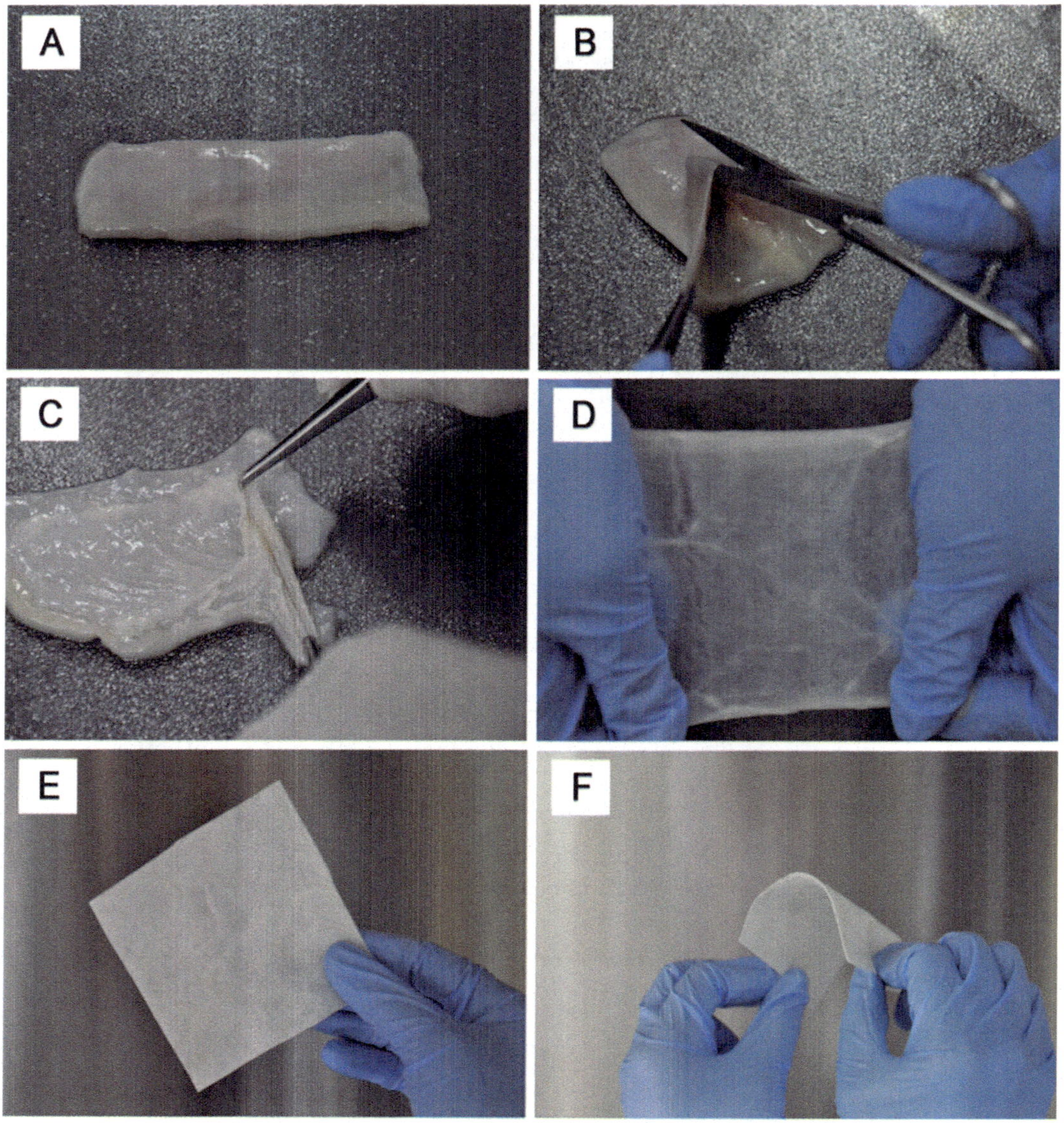

Fig. 1 Macroscopic images retrieved during isolation and after decellularization of bovine small intestinal submucosa (SIS): (**a**) Segment of fresh native bovine small intestine, (**b**) lengthwise cutting of the small intestine segment to bring into membrane form, (**c**) removal of layers from small intestine, (**d**) separated SIS layer , and (**e**, **f**) lyophilized form of decellularized bovine SIS membranes

4. Remove the muscle layer at the bottom of the intestine (under the serosa layer) in the same way.
5. Mucosa membrane is the innermost layer of the intestine, which can be removed by washing. Remove the mucosa by serial washing with 0.9 % NaCl solution (*see* **Note 10**). At the final stage, separate the SIS layer by mechanically removing the remaining muscle layer (Fig. 1d).

3.2 Decellularization of the Bovine Small Intestinal Submucosa Layer

1. Wash the separated bovine SIS layers with 0.9 % NaCl solution until the tissue debris is completely removed (*see* **Note 11**).
2. Following the washing procedure, apply two freeze–thaw cycles. For this, store bovine SIS layers at −86 °C (liquid nitrogen is another option) for 15–20 min. Immediately then incubate the specimens at +37 °C until completely thawed (*see* **Note 12**).
3. Upon thawing, wash the bovine SIS layers with 0.9 % NaCl solution at room temperature for about 10–15 min (*see* **Note 11**).
4. Subsequently, treat bovine SIS layers with 0.1 M NaOH solution at room temperature for 1 h (*see* **Note 13**).
5. In order to remove the remaining NaOH solution from the bovine SIS layers, wash the samples with 0.9 % NaCl solution at room temperature (*see* **Note 11**).
6. Incubate bovine SIS layers for 36 h at 37 °C with 0.1 M ascorbic acid and 0.15 % PAA (3:1) (v:v) solution, prepared in distilled water (*see* **Note 14**).
7. Wash again with 0.9 % NaCl solution at room temperature to remove the chemical residues (*see* **Note 15**).
8. Soak bovine SIS layers in 0.15 % M PAA solution prepared in 70 % ethanol for 20–24 h, at room temperature under constant shaking (*see* **Note 16**).
9. Wash again with 0.9 % NaCl at room temperature, to remove any remaining chemical residues (*see* **Note 11**).
10. Thereafter, treat bovine SIS layers with 0.15 % PAA solution prepared in 10–15 % hydrogen peroxide (H_2O_2) for 12–16 h at room temperature for bleaching and disinfection (*see* **Notes 4** and **5**).
11. At the final stage, repetitively wash bovine SIS layers with 0.9 % NaCl until the chemicals are completely removed. This is followed by lyophilization.
12. Place decellularized bovine SIS layers in specified molds and freeze at −86 °C for at least 12 h. Then, lyophilize overnight at −80 °C (*see* **Note 17**).
13. Finally, prepare lyophilized bovine SIS layers for sterilization (Fig. 1e, f) (*see* **Note 18**).

3.3 Spectrophotometric Evaluation of Decellularization Efficiency

1. Cut 1 cm^2 samples from decellularized and native bovine SIS membranes (*see* **Note 19**).
2. Weigh the samples and record initial weights (*see* **Note 20**).
3. Incubate samples in the digestion solution [10 mM Tris–HCl (pH 8.0), 50 mM KCl, 1.5 mM $MgCl_2$, 0.5 % Tween-20, and 20 mg/mL proteinase K] at 55 °C for 48 h or until the solution is clear, inside a water bath.
4. After incubation, centrifuge the final solution at +4 °C at 3,000 rpm for 15 min and collect the supernatant.
5. Add 100 μL mixture of phenol:chloroform:isoamyl alcohol (volume ratio of 25:24:1) onto the collected supernatant. Centrifuge for 15 min at +4 °C and 3,000 rpm (*see* **Note 7**).
6. Add 200 μL of 3 M sodium acetate (pH 5.5) and 500 μL of 95 % ethanol to the mixture, and incubate at −80 °C for 20 min.
7. Subsequently incubate the samples at 37 °C for 30 min.
8. Centrifuge the samples at 25 °C and 10,000 rpm for 10 min, and then discard the supernatant.
9. Add 100 μL of DNase/RNase-free water on the pellet, resuspend and perform spectrophotometric measurements at 260 and 280 nm. Calculate the DNA concentration for SIS samples using Eq. 1 (*see* **Note 21**) (Fig. 2).

$$\text{DNA concentration} = 50\,\mu\text{g/mL} \times \text{OD}_{260} \times \text{dilution factor} \quad (1)$$

$$\text{DNA purity} = \text{OD}_{260/280} = \begin{array}{ll} \sim 1.8 & \text{(ideal)} \\ \ll 1.8 & \text{contamination} \\ \gg 1.8 & \text{contamination} \end{array}$$

4 Notes

(1) Prepare the solutions in large volumes for liberal use during the procedure, and store at room temperature until use unless otherwise stated.

(2) The NaOH solution should be prepared in glass conical flask or beaker due to the corrosive nature of NaOH. Wear protective glasses and gloves to avoid skin and eye contact during preparation.

(3) Take note of the PAA stock solution, which is available in varying concentrations (e.g., 30–40 %) in the market. The quantities and volumes are specified for the product code in this protocol.

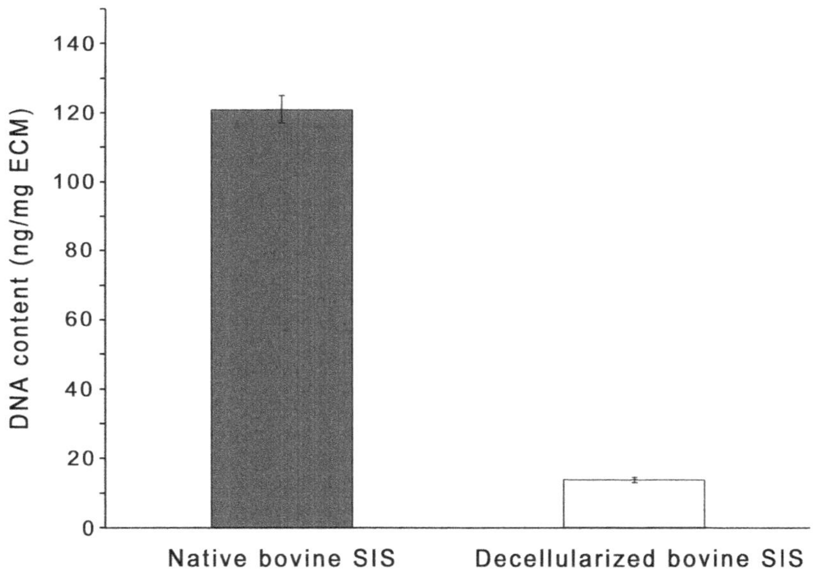

Fig. 2 Evaluation of protocol efficiency in terms of DNA content of decellularized bovine SIS membranes relative to native bovine SIS layers. Following decellularization process, DNA content was significantly reduced compared to the native bovine SIS (Control). Amounts of DNA content were determined as: 121 $\pm$ 4.02 ng per mg extracellular matrix (ECM) for native bovine SIS and 13.87 $\pm$ 0.86 ng per mg ECM for decellularized bovine SIS (the ideal DNA amount of decellularized tissue has not been defined in the literature; however, most studies suggest that there should be no more than 50 ng of DNA per mg ECM dry weight after decellularization [15])

Importantly, PAA stock solution is highly volatile, due to its high amount of acetic acid content, thus handle it under a fume hood and pay attention during pipetting. Add and solubilize ascorbic acid slowly, if high amounts are used. The final solution must be stored at 2–8 °C until use. Prepare the solutions in a glass conical flask or beaker and seal with parafilm for storage.

(4) Perform all steps under a fume hood. If possible, perform in a closed glass bottle under magnetic stirring.

(5) The calculations may change depending on the concentration of H_2O_2 stock solution used. Commercial H_2O_2 solutions are available with a range between 1 and 15 %. The present protocol uses 30 % stock solution of H_2O_2.

(6) Proteinase K solution should be added after the temperature of the main solution reaches 55 °C in the water bath, since proteinase K displays maximum activity at 55 °C. On the other hand, add Tween-20 solution at a later stage and mix at low speeds to prevent foaming during the preparation.

(7) The solution used in this step is corrosive and can cause severe toxic effects upon inhalation or skin contact. Perform all steps under a fume hood. All personnel must wear protective masks.

(8) Following the extraction, small intestine samples should be immediately transferred to the laboratory in cold 0.9 % NaCl solution. The main purpose of the washing step is to clean the inside of the intestine. Do not use any auxiliary equipment at this stage.

(9) The bovine intestine has significant amount of fat tissue, especially in the nodal areas. To avoid rupture of the intestines taken from the ice-cold solution, fat tissue should be eliminated during frozen state.

(10) Standard washing process removes most or entire mucosa layer since it is a thin epithelial structure. Further mechanical separation may be redundant.

(11) At this stage, the use of 0.9 % NaCl solution allows for general tissue cleansing as well as facilitating cell lysis and disruption of DNA–protein interaction through osmotic shock. It is therefore recommended to refresh the salt solution (every 10 min) depending on the size and number of tissue pieces. Perform washing procedure on a shaker.

(12) Perform to disrupt the cell membrane (by the formation of intracellular ice crystals) by the freeze–thaw process. Place the samples on aluminum sheets to freeze at −86 °C. Alternatively, implementation of liquid nitrogen may freeze tissues faster and increase decellularization efficiency.

(13) The alkaline NaOH solution dissolves the cytoplasmic components of the cells and breaks down the nucleic acids. At this stage, the solution may be refreshed depending on the amount of sample. Please refer to **Note 3** when working with NaOH solution.

(14) PAA allows the removal of nucleic acid residues, which can increase the effectiveness of the agents used in further steps, while minimally damaging the ECM content. It is important to refresh the solution at least three times depending on the number and size of samples. Check **Note 4** while working with PAA.

(15) In addition to **Note 11**, the washing duration may vary according to the type of chemical to be removed. For this purpose, unlike routine washes, use fresh 0.9 % NaCl solution until pH stabilizes.

(16) The dehydration effect of ethyl alcohol enhances the removal of lipids and cell lysis, besides the PAA effect described in **Note 14**. The use of ethyl alcohol makes the solution

very volatile. Keep the container covered during the decellularization process or work under a fume hood.

(17) Prior to lyophilization, samples must be frozen in membrane form on a flat surface, e.g., aluminum sheet. Avoid stretching the natural SIS layer, since freezing weakens the mechanical properties.

(18) All samples should be sealed in packages before sterilization (gamma, ethylene oxide, etc.). Upon sterilization, samples must be stored at +4 °C until use.

(19) Cut the samples with sterile scissors inside the biosafety cabinet to avoid possible environmental DNA contamination.

(20) It is important that the weights of the tissue pieces are close to each other. Reduce the size of tissue pieces to equalize the weights.

(21) The specified calculation is useful for data acquired by conventional spectrophotometric measurements or with bench-top small-volume spectrophotometers.

Competing Interests

Y.M.E. is the founder of Biovalda, Inc. (Ankara, Turkey) and has intellectual properties related to decellularized tissues.

References

1. Cornwell KG, Landsman A, James KS (2009) Extracellular matrix biomaterials for soft tissue repair. Clin Podiatr Med Surg 26(4):507–523
2. Parmaksiz M, Dogan A, Odabas S, Elcin AE, Elcin YM (2016) Clinical applications of decellularized extracellular matrices for tissue engineering and regenerative medicine. Topical review. Biomed Mater 11(2):022003
3. Voytik-Harbin SL, Brightman AO, Kraine MR, Waisner B, Badylak SF (1997) Identification of extractable growth factors from small intestinal submucosa. J Cell Biochem 67(4):478–491
4. McDevitt C, Wildey GM, Cutrone RM (2003) Transforming growth factorbeta1 in a sterilized tissue derived from the pig small intestine submucosa. J Biomed Mater Res A 67:637–640
5. Hodde JP, Record RD, Liang HA, Badylak SF (2001) Vascular endothelial growth factor in porcine-derived extracellular matrix. Endothelium 8(1):11–24
6. Londono R, Badylak SF (2015) Biologic scaffolds for regenerative medicine: mechanisms of in vivo remodeling. Ann Biomed Eng 43:577–592
7. Parmaksiz M, Elcin AE, Elcin YM (2015) Decellularization of bovine small intestinal submucosa and its use for the healing of a critical-sized full-thickness skin defect, alone and in combination with stem cells, in a small rodent model. J Tissue Eng Regen Med. doi:10.1002/term.2071 (Epub ahead of print)
8. Hoshiba T, Lu H, Kawazoe N, Chen G (2010) Decellularized matrices for tissue engineering. Expert Opin Biol Ther 10:1717–1728
9. Badylak SF, Lantz GC, Coffey A, Geddes LA (1989) Small intestinal submucosa as a large diameter vascular graft in the dog. J Surg Res 47(1):74–80
10. Badylak SF, Liang A, Record R, Tullius R, Hodde J (1999) Endothelial cell adherence to small intestinal submucosa: an acellular bioscaffold. Biomaterials 20(23–24):2257–2263
11. Lindberg K, Badylak SF (2001) Porcine small intestinal submucosa (SIS): a bioscaffold supporting in vitro primary human epidermal cell differentiation and synthesis of basement membrane proteins. Burns 27(3):254–266

12. Pribitkin EA, Ambro BT, Bloeden E, O'Hara BJ (2004) Rabbit ear cartilage regeneration with a small intestinal submucosa graft. Laryngoscope 114:1–19
13. Caione P, Capozza N, Zavaglia D, Palombaro G, Boldrini R (2006) In vivo bladder regeneration using small intestinal submucosa: experimental study. Pediatr Surg Int 22(7):593–599
14. Gilbert T, Sellaro TL, Badylak SF (2006) Decellularization of tissues and organs. Biomaterials 27:3675–3683
15. Crapo PM, Gilbert TW, Badylak SF (2011) An overview of tissue and whole organ decellularization processes. Biomaterials 32(12):3233–3243
16. Kawecki M, Łabuś W, Klama-Baryla A, Kitala D, Kraut M, Glik J, Misiuga M, Nowak M, Bielecki T, Kasperczyk A (2017) A review of decellurization methods caused by an urgent need for quality control of cell-free extracellular matrix scaffolds and their role in regenerative medicine. J Biomed Mater Res B Appl Biomater. doi:10.1002/jbm.b.33865 (Epub ahead of print)
17. Brown BN, Freund JM, Han L, Rubin JP, Reing JE, Jeffries EM, Wolf MT, Tottey S, Barnes CA, Ratner BD, Badylak SF (2011) Comparison of three methods for the derivation of a biologic scaffold composed of adipose tissue extracellular matrix. Tissue Eng Part C Methods 17:411–421
18. Keane TJ, Swinehart IT, Badylak SF (2015) Methods of tissue decellularization used for preparation of biologic scaffolds and in vivo relevance. Methods 85:25–34
19. Ott HC, Matthiesen TS, Goh SK, Black LD, Kren SM, Netoff TI, Taylor DA (2008) Perfusion decellularized matrix: using nature's platform to engineer a bioartificial heart. Nat Med 14:213–221
20. Xu H, Xu B, Yang Q, Li X, Ma X, Xia Q, Zhang Y, Zhang C, Wu Y, Zhang Y (2014) Comparison of decellularization protocols for preparing a decellularized porcine annulus fibrosus scaffold. PLoS One 9:1–13
21. Roosens A, Somers P, De Somer F, Carriel V, Van Nooten G, Cornelissen R (2016) Impact of detergent-based decellularization methods on porcine tissues for heart valve engineering. Ann Biomed Eng 44:2827–2839

Methods in Molecular Biology (2018) 1577: 139–146
DOI 10.1007/7651_2017_59

Published online: 11 August 2017

Decellularization and Recellularization of Cartilage

Catherine A. Bautista and Bahar Bilgen

Abstract

Decellularization of cartilage enables the use of cartilage allografts or xenografts as natural scaffolds for repair and regeneration of injured cartilage. The preservation of the extracellular matrix ultrastructure of the graft makes this a promising tool for cartilage tissue engineering. We have optimized the decellularization protocol by enzymatically digesting proteoglycans while preserving the native collagen architecture. Here we describe our methods for cartilage decellularization and cell labeling for the tracking of infiltration for recellularization in detail.

Keywords Allograft, Cartilage, Cell seeding, Decellularization, Recellularization, Transplantation, Viability

1 Introduction

The limited healing capacity of articular cartilage demonstrates a need for either reparative tissue grafts or regenerative biomaterials. Fresh osteochondral allografts have limited availability and often evoke an immune response, which has been associated with unfavorable clinical outcomes [1]. Frozen storage allows for increased availability of allograft tissue, but causes cell death, which may still promote an inflammatory immune response due to residual cell debris in the tissue [2, 3]. To solve these issues, researchers have sought to engineer scaffolds that mimic the extracellular matrix (ECM) composition and organization of native articular cartilage [4]. Decellularization is of great interest to cartilage tissue engineers because it ideally preserves the unique ultrastructure and bioactivity of the native ECM, while reducing immunogenicity of allogeneic and xenogeneic donor tissue [5, 6]. One approach is to use freeze-thaw cycles, detergent, and enzymatic DNA digestion to decellularize intact cartilage [7, 8]. The relatively dense proteoglycan and collagen matrix of intact cartilage presents a challenge for decellularizing and recellularizing the tissue [9–11]. Therefore, we optimized a protocol for enzymatically eliminating proteoglycans to increase removal of cell debris, while preserving the native collagen fiber network [12]. The decellularized scaffold should be characterized in terms of the remaining DNA content to verify the removal of DNA from the scaffold. Depending on the functional requirements of the decellularized scaffold, biochemical assays confirming the amount of remaining extracellular matrix proteins such as proteoglycans or collagens, or mechanical testing to assess mechanical function can be performed. The fluorescent labeling of the seeded cells enables the tracking of these cells to assess the infiltration capacity of these cells. Confocal microscopy or histological sectioning would be appropriate methods for the assessment of the cell infiltration in the decellularized scaffold. Here we describe our methods for cartilage decellularization and cell labeling for recellularization in detail. Figure 1 shows representative results of these decellularization methods applied on porcine articular cartilage.

2 Materials

2.1 Decellularization Solutions

Prepare all buffers using ultrapure water that has been sterilized by autoclaving. Sterilize final working solutions with 0.22 μm filters. Store all reagents at room temperature (RT), unless otherwise noted.

1. Phosphate-buffered saline (PBS).
2. Hypotonic buffer: 10 mM Tris–HCl, pH 8.0. To 1-l glass beaker, add 10 ml 1 M Tris–HCl stock and fill to 1 l with ultrapure water (see **Note 1**).
3. Protease Inhibitor Cocktail (Sigma P8340). Aseptically add 2.95 μl/ml to specified working solutions after they have been sterile-filtered in biological safety cabinet, immediately before use. Store stock at −20 °C.

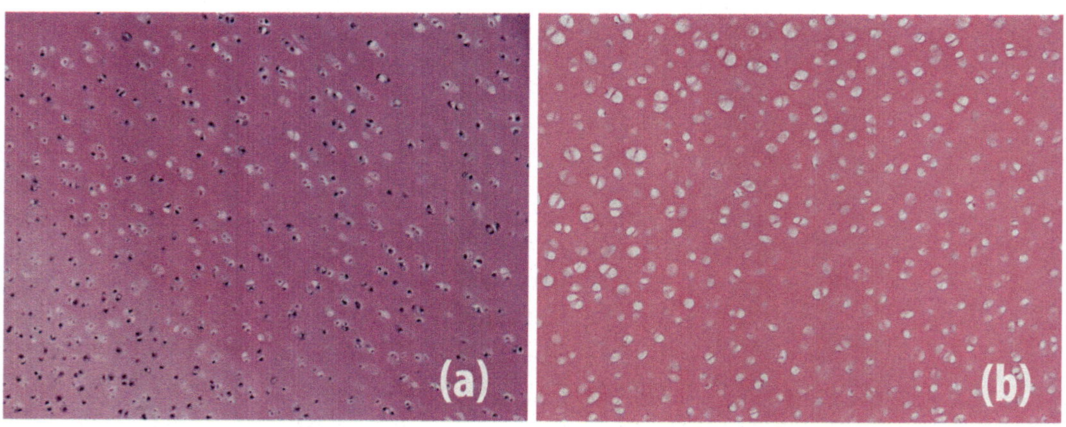

Fig. 1 Representative images of a decellularized scaffold. Porcine knee femur cartilage was decellularized using the methods described in this chapter. Hematoxylin & Eosin (H&E) staining of histological sections of (**a**) native articular cartilage and (**b**) decellularized cartilage scaffold

4. Chondroitinase ABC: 0.125 U/ml chondroitinase ABC, 50 mM Tris–HCl, 60 mM sodium acetate, 2 mM EDTA, pH 8.0. For a 100-ml solution: first add 80 ml water to a 250-ml glass beaker, then add the following (stirring after each addition): 0.394 g Trizma-HCl (Sigma T3253), 0.408 g sodium acetate (Sigma S2889), and 62.5 μl chondroitinase ABC stock (see **Note 2**). Adjust to pH 8.0 using concentrated NaOH. Add 0.037 g EDTA (Sigma E5134) and stir. Prepare working solution immediately before use.
5. Anionic detergent: 0.1% (w/v) sodium dodecyl sulfate (Thermo Fisher Scientific 15525017), 0.1% (w/v) EDTA in hypotonic buffer.
6. Nuclease solution: 100 U/ml DNase, 1 U/ml RNase, 10 mM magnesium chloride, 50 μg/ml BSA, 50 mM Trizma base. For a 100-ml solution: in a 250-ml glass beaker, add 304.75 mg Trizma base (Sigma T1503) and 80 ml water. Adjust to pH 7.5 using concentrated HCl. Then add the following (stirring after each addition): 101.65 mg magnesium chloride (Sigma M2670), 2.5 mg BSA, 1 ml RNase stock (see **Note 3**), and 2.500 ml DNase (Thermo Fisher Scientific AM2239). Prepare working solution immediately before use.
7. Peracetic acid: 0.1% (v/v) peracetic acid (Sigma 269336) in PBS.
8. Rinsing solution: 1% (v/v) penicillin streptomycin glutamine (Thermo Fisher Scientific 10378016), 1% (v/v) amphotericin B (Thermo Fisher Scientific 15290018) in PBS. Store at 4 °C.

2.2 Decellularization Materials

1. 24-Well non-treated cell culture plates (Corning 3738). See **Note 4**.
2. Nunc™ Biobanking and Cell Culture Cryogenic Tubes (Thermo Fisher Scientific 377267).
3. Nunc™ CryoColor Vial Closure Coder, PS (Thermo Fisher Scientific 375930).
4. Whatman® Qualitative Filter Paper, Grade 1 (Sigma Z240079).
5. Orbital shaker (New Brunswick Innova 2000).

2.3 Recellularization Solutions

1. 0.25% trypsin-EDTA (Thermo Fisher Scientific 25200056; see **Note 5**). Aliquot and store at −20 °C.
2. Serum-free medium: 1% (v/v) penicillin streptomycin glutamine, 1% (v/v) amphotericin B, and 1% (v/v) ITS+ Premix (Corning 354352) in DMEM, high glucose, pyruvate (Thermo Fisher Scientific 11995) (see **Note 6**). Store at 4 °C.
3. 10% FBS medium: 10% (v/v) fetal bovine serum (FBS; Atlanta Biologicals S11150) in serum-free medium. Store at 4 °C.

4. Vybrant DiI Cell-Labeling Solution (Thermo Fisher Scientific V22885; see **Note 7**). Store at RT, protected from light.

3 Methods

Perform all steps in a biological safety cabinet to minimize contamination.

3.1 Freeze-Thaw Cycles

1. Gently transfer each cartilage plug to an individual cryovial using a sterile micro spatula (see **Note 8**).
2. Using sterile scissors, cut one strip (approximately 5 mm × 20 mm) of filter paper per cartilage plug. Using sterile tweezers, immerse each strip in PBS and place in each cryovial.
3. Place cryovials in a −20 °C freezer for at least 3 h (or until crystal formation).
4. Thaw samples on bench for 2–4 h.
5. Repeat Steps 3 and 4.
6. Using sterile tweezers, remove filter paper from each cryovial.
7. Add 1 ml hypotonic buffer to each cryovial (see **Note 9**). Ensure the entire cartilage plug is submerged in the solution.
8. Place cryovials in a −20 °C freezer for at least 3 h.
9. Thaw samples on bench for 2–4 h.

3.2 Chondroitinase ABC Treatment

1. Pour contents of each cryovial into one well of a 24-well plate.
2. Aspirate old solution. Add 1 ml chondroitinase ABC solution with 2.95 μl/ml protease inhibitor cocktail to each well.
3. Agitate on orbital shaker set at 180 RPM at 37 °C incubator for 24 h (see **Note 10**).

3.3 SDS Application

1. Aspirate old solution. Add 1 ml hypotonic buffer with protease inhibitor cocktail to each well.
2. Agitate on an orbital shaker set at 180 RPM at 45 °C for 24 h.
3. Aspirate old solution. Add 1 ml anionic detergent with protease inhibitor cocktail to each well.
4. Agitate on an orbital shaker set at 180 RPM at 45 °C for 24 h.
5. Repeat Steps 1–4 twice for a total of three detergent cycles.
6. Aspirate old solution. Add 1 ml PBS with protease inhibitor cocktail to each well.
7. Agitate on an orbital shaker at 180 RPM at room temperature for 30 m.
8. Repeat Steps 6–7.
9. Repeat Step 6.

10. Agitate on an orbital shaker at 180 RPM at 45 °C for 24 h.

3.4 Nuclease Digestion

1. Aspirate old solution. Add 1 ml nuclease solution to each well.
2. Agitate on an orbital shaker set at 180 RPM at 37 °C for 3 h.
3. Aspirate old solution. Add 1 ml PBS with protease inhibitor cocktail to each well.
4. Agitate on orbital shaker set at 180 RPM at RT for 1 h.

3.5 Peracetic Acid Decontamination

1. Aspirate old solution. Add 1 ml peracetic acid solution to each well.
2. Agitate on an orbital shaker at 180 RPM at RT for 3 h.
3. Aspirate old solution. Add 1 ml rinsing solution.
4. Agitate on orbital shaker set at 180 RPM at RT for 30 m.
5. Repeat Steps 3–4.
6. Repeat Step 3.
7. Agitate on orbital shaker set at 180 RPM at RT for 24 h.
8. Repeat Step 3.
9. Samples are now ready for recellularization. Store at 4 °C until use (see **Note 11**).

3.6 Seeded Cell Labeling

1. Trypsinize mesenchymal stem cells intended for recellularization (see **Note 12**).
2. Resuspend cells in cold serum-free medium at a density of 2×10^6 cells/ml in a 50-ml centrifuge tube (see **Note 13**).
3. Rinse cells by centrifuging at 300 × *g* and resuspending in warm serum-free medium at a density of 1×10^6 cells/ml.
4. Add 5 μl DiI per ml cell suspension and mix well by gentle pipetting.
5. Tape tube with cells on its side on an orbital shaker set at 50 RPM in a 37 °C, 5% CO_2 incubator. Incubate for 40 min (see **Note 14**).
6. Rinse cells by centrifuging at 300 × *g* for 5 m, discarding supernatant, and resuspending in 10 ml warm serum-free medium (see **Note 15**).
7. Repeat Step 5 twice.
8. Labeled cells can now be used for seeding the decellularized scaffolds.

4 Notes

1. To make 1 M Tris–HCl stock: in a 1-l glass beaker with magnetic stir bar, add 121.1 g Trizma base (Sigma T1503) and

800 ml water. Dissolve using a hotplate with magnetic stirrer. Allow solution to reach room temperature (RT), then adjust to pH 8.0 with concentrated HCl. Fill to final volume of 1000 ml and sterilize by autoclaving.

2. To make 100 U/ml chondroitinase ABC stock: reconstitute 10 U chondroitinase ABC (Sigma C3667) in 100 μl 0.1% BSA. Aliquot and store at −20 °C.
3. To make RNase stock: reconstitute Ribonuclease A (Sigma R5000) in PBS to a final concentration of 100 U/ml. Aliquot and store at −20 °C.
4. Well size, volume per well, and orbital shaker RPM have been optimized for cylindrical porcine articular cartilage plugs with a diameter of 4 mm and height of 2–4 mm. These parameters may need to be adjusted for plugs of different sizes and from different sources.
5. We use 0.25% trypsin-EDTA for synovium-derived mesenchymal stem cells. If using other cells, we recommend using the typical cell dissociation reagent for the specific cell type used.
6. Medium composition is based on that of serum-free medium for standard synovial membrane cell culture. As in **Note 5**, composition may need to be adapted based on the final cell type used for recellularization.
7. Other Vybrant dyes (e.g., DiO, DiD) are available if multiple batches or types of cells are to be used for recellularization or if secondary antibodies for immunostaining overlap with DiI fluorescence spectra. The absorption and emission maxima for DiI are 549 and 565 nm, respectively.
8. We typically weigh the labeled cryovials when they are empty, before decellularization, and after decellularization to normalize water and extracellular matrix content to sample weight.
9. We found that use of a repeater pipet, such as the Eppendorf Repeater M4 (with 25-ml Combitips), saves time when decellularizing larger batches of samples (≥24 plugs). When using a repeater pipet, take care not to dispense the solution with too much force so that the plug does not "jump" out of its cryovial or well.
10. To prevent solution evaporation, we recommend sealing the edges of the plates with Parafilm. Additionally, to secure the plate on the orbital shaker, use one or two strips of lab tape across the width of the plate.
11. We recommend starting the recellularization process as soon as possible because degradation of the collagen network over time in 4 °C storage has not yet been investigated.

12. Typical trypsinization protocol for one 150-cm^2 flask of synovium-derived mesenchymal stem cells: discard medium. Gently rinse cells with 20 ml PBS and discard. Add 10 ml trypsin-EDTA and incubate at 37 °C for 2–3 m. Check to see if cells are detached, gently tapping the slide of the flask to facilitate detachment; return flask to 37 °C for a maximum of 2–3 m if most cells are still attached. Once most cells are detached, add 20 ml cold 10% FBS medium to inactivate the trypsin and transfer cell suspension to 50-ml centrifuge tube. Centrifuge at 300 × *g* for 5 m (preferably at 4 °C) and discard supernatant. Resuspend cells in 10 ml cold serum-free medium. Count cells and transfer the desired number of cells into a new 50-ml centrifuge tube. Centrifuge at 300 × *g* for 5 m and discard supernatant.
13. BSA in FBS inhibits DiI binding to cells; therefore serum-free medium should be used.
14. Incubation timing should be optimized for each new cell type used. Ideally, a range of incubation times (e.g., 20, 30, 40, 50 m) should be tested on a small population of cells for every new cell type and donor.
15. Centrifugation at this stage should preferably be conducted at 37 °C.

Acknowledgments

We would like to acknowledge the assistance of Courtney M. Mazur in protocol development. This work was supported by Career Development Award # IK2 RX000760 from the United States (U.S.) Department of Veterans Affairs Rehabilitation Research and Development Service.

References

1. Sirlin CB, Brossmann J, Boutin RD, Pathria MN, Convery FR, Bugbee W, Deutsch R, Lebeck LK, Resnick D (2001) Shell osteochondral allografts of the knee: comparison of mr imaging findings and immunologic responses. Radiology 219(1):35–43
2. Pallante AL, Görtz S, Chen AC, Healey RM, Chase DC, Ball ST, Amiel D, Sah RL, Bugbee WD (2012) Treatment of articular cartilage defects in the goat with frozen versus fresh osteochondral allografts: effects on cartilage stiffness, zonal composition, and structure at six months. J Bone Joint Surg Am 94 (21):1984–1995. doi:10.2106/JBJS.K.00439
3. Londono R, Dziki JL, Haljasmaa E, Turner N, Leifer C, Badylak SF (2017) The effect of cell debris within biologic scaffolds upon the macrophage response. J Biomed Mater Res A. doi:10.1002/jbm.a.36055
4. Camarero-Espinosa S, Rothen-Rutishauser B, Foster EJ, Weder C (2016) Articular cartilage: from formation to tissue engineering. Biomater Sci 4(5):734–767. doi:10.1039/c6bm00068a
5. Sutherland AJ, Converse GL, Hopkins RA, Detamore MS (2015) The bioactivity of cartilage extracellular matrix in articular cartilage regeneration. Adv Healthc Mater 4(1):29–39. doi:10.1002/adhm.201400165

6. Cheng CW, Solorio LD, Alsberg E (2014) Decellularized tissue and cell-derived extracellular matrices as scaffolds for orthopaedic tissue engineering. Biotechnol Adv 32(2):462–484. doi:10.1016/j.biotechadv.2013.12.012
7. Fermor HL, Russell SL, Williams S, Fisher J, Ingham E (2015) Development and characterisation of a decellularised bovine osteochondral biomaterial for cartilage repair. J Mater Sci Mater Med 26(5):186. doi:10.1007/s10856-015-5517-0
8. Elder BD, Kim DH, Athanasiou KA (2010) Developing an articular cartilage decellularization process toward facet joint cartilage replacement. Neurosurgery 66(4):722–727.; discussion 727. doi:10.1227/01.NEU.0000367616.49291.9F
9. Luo L, Chu JY, Eswaramoorthy R, Mulhall KJ, Kelly DJ (2016) Engineering tissues that mimic the zonal nature of articular cartilage using decellularized cartilage explants seeded with adult stem cells. ACS Biomater Sci Eng. doi:10.1021/acsbiomaterials.6b00020
10. Utomo L, Pleumeekers MM, Nimeskern L, Nürnberger S, Stok KS, Hildner F et al (2015) Preparation and characterization of a decellularized cartilage scaffold for ear cartilage reconstruction. Biomed Mater 10(1):015010. doi:10.1088/1748-6041/10/1/01501
11. Schwarz S, Elsaesser AF, Koerber L, Goldberg-Bockhorn E, Seitz AM, Bermueller C et al (2015) Processed xenogenic cartilage as innovative biomatrix for cartilage tissue engineering: effects on chondrocyte differentiation and function. J Tissue Eng Regen Med 9(12): E239–E251. doi:10.1002/term.1650
12. Bautista CA, Park HJ, Mazur CM, Aaron RK, Bilgen B (2016) Effects of chondroitinase ABC-mediated proteoglycan digestion on decellularization and recellularization of articular cartilage. PLoS One 11(7):e0158976. doi:10.1371/journal.pone.0158976

Methods in Molecular Biology (2018) 1577: 147–154
DOI 10.1007/7651_2017_27

Published online: 28 April 2017

A Protocol to Prepare Decellularized Stem Cell Matrix for Rejuvenation of Cell Expansion and Cartilage Regeneration

Jingting Li and Ming Pei

Abstract

Traditional ex vivo expansion of adult stem cells yields an insufficient quantity of less potent cells. Here we describe the fabrication of decellularized matrix deposited by synovium-derived stem cells (SDSCs). This matrix could serve as a three-dimensional expansion system to rejuvenate cells for proliferation and tissue-specific differentiation potential, which could benefit cartilage regeneration. The decellularized stem cell matrix (DSCM) might be a powerful system for tissue engineering and regeneration.

Keywords Decellularized stem cell matrix, Cell expansion, Synovium-derived stem cells, Chondrogenesis, Cartilage tissue engineering, Cartilage regeneration

1 Introduction

Injury and degenerative diseases such as osteoarthritis are the main causes of cartilage defects. Cartilage defects rarely heal spontaneously due to their intrinsic avascular nature. Despite the fact that both autologous chondrocyte transplantation and stem cell based therapy are promising approaches for the repair of cartilage defects [1], unstable lineage differentiation and an insufficient cell number after traditional two-dimensional cell culture and expansion are the two major challenges for cartilage engineering and regeneration [2].

Physiologically, stem cells reside in the so-called niche structure in vivo, which is mainly composed of extracellular matrix and soluble factors. Thus, the decellularized extracellular matrix not only can mimic the three-dimensional structure of the stem cell niche but it can also retain the biochemical signals that direct the stem cells to undergo lineage differentiation [3]. Synovium-derived stem cells (SDSCs) have been identified as a tissue-specific stem cell for chondrogenesis [4, 5]. To reconstruct an optimal microenvironment for ex vivo cell expansion to supplement the shortage of cells for cartilage tissue engineering [6], we developed a decellularized stem cell matrix (DSCM) deposited by SDSCs as an expansion substrate with rejuvenation capacities. DSCM expanded cells demonstrated significantly enhanced proliferation and chondrogenic capacity in both porcine and human SDSCs in vitro [7–10]. Repair of cartilage defects using DSCM pretreated SDSCs also showed better in vivo outcomes when compared to plastic flask expanded SDSCs [11]. So far, we and other labs have performed this technique using bone marrow- [12, 13] and adipose-derived mesenchymal stem cells (MSCs) [14], chondrocytes [15], and nucleus pulposus cells [16, 17] for tissue-specific regeneration purposes. This intriguing technique would help us further understand the biology of stem cells and their niches.

Here we describe the fabrication of the SDSC deposited DSCM with steps indicated below (Fig. 1). After plating on DSCM, SDSCs had their morphology changed as well as cell proliferation and chondrogenic capacity rejuvenated (Fig. 2). While the protocol presented here primarily focuses on the procedures for preparing DSCM and expanding cells, it is important to note that the mechanisms underlying the rejuvenating effect are exciting [6]. Full characterization of DSCM deposited by SDSCs, including analyses of components of

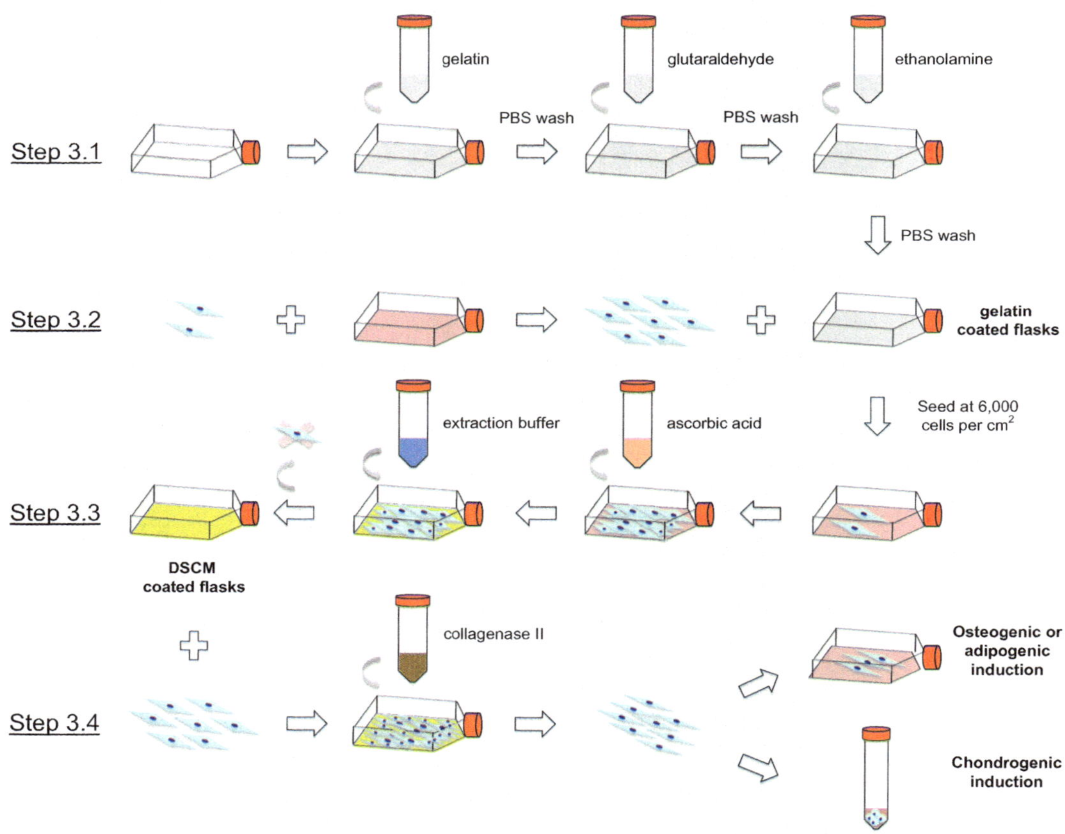

Fig. 1 Stepwise preparation of decellularized stem cell matrix (DSCM) and synovium-derived stem cell (SDSC) expansion. Step 3.1. Surface preparation for DSCM deposition. Step 3.2. Culture of SDSCs. Step 3.3. DSCM fabrication using SDSCs. Step 3.4. Cell seeding and harvesting after DSCM expansion and preparation of multi-lineage differentiation

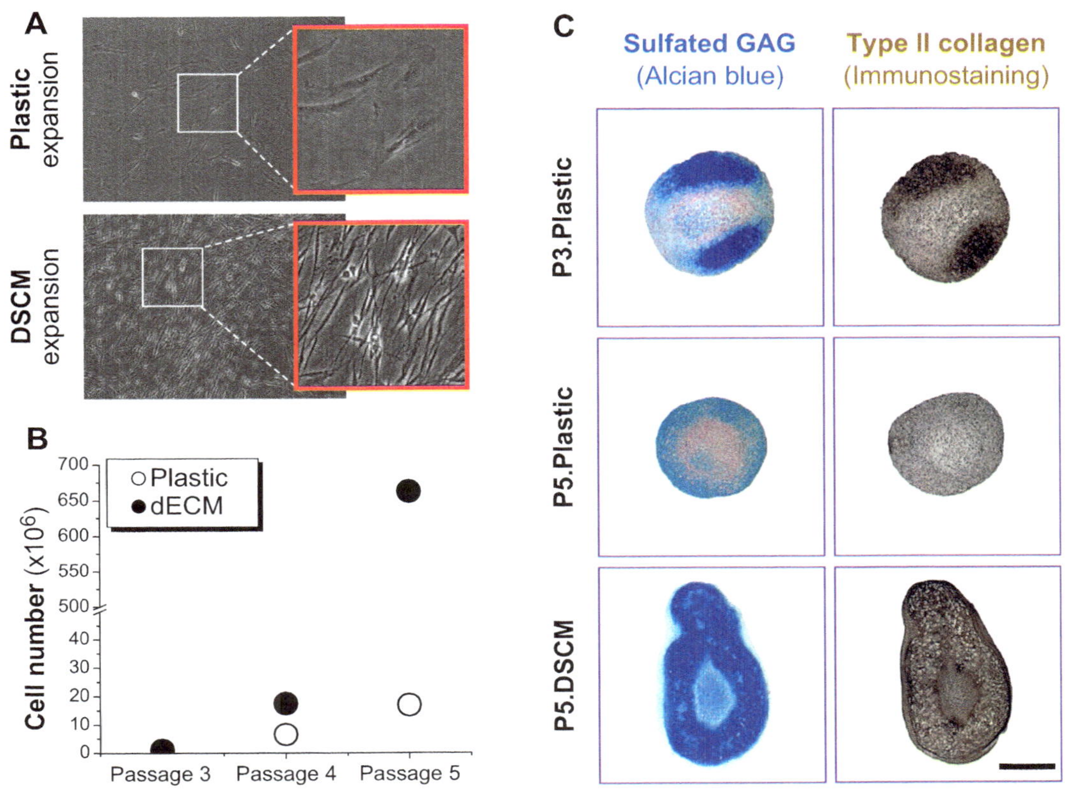

Fig. 2 Effect of DSCM deposited by SDSCs on porcine SDSCs' proliferation and chondrogenic differentiation. (**a**) Cell morphology 5 days after expansion on DSCM and plastic flasks. (**b**) Cell proliferation from passage 3 (P3) SDSCs grown on either DSCM or plastic flasks for two consecutive passages. (**c**) *Alcian blue staining* for sulfated glycosaminoglycan (GAG) and immunostaining for type II collagen (scale bar: 800 mm) of 2-week chondrogenically induced SDSCs in a pellet culture system after two passages on DSCM (P5.DSCM) or plastic flasks (P5.Plastic) with pre-expansion SDSCs (P3.Plastic) as a control. Reproduced from Ref. [7] with permission from Mary Ann Liebert, Inc. Publications

DECM by proteomics, has been attempted [18]. More in vivo animal models and human studies about its efficiency and immunity issues are needed before it can be applied clinically.

2 Materials

2.1 Solutions for Matrix Deposition

1. Gelatin: Weigh 0.2 g of gelatin and dissolve in 100 mL of phosphate buffered saline (PBS) while heating to 37 °C. Cool and filter through a 0.2 μm filter. Store at 4 °C
2. Glutaraldehyde: 1 % (v/v) solution in PBS and filter through a 0.2 μm filter. Store at 4 °C
3. Ethanolamine: 1 M solution in water and filter through a 0.2 μm filter. Store at 4 °C

4. 1× PBS with Ca^{2+} and Mg^{2+}
5. T175 flasks/plastic container

2.2 Culturing Synovium-Derived Stem Cells

1. Growth medium: Alpha Minimum Essential Medium Eagle—Alpha Modifications (alpha MEM) containing 10 % fetal bovine serum (FBS), 100 U/mL penicillin, 100 μg/mL streptomycin, and 0.25 μg/mL Fungizone
2. 0.5 % trypsin–Ethylenediaminetetraacetic acid (EDTA)
3. 1× PBS without Ca^{2+} and Mg^{2+}

2.3 Decellularized Stem Cell Matrix Fabrication

1. Matrix deposition medium: Ascorbic acid should be freshly prepared prior to use as a 1,000× stock concentration of 250 mM in growth medium (from Step 2.2.1) in order to yield a final concentration of 250 μM [19]. Weigh 7.24 g of L-Ascorbic acid phosphate magnesium salt and dissolve in 100 mL growth medium. Filter sterilize with a 0.2 μm filter, aliquot, and store at −20 °C.
2. Extraction buffer: 0.5 % Triton X-100 and 20 mM NH_4OH in PBS. Filter sterilize and store at 4 °C.
3. PBS supplement with antibiotics: 1× PBS with Ca^{2+} and Mg^{2+} containing 100 U/mL penicillin, 100 μg/mL streptomycin, and 0.25 μg/mL Fungizone. Store at 4 °C.

2.4 Cell Expansion on Decellularized Stem Cell Matrix and Multi-lineage Differentiation

1. Growth medium
2. 0.2 % collagenase type II: Dissolve in 1× PBS and 50 mM $CaCl_2$. Filter to sterilize and store at −20 °C
3. Cell strainer (70 μm Nylon)
4. Lineage differentiation media: **Chondrogenic medium** consists of high-glucose Dulbecco's modified Eagle's medium (DMEM), 40 mg/mL proline, 0.1 μM dexamethasone, 100 U/mL penicillin, 100 μg/mL streptomycin, 0.1 mM L-Ascorbic acid phosphate magnesium salt, and 1× ITS Premix (6.25 μg/mL insulin, 6.25 μg/mL transferrin, 6.25 μg/mL selenous acid, 5.35 μg/mL linoleic acid, and 1.25 μg/mL bovine serum albumin) supplemented with 10 ng/mL transforming growth factor β3. **Adipogenic medium** consists of growth medium supplemented with 1 mM dexamethasone, 0.5 mM isobutyl-1-methyxanthine, 200 μM indomethacin, 10 μM insulin, and 1 nM 3,3′5′-triiodo-L-thyronine. **Osteogenic medium** consists of growth medium supplemented with 0.01 μM dexamethasone, 10 mM β-glycerol phosphate, 50 μM L-Ascorbic acid phosphate magnesium salt, and 0.01 μM 1,25-dihydroxyvitamin D3

3 Methods

3.1 Surface Preparation for Decellularized Stem Cell Matrix Deposition

1. To prepare surface for matrix deposition, add enough 0.2 % gelatin solution to cover the bottom of a tissue culture plastic flask or dish and incubate at 37 °C for 1 h (scale up the volume of all reagents according to container size).
2. Aspirate gelatin solution and rinse once with PBS for 5 min.
3. Add 1 % glutaraldehyde to each flask or dish and incubate at room temperature (RT) for 30 min.
4. Remove glutaraldehyde and wash three times with PBS for 5 min each.
5. Incubate flask or dish with 1 M ethanolamine at RT for 30 min and aspirate ethanolamine.
6. Repeat PBS wash three times for 5 min each.
7. Aspirate PBS and add growth medium to the flask or dish. If medium turns purple, repeat Steps 3.1.6 and 3.1.7 to remove any trace amounts of ethanolamine.
8. At this point, the surface is prepared and ready to seed cells. Another option is to keep the flask or dish with coated surface in PBS at RT for days (less than 1 week) under sterile conditions.

3.2 Culturing Synovium-Derived Stem Cells

1. Human SDSCs (primary cells) can be isolated from synovial tissue or obtained from companies. The isolated synovial cells need to be characterized for multi-lineage differentiation capacity as described previously [9] before use in the following experiments.
2. The characterized SDSCs are cultured in growth medium until 95 % confluence. Replace growth medium every 2 days.
3. Rinse the cells with pre-warmed PBS without Ca^{2+} and Mg^{2+} briefly to remove any traces of serum that may inhibit trypsin activity.
4. Add warm 0.5 % trypsin/EDTA to cover the cell layer and let sit for 1 min. Observe under an inverted microscope at RT until the cells shrink and become spherical (about 2 min). Slap sides of the flask or dish to detach cells. Trypsin exposure to the cells should not exceed 4 min.
5. Neutralize the trypsin/EDTA with proportional growth medium (twofold volume of trypsin/EDTA) and collect the cell suspension into centrifuge tube.
6. Centrifuge the cell suspension at 1,200 rpm at 4 °C for 7 min and discard the supernatant carefully without disturbing the cell pellet.
7. Resuspend the cell pellet with fresh growth medium and count the cells with a hemocytometer.

3.3 Decellularized Stem Cell Matrix Fabrication Using Synovium-Derived Stem Cells

1. Dilute detached SDSCs to appropriate concentration from Step 3.2 and seed them into gelatin-coated surfaces (from Step 3.1) at 6,000 cells per cm^2 (according to the surface area of the cell containers).
2. Culture SDSCs with growth medium until 100 % confluence.
3. Switch to matrix deposition medium (growth medium containing 250 μM L-Ascorbic acid phosphate magnesium salt) and continue culture for another 10 days with medium change every 2 days [19].
4. Aspirate the medium and rinse flasks carefully with PBS without disturbing the cell layer with matrix.
5. Gently add pre-warmed extraction buffer (*see* Step 2.3) with the matrix surface covered.
6. Immediately observe cell lysis under an inverted microscope until no intact cells are visualized (about 5 min).
7. Slowly add PBS (the same volume as extraction buffer) to dilute cell debris with extraction buffer; avoid touching the matrix with pipet. Gently move the flasks or dishes to 4 °C refrigerator and leave them to incubate overnight.
8. Carefully aspirate the diluted cell debris and rinse the matrix with PBS three times.
9. To store the matrix, cover the flasks or dishes with PBS supplemented with 100 U/mL penicillin, 100 μg/mL streptomycin, and 0.25 μg/mL Fungizone. Seal with Parafilm and store at 4 °C for up to 4 weeks (*see* **Note 4.1**).

3.4 Cell Seeding and Harvesting After Decellularized Stem Cell Matrix Expansion and Preparation of Multi-lineage Differentiation

1. Warm the DSCM and growth medium to RT.
2. Obtain the cell suspension as described in Step 3.2 and count cell number.
3. Seed SDSCs at about 3,000 cells per cm^2 into flask or dish coated with matrix (prepared in Step 3.3) or plastic flasks as a control. Monitor the cell morphology and status of confluence under an inverted microscope. Replace growth medium every 2 days.
4. After about 8-day expansion or when the cells in the plastic flask reach 95 % confluence, aspirate medium and rinse with PBS.
5. Incubate the cells with pre-warmed 0.2 % collagenase type II for 20 min until the matrix dissolves.
6. Pipet up and down the mixture of cells and matrices for 2 min until there are no visible pieces of matrix.
7. Filter the cell matrix suspension with 70 μm cell strainer and centrifuge at 1,200 rpm at 4 °C for 7 min. Discard the supernatant without disturbing the pellet.

8. Resuspend the cell pellet with fresh medium and count cell number with a hemocytometer.
9. Cells are ready to undergo multi-lineage differentiation induction, such as chondrogenic, osteogenic, and adipogenic differentiation.

4 Notes

1. The most critical step of this protocol is to avoid detaching the matrix, especially after decellularization. Special care should be taken when transferring flasks with a large volume of fluids or when the matrix is left to dry. Failing to ensure these conditions can lead to detachment of matrix. The DSCM coated flasks can be stored at 4 °C in sterile conditions for up to 4 weeks.

Acknowledgements

We thank Suzanne Danley for editing the manuscript. This project was partially supported by Research Grants from the Musculoskeletal Transplant Foundation (MTF) and the National Institutes of Health (AR062763-01A1 and AR067747-01A1).

References

1. Karnes J, Zhang Y, Pei M (2014) Cell therapy for the creation of cartilage and related clinical trials. In: Templeton NS (ed) Gene and cell therapy: therapeutic mechanisms and strategies, 4th edn. Taylor & Francis, CRC Press, pp 1123–1135
2. Li JT, Pei M (2012) Cell senescence: a challenge in cartilage engineering and regeneration. Tissue Eng Part B Rev 18:270–287
3. Pei M, Li JT, Shoukry M, Zhang Y (2011) A review of decellularized stem cell matrix: a novel cell expansion system for cartilage tissue engineering. Eur Cell Mater 22:333–343
4. Jones B, Pei M (2012) Synovium-derived stem cells: a tissue-specific stem cell for cartilage engineering and regeneration. Tissue Eng Part B Rev 18:301–311
5. Pizzute T, Lynch K, Pei M (2015) Impact of tissue-specific stem cells on lineage specific differentiation: a focus on musculoskeletal system. Stem Cell Rev 11:119–132
6. Pei M (2017) Environmental preconditioning rejuvenates stem cells' chondrogenic potential. Biomaterials 117:10–23
7. He F, Chen X, Pei M (2009) Reconstruction of an in vitro tissue-specific microenvironment to rejuvenate synovium-derived stem cells for cartilage tissue engineering. Tissue Eng Part A 15:3809–3821
8. Li JT, Pei M (2011) Optimization of an in vitro three-dimensional microenvironment to reprogram synovium-derived stem cells for cartilage tissue engineering. Tissue Eng Part A 17:703–712
9. Li J, He F, Pei M (2011) Creation of an in vitro microenvironment to enhance human fetal synovium-derived stem cell chondrogenesis. Cell Tissue Res 345:357–365
10. Pei M, Zhang Y, Li JT, Chen DQ (2013) Antioxidation of decellularized stem cell matrix promotes human synovium-derived stem cell-based chondrogenesis. Stem Cells Dev 22:889–900
11. Pei M, He F, Li J, Tidwell JE, Jones AC, McDonough EB (2013) Repair of large animal partial-thickness cartilage defects through intraarticular injection matrix-rejuvenated synovium-derived stem cells. Tissue Eng Part A 19:1144–1154
12. Lai Y, Sun Y, Skinner CM, Son EL, Lu Z, Tuan RS, Jilka RL, Ling J, Chen XD (2010) Reconstitution of marrow-derived extracellular matrix ex vivo: a robust culture system for

expanding large-scale highly functional human mesenchymal stem cells. Stem Cells Dev 19:1095–1107

13. Pei M, He F, Kish VL (2011) Expansion on extracellular matrix deposited by human bone marrow stromal cells facilitates stem cell proliferation and tissue-specific lineage potential. Tissue Eng Part A 17:3067–3076
14. He F, Pei M (2013) Extracellular matrix enhances differentiation of adipose stem cells from infrapatellar fat pad toward chondrogenesis. J Tissue Eng Regen Med 7:73–84
15. Pei M, He F (2012) Extracellular matrix deposited by synovium-derived stem cells delays replicative senescent chondrocyte dedifferentiation and enhances redifferentiation. J Cell Physiol 227:2163–2174
16. He F, Pei M (2012) Rejuvenation of nucleus pulposus cells using extracellular matrix deposited by synovium-derived stem cells. Spine (Phila Pa 1976) 37:459–469
17. Pei M, Shoukry M, Li JT, Daffner S, France J, Emery SE (2012) Modulation of in vitro microenvironment facilitates synovium-derived stem cell-based nucleus pulposus tissue regeneration. Spine 37:1538–1547
18. Li J, Hansen KC, Zhang Y, Dong C, Dinu CZ, Dzieciatkowska M, Pei M (2014) Rejuvenation of chondrogenic potential in a young stem cell microenvironment. Biomaterials 35:642–653
19. Pizzute T, Zhang Y, He F, Pei M (2016) Ascorbate-dependent impact on cell-derived matrix in modulation of stiffness and rejuvenation of infrapatellar fat derived stem cells toward chondrogenesis. Biomed Mater 11:045009

Methods in Molecular Biology (2018) 1577: 155–160
DOI 10.1007/7651_2017_52

Published online: 21 July 2017

Tracheal Cartilage Isolation and Decellularization

Zachary Galliger and Angela Panoskaltsis-Mortari

Abstract

Tissue decellularization allows isolation of tissue extracellular matrix components, enabling bioengineering of native tissue microenvironments with minimal antigenic components. Here we describe a method to harvest decellularized cartilage tissue from donor trachea using a series of chemical and enzymatic washes and incorporating the extracellular matrix in gelatin methacrylate hydrogels. This decellularized cartilage tissue is easily incorporated into a variety of hydrogels to create a cartilage tissue scaffold.

Keywords Cartilage, Decellularization, Extracellular matrix, Hydrogel, Trachea

1 Introduction

Decellularized tissues have gained significant popularity as scaffolds for tissue engineering, especially for tracheal regeneration [1]. Since these scaffolds are derived from native tissues, they may retain their microscopic structure and the extracellular matrix (ECM), allowing them to better recapitulate the microenvironment of living tissue when compared to synthetic materials alone [2]. One of the most attractive features of decellularized scaffolds is their minimal antigenic potential, thereby increasing the versatility of allografts and xenografts [3].

Previously our lab has developed a method for whole organ level decellularization [4]. However, not every application requires a whole decellularized organ. There are many cases where isolated tissue-specific ECM is beneficial, such as bioprinting, chemical modification of ECM proteins, and stem cell differentiation [5, 6]. The following protocol illustrates a method to isolate and decellularize tracheal cartilage ECM and prepare it for hydrogel incorporation, using porcine trachea and gelatin methacrylate as an example.

2 Materials

2.1 Tissue Isolation

Deionized water (DI H_2O).

2.2 Decellularization (All Solutions in DI H_2O)

1. Triton solution.

   Triton X-100 (0.1% v/v).
2. Sodium deoxycholate solution.

   Sodium deoxycholate (2% w/v).
3. Sodium chloride solution.

   Sodium chloride (1 M).
4. DNase solution.

   Porcine pancreatic DNase, (1 μg/ml).
   Magnesium sulfate (1.3 mM).
   Calcium chloride (2 mM).
5. Phosphate buffered saline (PBS).

2.3 Tissue Digestion (All Solutions in DI H_2O)

1. Digestion solution.

   Hydrochloric acid (0.01 N).
   Porcine pepsin (1 mg/ml).
2. Neutralizing solution.
   Sodium hydroxide (0.1 N).

2.4 Gel Incorporation

1. Hydrogel solution (in PBS or cell culture media).
   Gelatin methacrylate (GelMA) (10% wt/v%).
   Lithium phenyl-2,4,6-trimethylbenzoylphosphinate (LAP) (1% wt/v%).

2.5 Equipment

1. Size 7-sa tweezers.
2. SU 1710 scissors or no. 22 scalpel.
3. 250 ml beaker.
4. Stir bar.
5. Stir plate.
6. Cold room.
7. VirTis automatic freeze-dryer.
8. Mill 1,700 rpm 1 mm sieve, or mortar.
9. Vortex mixer.
10. Water bath.
11. 405 nm wavelength light.

3 Methods

3.1 Tissue Isolation

1. Rinse the trachea with deionized water to removed non-adhered cells and blood. See Fig. 1a.
2. Using tweezers and scissors, remove the connective tissue surrounding the exterior of the trachea by pulling and snipping the outer layers. See Fig. 1b and c.

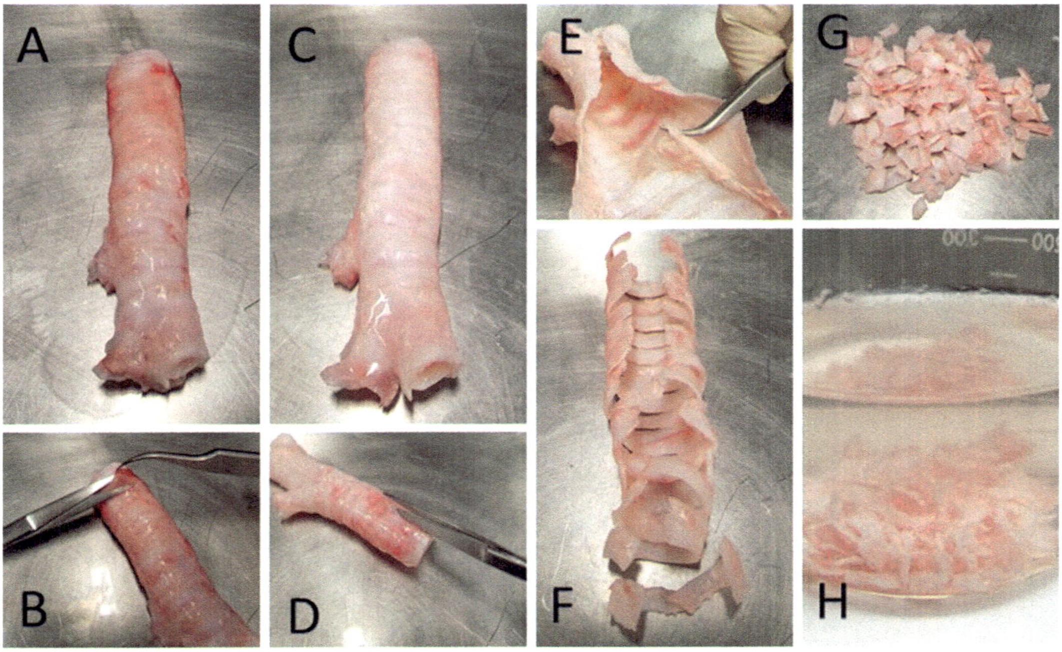

Fig. 1 Trachea dissection overview. (**a**) Trachea rinsed with deionized water. (**b**) Removal of exterior tissue. (**c**) Cleaned trachea. (**d**) Cutting along the dorsal side. (**e**) Removing epithelial and muscle layers. (**f**) Separated individual cartilage rings. (**g**) Cartilage sections. (**h**) Cartilage in deionized water

3. Open the trachea by making a full-length incision on the dorsal side. See Fig. 1d.
4. Remove the epithelial layers and smooth muscle by inserting the tip of the tweezers under the superficial layer and sliding it along the full length of the trachea. See Fig. 1e.
5. Separate each cartilage ring by cutting them apart. See Fig. 1f.
6. Cut each ring into 5–10 mm sections and place in deionized water. See Fig. 1g and h.

3.2 Decellularization

1. Combine the tissue with 100 ml of deionized water in a beaker. Allow tissue to soak for an hour while stirring. See Notes 1 and 2. See Fig. 2a and b.
2. Drain deionized water from tissue and add 100 ml triton X-100 (0.1% v/v). Cover the beaker and allow to spin for 18 h. See Fig. 2c and d.
3. Replace the triton solution with 100 ml of sodium deoxycholate solution (2% w/v) and allow it to spin for another 18 h. See Fig. 2e and f.

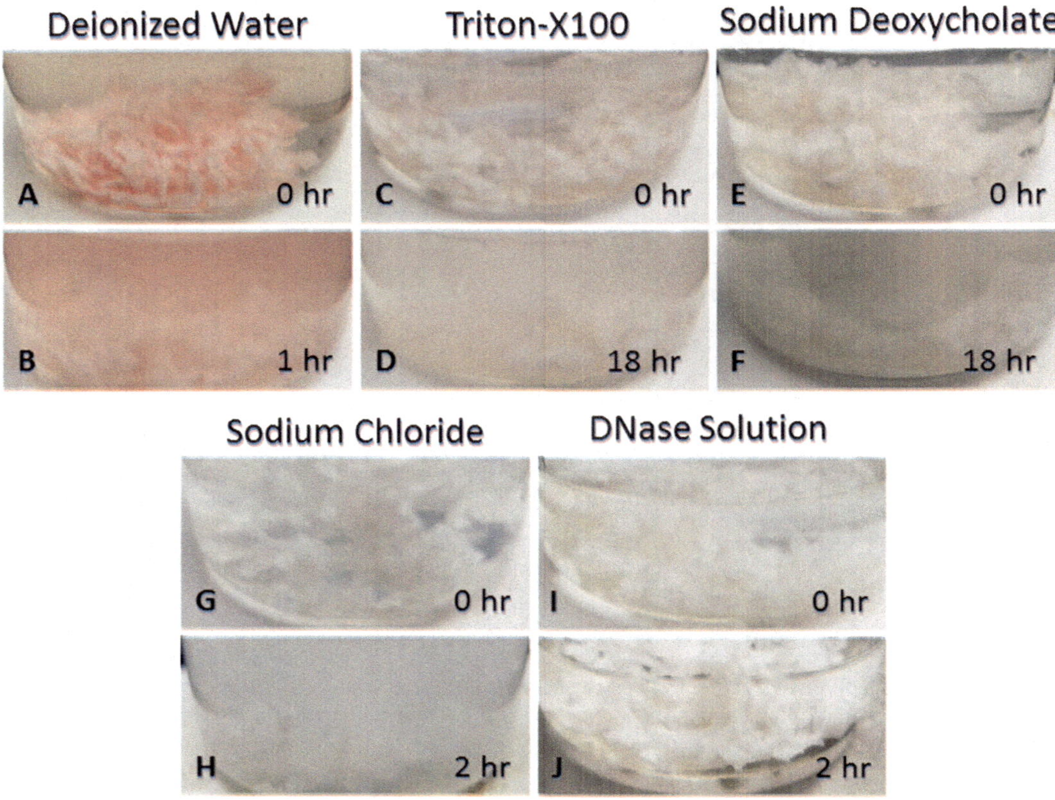

Fig. 2 Cartilage decellularization overview. (**a**) Initial tissue in deionized water. (**b**) Tissue in deionized water after 1 h. (**c**) Tissue in Triton-X100 after first wash. (**d**) Tissue in Triton-X100 after 18 h. (**e**) Tissue in sodium deoxycholate after second wash. (**f**) Tissue in sodium deoxycholate after 18 h. (**g**) Tissue in sodium chloride after third wash. (**h**) Tissue in sodium chloride after 2 h. (**i**) Tissue in DNase solution after fourth wash. (**j**) Tissue in DNase solution after 2 h

4. Remove the sodium deoxycholate solution and add 100 ml of sodium chloride (1 M) and stir for 2 h. See Fig. 2g and h.
5. Drain the sodium chloride solution and replace it with 100 ml of DNase solution (1 μg/ml porcine pancreatic DNase, 1.3 mM magnesium sulfate, and 2 mM calcium chloride) for another 2 h. See Fig. 2i and j.
6. Lastly, the tissue is rinsed with PBS to remove any residual detergents.
7. Aliquot the decellularized tissue into a freezer safe container and freeze the tissue overnight at −80 °C.
8. Lyophilize the frozen matrix until the water has been removed.

3.3 Tissue Digestion

1. Using a mill or mortar, grind the dried ECM into a powder.
2. Combine this powdered ECM at 10 mg/ml with digestion solution (pepsin 1 mg/ml in 0.01 N hydrochloric acid).
3. Allow solution to stir at room temperature for 2 days.
4. Neutralize the pH of the solution using sodium hydroxide (0.1 N).
5. Aliquot and freeze the digested ECM solution.
6. Lyophilize the frozen ECM again and grind into a powder.
7. Store at −80 °C. See Note 3.

3.4 Gel Incorporation

Acellular hydrogel scaffolds.

1. Prepare hydrogel solution in PBS at 50 °C. See Note 4.
2. Combine the powdered ECM with the hydrogel solution at 50 mg/ml.
3. Vortex to dissolve.
4. Add the ECM hydrogel solution to a tissue culture well-plate, size and volume determined by your application.
5. Cross-link the solution by exposing it to 405 nm light for 10 s.

3.5 Cell Laden Hydrogel Scaffolds

1. Prepare hydrogel solution in cell culture media with half the final volume and cool to 37 °C.
2. Combine the powdered ECM with the hydrogel solution at 50 mg/ml.
3. Vortex to dissolve.
4. Harvest cells using preferred lab method and create a cell suspension of half the final volume. See Note 5.
5. Combine ECM hydrogel solution with cell suspension to reach final fluid volume.
6. Add the cell laden hydrogel solution to a tissue culture well-plate, size and volume determined by your application.
7. Cross-link the solution by exposing it to 405 nm for 10 s.

4 Notes

1. The decellularization process is best done at 4 °C to prevent the growth of bacteria during the initial wash; however, room temperature also works.
2. Make sure the entire tissue sample is immersed in fluid and moving to ensure proper mixing during the decellularization process.

3. Once the digested ECM is powdered and stored at −80 °C, it can be used for a variety of applications.
4. In preparing the gel solution it is generally easiest to first dissolve the LAP, then dissolve the GelMA while heating the solution to 50 °C in a water bath.
5. Concentration of cell suspension and final concentration of cell seeding depend on the cell type used and your application.

Acknowledgments

We would like to thank Daniel Sorby for his help procuring animal tissue and processing. We would also like to thank the Hormel company (Austin, MN) for their generosity, donating porcine trachea for research.

References

1. Weiss DJ, Elliott M, Yang Q et al (2014) Tracheal bioengineering: the next steps: proceeds of an international society of cell therapy pulmonary cellular therapy signature series workshop, Paris, France, April 22, 2014. Cytotherapy 16:1601–1613. doi:10.1016/j.jcyt.2014.10.012
2. Saldin LT, Cramer MC, Velankar SS et al (2016) Extracellular matrix hydrogels from decellularized tissues: structure and function. Acta Biomater 49:1–15. doi:10.1016/j.actbio.2016.11.068
3. Badylak SF, Freytes DO, Gilbert TW (2009) Extracellular matrix as a biological scaffold material: structure and function. Acta Biomater 5:1–13. doi:10.1016/j.actbio.2008.09.013
4. Price AP, England KA, Matson AM et al (2010) Development of a decellularized lung bioreactor system for bioengineering the lung: the matrix reloaded. Tissue Eng Part A 16:2581–2591. doi:10.1089/ten.tea.2009.0659
5. Pati F, Jang J, Ha D-H et al (2014) Printing three-dimensional tissue analogues with decellularized extracellular matrix bioink. Nat Commun 5:3935. doi:10.1038/ncomms4935
6. Visser J, Levett PA, te Moller NCR et al (2015) Crosslinkable hydrogels derived from cartilage, meniscus, and tendon tissue. Tissue Eng Part A 21:1195–1206. doi:10.1089/ten.TEA.2014.0362

Methods in Molecular Biology (2018) 1577: 161–175
DOI 10.1007/7651_2017_60

Published online: 04 August 2017

Protocols for Rat Uterus Isolation and Decellularization: Applications for Uterus Tissue Engineering and 3D Cell Culturing

Arvind Manikantan Padma, Tom Tristan Tiemann, Ahmed Baker Alshaikh, Randa Akouri, Min Jong Song, and Mats Hellström

Abstract

Sophisticated culturing conditions are required to grow cells in a three-dimensional (3D) environment. Cells then require a type of scaffold rich in proteins, growth factors, and signaling molecules that simulates their natural environment. Tissues from all species of animals have an organ-specific extracellular matrix (ECM) structure that plays a key role in cell proliferation and migration. Hence, the scaffold composition plays a significant role for any successful 3D cell culturing system. We developed a whole rat uterus ECM scaffold by the perfusion of detergents and ionic solutions through the vascular system of an isolated normal rat uterus in a process termed "decellularization." The generated rat uterus scaffolds consist of a cell-free ECM structure similar to that of the normal rat uterus, and are thus excellent platforms on to which new cells can be added. Rat uterus 3D cell culturing systems based on these scaffolds could become valuable to decidual differentiation- and embryo implantation studies, or for investigating invasion mechanisms of endometrial cancer cells. They could also be used for the creation of tissue engineered uterine tissue, for partial or whole organogenesis developed for transplantation applications to treat absolute uterine infertility. This is a condition affecting about 1 in 500 women, and is only treatable by a uterus transplantation. This article provides valuable troubleshooting notes and describes in detail how to generate rat uterus scaffolds, including the delicate surgery required to isolate the uterus with an intact vascular tree which facilitates vascular perfusion and re-transplantation.

Keywords Bioengineering, ECM, Infertility, Microsurgery, Reproduction, Scaffold, Uterus transplantation

1 Introduction

Decellularization is a method used to generate scaffolds for tissue engineering applications where immunogenic cells and nuclear material have been removed from a donor organ by physical, chemical, or enzymatic methods. The aim is to produce a viable and healthy scaffold that can support the growth and migration of added cells. When transplanted, it should also be completely accepted by the recipient's immune system [1]. If successful, tissue engineered organs should reduce the need for donors and the use of immunosuppressive treatment following organ transplantation, including in uterus transplanted patients [2–4]. Since the extracellular matrix (ECM) provides cells with important signals, the decellularization process should preserve the extracellular proteins and glycoproteins and, if possible, limit growth factor loss so that the recipient's cells used in the reconstruction phase can adhere, grow, and proliferate [5–7]. Hence, a successful decellularization procedure should have minimal effects on the ECM, but should be aggressive enough to remove all cells and nuclear material from the organ. The outcome depends on multiple factors, including organ pressure exposure during decellularization, chemicals used, tissue lipid content, cell and organ density, and properties of the decellularization agents used. Many studies describe the decellularization of varied tissues and organs, including blood vessels, trachea, neuronal tissues, liver, lung, and heart [8–14]. Thin-cell layered organs such as a blood vessel or an amniotic membrane can be decellularized using agitation, while bigger and more complex organs need dynamic decellularization techniques usually performed by vascular perfusion [10, 11, 13, 15, 16].

Most of the decellularization protocols use ionic reagents like sodium dodecyl sulfate (SDS) and sodium deoxycholate (SDC) that break bonds between cytoplasmic proteins and cellular membranes. These are considered aggressive reagents since they simultaneously denature ECM proteins by disrupting protein–protein bonds [3, 10, 11]. Non-ionic detergents such as Triton X-100 are milder to the ECM than the ionic detergents since they more specifically break lipid–lipid and lipid–protein bonds (while leaving protein–protein bonds more intact) [3]. DMSO is a polar amphipathic solvent that is known to interact with ECM proteins by unfolding them [1, 3, 17–19].

Many published review papers discussed how the evaluation of the decellularized product should be performed [2, 10, 11, 20]. We recommend to qualitatively assess the scaffold by histology (hematoxylin and eosin, and 4′,6-Diamidin-2-phenylindol; DAPI) and by immunohistochemistry (for ECM-proteins like collagen, fibronectin, and others), and then also quantitatively assess the scaffold composition for collagen, elastin, glycosaminoglycans, growth factors (Luminex/Proteomics), and DNA. Its physical attributes (pore size and mechanistic properties) are also significant [21]. Based on a highly cited publication, the recommended levels of double stranded DNA residues should be less than 50 ng per milligram dry tissue weight, and the DNA fragments should be smaller than 200 bp to reduce immune rejection risks following transplantation [1]. However, these levels may be organ and tissue specific and should therefore be tested and confirmed in each system [4, 22].

The two decellularization protocols for whole rat uterus described herein remove on average 96–99% of the total DNA [20]. Patches of recellularized constructs were still viable 9 weeks post-transplantation without immunosuppression, which suggests efficient scaffold resilience towards the host's immune system [2]. Furthermore, these tissue engineered patches improved uterine repair after a full-thickness uterine wall injury, and gave strength--support to the uterus throughout the whole pregnancy [2]. After recellularization, these scaffolds thus proved to be efficient in vivo, and are interesting materials to use for rat uterus 3D cell culturing applications that may include decidual differentiation- and embryo implantation studies, studies investigating invasion mechanisms of endometrial cancer cells, or for the construction of a bioengineered uterus. A bioengineered patient-specific uterus may benefit patients diagnosed with absolute uterine infertility, a condition that can be treated only by a successful uterus transplantation [4, 23, 24]. This article provides valuable troubleshooting notes and describes in detail how to generate such scaffolds, including the delicate surgery required to isolate the rat uterus with an intact vascular tree to facilitate vascular perfusion and re-transplantation.

2 Materials

2.1 Rat Uterus Isolation

- Female rats between 140 and 180 g.
- Surgical microscope.
- Anesthesia machine.
- Isoflurane.
- Lab bench with exhaust ventilator.
- Heating pad.
- Gauze.
- Macro and micro scissors.
- Macro and micro straight forceps.
- 8-0; 6-0; 4-0 silk sutures.
- Titanium clips.
- Cannula, e.g., polythene tubing (PE-50) with an inner diameter of 0.58 mm and an outer diameter of 0.96 mm attached to a 23G needle.
- 3 ml syringe.
- 50 ml Falcon tube.
- Perfadex (+4 °C) solution, 30 ml/uterus, Vitrolife Sweden AB, Kungsbacka, Sweden.
- Xylocaine.
- Heparin.

2.2 Rat Uterus Decellularization

- Silicone tubing: inner diameter = 5 mm, outer diameter = 8 mm (about 1.5 m in total).
- Peristaltic perfusion pump (1–10 ml/min).
- Appropriate tube for pump (Note 1).
- Bubble trap (Note 2).
- 4-Way tubing connection (Note 3).
- 50 ml Falcon tubes.
- 23G syringe needles.
- 14G syringe needles.
- 1 ml syringes.
- 10 ml pipettes.
- Silicon glue.
- Tape.
- Phosphate buffered saline (pH: 7.4).
- Dimethyl sulfoxide.
- Triton X-100.
- Sodium deoxycholate.
- Peracetic acid.
- Penicillin–Streptomycin–Amphotericin B.
- Mucasol.
- 7 cm silicon tubing with a 2 mm inner diameter and a 3 mm outer diameter.
- 3× 5 cm polythene tubing with an inner diameter of 0.58 mm and an outer diameter of 0.96 mm.
- File.
- Cutting device/small saw.

2.3 Solutions

Solution 0: Perfadex solution; Perfadex with Xylocaine (0.4 mg/ml) and Heparin (50 IU/ml).

Solution 1: Mucasol 3% diluted in deionized water.

Solution 2: Phosphate buffered saline (pH: 7.4: PBS).

Solution 3: Distilled water.

Solution 4: Dimethyl sulfoxide (4% in solution 3).

Solution 5: Triton X-100 (1% in solution 2).

Solution 6: Sodium deoxycholate (2% in solution 3) (Note 4).

Solution 7: Peracetic acid (0.1% in sterile phosphate buffered saline).

Solution 8: Sterile phosphate buffered saline +100 U/ml Penicillin +100 μg/ml Streptomycin + 25 μg/ml Amphotericin B.

3 Methods

3.1 Rat Uterus Isolation

The surgical procedure described below will result in an isolated whole uterus which includes the upper part of the cervix and a complete vascular tree from the aorta to the vena cava (Figs. 1, 2, and 3). All vascular branches have been ligated and cut off except the uterine vessels. This enables future uterus re-transplantation with an end-to-side anastomosis to the donor aorta and vena cava. The work described was approved by the Animal Ethics Committee at Gothenburg University, Sweden.

1. Administer an anticoagulant to the rat, e.g., low molecular weight Heparin (100 IU/kg).
2. Anesthetize the rat using 2.5% Isoflurane with 2 l/min air, supplemented with 0.5 l/min O_2.
3. Shave and clean the rat's abdomen with 70% ethanol.
4. Place the rat on a heating pad under an operating stereomicroscope.
5. Perform a laparotomy via a midline incision from the symphysis pubis to the xiphoid process (Fig. 1a).
6. Retract the small intestines to the left side of the animal and cover them in 37 °C moistened gauze (Fig. 1b).
7. Cauterize the inferior mesenteric artery by using a bipolar diathermy at a maximum effect of 2 W.
8. Cut the colon just above the rectum and retract it to the side to acquire a better operating field (Fig. 1c).
9. Cauterize and cut the ureters (Fig. 1d).
10. Cauterize the inferior epigastric vessels to fully be able to expose the external iliac vessels on both sides.
11. Cut just above the ilioinguinal ligament.
12. Place a double ligature (8-0, silk) around the external iliac vessels 3–4 mm distal to the branching of the inferior epigastric vessels, then cauterize and cut between the ligatures.
13. Make a thorough dissection and remove any extravascular tissue.
14. Cauterize the small vessels surrounding the external iliac vessels up to a point just distal to the branching of the external pudendal vessels.
15. Cauterize and cut the pudendal vessels.
16. Expose the vessels supplying the lower portion of the cervix and the bladder (superior vesical, inferior vesical, cervical and vaginal vessels), then cauterize these to prevent bleeding.

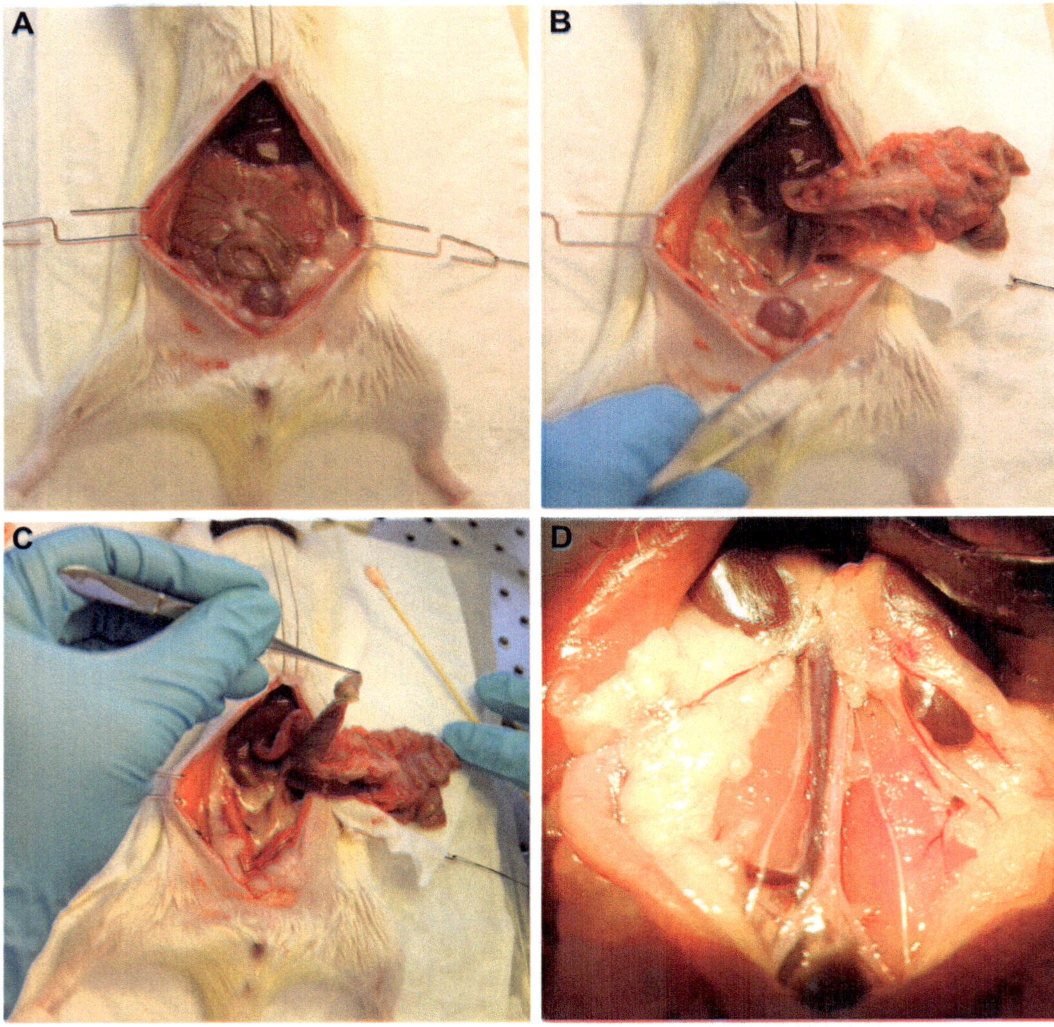

Fig. 1 Preparation for the isolation of the rat uterus. Midline vertical laparotomy from the symphysis pubis to the xiphoid process (**a**); Retraction of the small intestines to the left side of the animal (**b**); Cutting and retraction the colon to acquire a better operating field (**c**); Cutting the ureters (**d**)

17. Remove the bladder.
18. Transect the cervix at the level just below the attachment to the bladder (Note 5).
19. Gently lift up the cervix to identify the uterine artery and veins, then attach a titanium clip on the superior gluteal vessels.
20. Dissect these vessels free from underlying tissue up to the caudal vessels by cauterizing the small dorsal vessels.
21. Cauterize and cut the lumbar vessels [2–4] which branch dorsally from the aorta and vena cava.

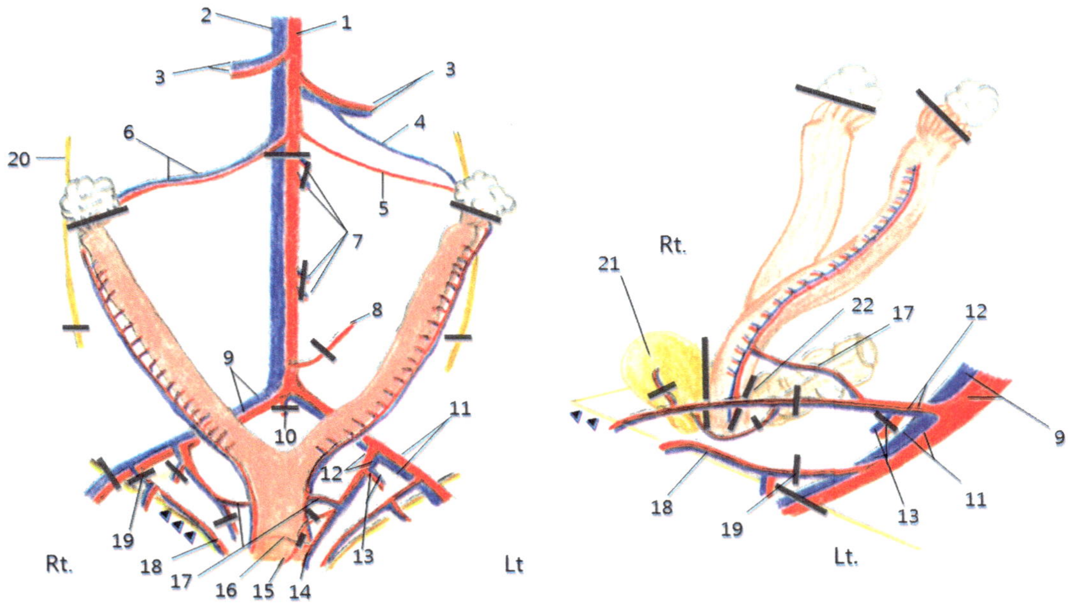

Fig. 2 Schematic drawing of vascular anatomy of the abdomen and the pelvis of the rat. Ligations are indicated by *bold black/gray bars*. 1 = abdominal aorta; 2 = vena cava; 3 = renal vessels; 4 = left renal vein; 5 = left renal artery; 6 = right renal vessels; 7 = lumbar vessels; 8 = inferior (caudal) mesenteric artery; 9 = common iliac vessels; 10 = caudal vessels; 11 = external iliac vessels; 12 = internal iliac vessels; 13 = superior gluteal vessels; 14 = umbilical vessels; 15 = inferior vesical vessels; 16 = superior vesical vessels; 17 = uterine vessels; 18 = external pudendal vessels; 19 = inferior epigastric vessels; 20 = ureter; 21 = urinary bladder; 22 = recto-sigmoid colon (*Rt* right side, *Lt* left side)

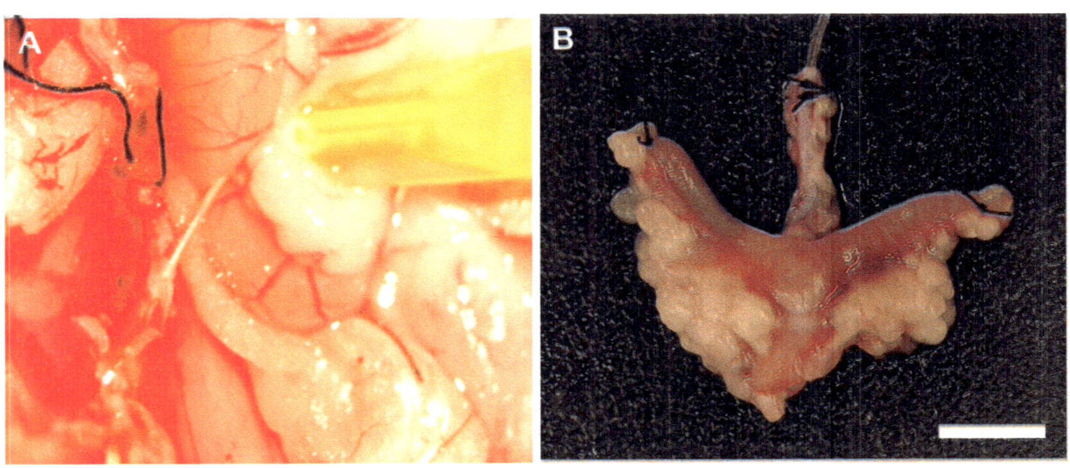

Fig. 3 Perfusion of the specimen with Perfadex solution to prevent thrombosis (**a**); Insertion of a cannula for perfusion and decellularization (**b**)

22. Separate the aorta and the vena cava from each other by gentle upward traction on the aorta and use blunt dissection in

between from a point just below the branching of the renal vessels and 3–4 mm distally (Notes 6 and 7).

23. Use 6-0 double ligature, or titanium clips on both uterine horns to close the site between the tip of the uterine horn and the oviduct, and then cut between the ligatures.
24. Dissect the uterine horns free from their attachments to the dorsal peritoneum, and then fold the uterus over to expose the hypogastric trunk.
25. Use an 8-0 suture to tie off the aorta and the vena cava (inferior to the ovarian vessels; exitus letalis of rat couple min later). This creates a complete segment, including the inferior aorta and the inferior vena cava with the vascular connection to both uterine horns.
26. Make a small incision in the vena cava and in the aorta, just below the ligations. Then cannulate the aorta (you may use a G23 gauge cannula and a PE-50 tubing) connected to a 3 ml syringe prefilled with cold (+4 °C) Perfadex solution supplemented with xylocaine (0.4 mg/ml) and heparin (50 IU/ml) (Fig. 3a).
27. Gently perfuse the specimen with enough Perfadex solution to remove all blood remnants to prevent thrombosis (Note 8), then cut the aorta and vena cava (Fig. 3b).
28. Then place the isolated uterus specimen in a prefilled sterile 50 ml Falcon tube containing cold Perfadex solution (Solution 0) and freeze in −20 °C, and then transfer to −80 °C for long-term storage.

3.2 Rat Uterus Decellularization

Start by assembling the perfusion system as outlined below and as shown in Fig. 4. The system is based on using cheap, sterile, and single packed plastic consumables together with other durable materials that enable the use of most detergents and chemicals without a great risk for leaks or other negative events.

3.2.1 Assemble and Mold the Lid for the Decellularization Chamber

1. Pierce three 23G needles through the lid of a 50 ml Falcon tube in a straight row.
2. Just above the middle 23G needle, pierce a 14G needle through the lid (Note 9).
3. File the sharp ends of the needles to make them blunt and suitable for cannulas.
4. Remove the piston from four 1 ml syringes and then cut off the ends of the syringes using a clean cutting tool.
5. Attach the modified syringes to the needles that penetrate the 50 ml Falcon tube.

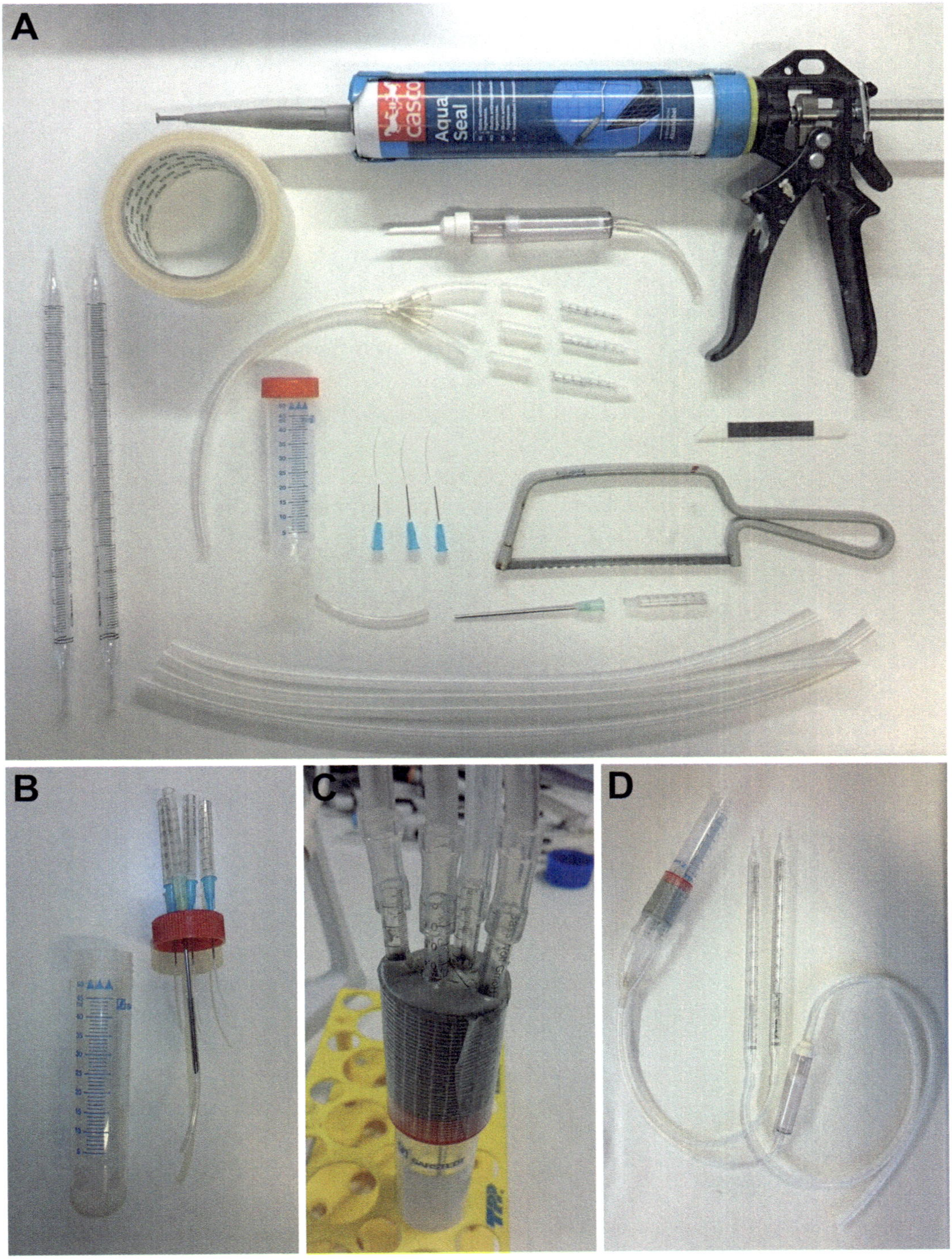

Fig. 4 All materials required to build the decellularization set-up (**a**). Needle penetrations through the lid. Cannulas inside the Falcon tube protect the organ. Only one cannula is attached to the organ, the others act as pressure valves to prevent overpressure in the organ (**b**). The mold with hardened silicone glue prevents leaks (**c**). The final decellularization set-up excluding the pump (**d**)

6. Use tape (4–5 cm wide) to make a mold around the lid, the lid penetrations and the syringes on top of the Falcon tube (Note 10).
7. Using silicon glue which hardens, fill the mold and secure the needle penetrations on the lid top, cover also the syringe parts (Note 11).
8. Wait >2 days to let the silicon glue harden.

3.2.2 Prepare the Perfusion System

1. Make sure the peristaltic pump is in order and has the appropriate tubing mounted.
2. Prepare three 40 cm long silicon tubings (inner diameter 5 mm, outer diameter 8 mm), and one 5 cm long tubing of the same sort.
3. Attach one of the long silicon tubings to each of the two pump tubings.
4. Attach a 10 ml pipette to the silicon tubing on the suction side of the pump, then place the pipette in the reservoir flask containing 500 ml of Mucasol 3% (Solution 1).
5. Attach a bubble trap at the second tubing leaving the pump (Note 3).
6. Attach the short (5 cm) piece of silicon tubing after the bubble trap.
7. Attach a 4-way connector of appropriate size to the other end of the silicon tubing.
8. On each end of the remaining three ends, attach a 3 cm segment of silicon tubing to the 4-way connector.
9. Connect these ends to the molded syringes attached to the 23G needle penetrations in the 50 ml Falcon tube lid to the decellularization chamber.
10. Attach an appropriate length silicon tubing to the syringe attached to the 14G needle penetration and attach a 10 ml pipette at the other end and insert this into the same reservoir flask containing 500 ml of Mucasol 3% (Solution 1) as mentioned in 4 (Note 12).
11. Start the pump and re-circulate Mucasol 3% (Solution 1) to clean the loop. Adjust connections and repair possible leaks (Note 13).
12. After 30 min, replace the Mucasol solution with dH_2O and rinse out all the cleaning solution by a non-recirculating loop for 60 min.
13. Replace the dH_2O in the loop with PBS (Solution 3) and remove all air bubbles.
14. Adjust perfusion speed to 6 ml per min (Note 14).

3.2.3 Start the Rat Uterus Decellularization

1. Locate the isolated uterus kept in long-term storage (−80 °C) and thaw the organ slowly in air at room temperature (Note 15).
2. Once thawed, cannulate the aorta of the isolated rat uterus using a prefilled (PBS; Solution 2) tubing (polythene tubing with an inner diameter of 0.58 mm and an outer diameter of 0.96 mm).
3. Secure the cannula with two ligatures around the aorta with 4-0 silk sutures.
4. Open the decellularization chamber and attach the cannulated rat uterus to one of the three 23G needles in the lid, then put the rat uterus inside the 50 ml Falcon tube and gently close the lid.
5. Turn on the pump and perfuse PBS (Solution 2) through the system with a speed of 6 ml/min for 30 min to remove blood remnants (Notes 14 and 16).
6. Select a decellularization protocol from Table 1, or optimize your own new protocol (here we give example of two efficient protocols previously published and characterized; Table 1) [17].
7. After completed protocol, carefully transfer the decellularization set-up into a sterile LAF bench (Note 17).
   (a) Perfuse with Peracetic acid (Solution 7) inside the LAF bench at 6 ml/min for 30 min to sterilize the tissue (Note 18).
   (b) Perfuse with sterile PBS (Solution 2) at 6 ml/min for 1 h to remove the remnants of Solution 7 (Note 19).
8. Disconnect all tubings and carefully transfer the sterile uterus scaffold into a 50 ml Falcon tube containing solution 8 and leave it overnight at +4 °C for 5 days to leach out possible remnants of any solutions used during the decellularization or sterilization steps.

Table 1
Two suggested protocols for an efficient decellularization of the rat uterus

Protocol 1	Protocol 2
1. Perfuse with DMSO 4% (Solution 4) at 6 ml/min for 4 h to disrupt cell walls	1. Perfuse with SDC 2% (Solution 6) at 6 ml/min for 4 h to disrupt cell walls and remove cells
2. Perfuse with Triton X-100 1% (Solution 5) at 6 ml/min for 4 h to remove cells and cell debris	2. Perfuse with distilled water (Solution 3) at 6 ml/min overnight to remove remnants of Solution 6
3. Perfuse with distilled water (Solution 3) at 6 ml/min overnight to remove remnants of solutions 4 and 5	3. Repeat step 1–2 four more times
4. Repeat step 1–3 four more times	

9. Freeze in −20 °C, and then transfer to −80 °C for long-term storage until use for downstream applications.

4 Notes

1. A Masterflex precision pump and tygon tubing with an inner diameter of 3 mm and an outer diameter of 6 mm that fits well inside the larger silicon tubing were used for the loop.
2. This item was obtained from a "BIEGLER Heating bag system" commonly used for blood and infusions in the clinic (Art no. FP1003001).
3. This is important to prevent air perfusion or air embolisms.
4. SDC causes irritation of airways. Always prepare this in a fume hood, wear a mask, and avoid spilling the powder. A good practice is to prepare a 5× stock solution and store at room temperature for the whole decellularization procedure.
5. Special caution needs to be taken to avoid the vascular branches from the hypogastric trunk on both sides.
6. Always angle sharp tips from instruments away from the vena cava to minimize unintentional damage to the fragile vein wall.
7. Again, special caution is necessary to avoid the vascular branches from the hypogastric trunk on both sides.
8. Flush until the uterus is blanched and a clear perfusate can be observed from the incision in the vena cava. Perfuse the 3 ml Perfadex solution slowly for the duration of at least 1–2 min, without inducing too much pressure in the organ. Refill the syringe with additional Perfadex solution if necessary. This step is important to prevent blood clots that will interfere with the downstream decellularization procedure.
9. Replace the needles if the needles are blocked due to plastic remnants from the lid. Use the same passages for new needles.
10. Make sure to leave >1 cm of the syringes over the mold edge so that silicon tubing can be attached later.
11. Construction-grade silicon glue was used that is usually used for building aquariums or for tiling in wet-room facilities.
12. The perfusion loop is now complete but the rat uterus is not yet connected.
13. Twist and lift tubing, including bubble trap and the decellularization chamber to remove all air bubbles. Reduce pump speed since a high speed may cause foaming of the detergent.

14. 6 ml per min equals the total amount of solution entering the 50 ml Falcon tube from all three lid penetrations. This amount can conveniently be calculated by weighing the amount of fluid that was circulated during a specific time.
15. Thawing it too fast increases risks for thawing damages to the tissue.
16. The efficiency may be improved by the addition of Xylocaine (0.4 mg/ml) and Heparin (50 IU/ml) to the PBS.
17. While transferring the set-up, make sure to keep the whole set-up on a trolley while transferring it into the LAF bench. Be careful and do not disconnect the tubes to prevent bubbles from being formed which could enter the organ. Bubbles may occlude the vasculature resulting in improper decellularization and sterilization.
18. Discard all solutions according to biosafety rules.
19. Discard all solutions containing antibiotics according to safety rules to prevent superbug formation.

General notes to improve decellularization

1. You may use an electric timer switch connected to the pumps to minimize risk of missing time points and the perfusion solution to run dry.
2. Using 1.86 g/l Ethylenediaminetetraacetic acid (EDTA) with solution 3 before starting decellularization for 2–4 h could help break already existing blood clots. However, note that you must wash away all EDTA before perfusing with SDC. Wash with PBS for at least 2 h to remove remnants of EDTA as they react and form turbid clumps that could occlude the vasculature.
3. Attaching a pressure meter to the inlet cannulas would allow for pressure-controlled perfusion during the decellularization.

Acknowledgements

The authors report no conflict of interest. The authors also thank Prof. Miran Kim and Prof. Joohee Yoon in the Uterine Transplantation Team, College of Medicine, The Catholic University of Korea. This work was supported by Hjalmar Svensson Research Foundation, Adlerbertska Research Foundation, the Swedish Government LUA grant, and the Swedish Science Research Council (Vetenskapsrådet; Grant No. 116008).

References

1. Crapo PM, Gilbert TW, Badylak SF (2011) An overview of tissue and whole organ decellularization processes. Biomaterials 32(12):3233–3243
2. Hellström M, Moreno-Moya JM, Bandstein S, Bom E, Akouri RR, Miyazaki K et al (2016) Bioengineered uterine tissue supports pregnancy in a rat model. Fertil Steril 106(2):487–496
3. Gilbert TW, Sellaro TL, Badylak SF (2006) Decellularization of tissues and organs. Biomaterials 27(19):3675–3683
4. Hellström M, Bandstein S, Brännström M (2016) Uterine tissue engineering and the future of uterus transplantation. Ann Biomed Eng. s10439-016-1776-2
5. Arenas-Herrera JE, Ko IK, Atala A, Yoo JJ (2013) Decellularization for whole organ bioengineering. Biomed Mater 8(1):014106
6. Hoshiba T, Lu H, Kawazoe N, Chen G (2010) Decellularized matrices for tissue engineering. Expert Opin Biol Ther 10(12):1717–1728
7. Riau AK, Beuerman RW, Lim LS, Mehta JS (2010) Preservation, sterilization and de-epithelialization of human amniotic membrane for use in ocular surface reconstruction. Biomaterials 31(2):216–225
8. Godinho MJ, Teh L, Pollett MA, Goodman D, Hodgetts SI, Sweetman I et al (2013) Immunohistochemical, ultrastructural and functional analysis of axonal regeneration through peripheral nerve grafts containing Schwann cells expressing BDNF, CNTF or NT3. PLoS One 8(8):e69987
9. Hu Y, Leaver SG, Plant GW, Hendriks WT, Niclou SP, Verhaagen J et al (2005) Lentiviral-mediated transfer of CNTF to schwann cells within reconstructed peripheral nerve grafts enhances adult retinal ganglion cell survival and axonal regeneration. Mol Ther 11(6):906–915
10. Macchiarini P, Jungebluth P, Go T, Asnaghi MA, Rees LE, Cogan TA et al (2008) Clinical transplantation of a tissue-engineered airway. Lancet 372(9655):2023–2030
11. Olausson M, Patil PB, Kuna VK, Chougule P, Hernandez N, Methe K et al (2012) Transplantation of an allogeneic vein bioengineered with autologous stem cells: a proof-of-concept study. Lancet 380(9838):230–237
12. Ott HC, Clippinger B, Conrad C, Schuetz C, Pomerantseva I, Ikonomou L et al (2010) Regeneration and orthotopic transplantation of a bioartificial lung. Nat Med 16(8):927–933
13. Ott HC, Matthiesen TS, Goh SK, Black LD, Kren SM, Netoff TI et al (2008) Perfusion-decellularized matrix: using nature's platform to engineer a bioartificial heart. Nat Med 14(2):213–221
14. Uygun BE, Soto-Gutierrez A, Yagi H, Izamis ML, Guzzardi MA, Shulman C et al (2010) Organ reengineering through development of a transplantable recellularized liver graft using decellularized liver matrix. Nat Med 16(7):814–820
15. Badylak SF, Weiss DJ, Caplan A, Macchiarini P (2012) Engineered whole organs and complex tissues. Lancet 379(9819):943–952
16. Akhyari P, Aubin H, Gwanmesia P, Barth M, Hoffmann S, Huelsmann J et al (2011) The quest for an optimized protocol for whole-heart decellularization: a comparison of three popular and a novel decellularization technique and their diverse effects on crucial extracellular matrix qualities. Tissue Eng Part C Methods 17(9):915–926
17. de Castro Bras LE, Ramirez TA, DeLeon-Pennell KY, Chiao YA, Ma Y, Dai Q et al (2013) Texas 3-step decellularization protocol: looking at the cardiac extracellular matrix. J Proteome 86:43–52
18. Lumpkins SB, Pierre N, McFetridge PS (2008) A mechanical evaluation of three decellularization methods in the design of a xenogeneic scaffold for tissue engineering the temporomandibular joint disc. Acta Biomater 4(4):808–816
19. Xu H, Xu B, Yang Q, Li X, Ma X, Xia Q et al (2014) Comparison of decellularization protocols for preparing a decellularized porcine annulus fibrosus scaffold. PLoS One 9(1):e86723
20. Hellström M, El-Akouri RR, Sihlbom C, Olsson BM, Lengqvist J, Backdahl H et al (2014) Towards the development of a bioengineered uterus: comparison of different protocols for

rat uterus decellularization. Acta Biomater 10 (12):5034–5042

21. Watt FM, Huck WT (2013) Role of the extracellular matrix in regulating stem cell fate. Nat Rev Mol Cell Biol 14(8):467–473
22. Friedrich LH, Jungebluth P, Sjoqvist S, Lundin V, Haag JC, Lemon G et al (2014) Preservation of aortic root architecture and properties using a detergent-enzymatic perfusion protocol. Biomaterials 35(6):1907–1913
23. Brännström M (2015) Uterus transplantation. Curr Opin Organ Transplant 20(6):621–628
24. Brännström M, Johannesson L, Bokström H, Kvarnström N, Mölne J, Dahm-Kahler P et al (2015) Livebirth after uterus transplantation. Lancet 385(9968):607–616

Methods in Molecular Biology (2018) 1577: 177–181
DOI 10.1007/7651_2017_35

Published online: 17 May 2017

Application of Decellularized Tissue Scaffolds in Ovarian Tissue Transplantation

Enes Taylan and Kutluk Oktay

Abstract

In tissue engineering, decellularized scaffolds have been proved to have remarkable capacity to promote regeneration in various organs such as kidney, heart, lung, and liver. Marrying the field of cryobiology and reproductive medicine resulted in considerable progress and breakthroughs, which led to the emergence of ovarian tissue cryopreservation and transplantation as a promising option for fertility preservation. Here we describe an innovative application of decellularized tissue scaffolds as a regenerative platform for reconstruction of ovarian grafts for auto-transplantation.

Keywords: Cryopreservation, Decellularized tissue scaffold, Extracellular tissue matrix, Fertility preservation, Ovarian tissue transplantation

1 Introduction

One of the most remarkable achievements in reproductive surgery and fertility preservation is ovarian tissue cryopreservation and transplantation [1]. In 2000, Oktay and Karlikaya reported the first successful orthotopic transplantation of cryopreserved ovarian tissue and over 80 live births ensued in the next two decades [2, 3]. This procedure is currently the only practical option for fertility preservation in female cancer patients of prepubertal age, as well as the reproductive age women who cannot delay chemotherapy to resort to more established methods of fertility preservation [4]. The main purpose of ovarian cryopreservation is to preserve a large dormant population of primordial follicle oocytes, which make up the ovarian reserve. Because many cancer treatments including chemotherapy and radiotherapy will result in the irreversible loss of this reserve and as a consequence premature menopause and infertility, safeguarding these primordial follicles by cryopreservation before such toxic treatments offers chance for restoring fertility and ovarian endocrine function upon auto-transplantation once the patient is cured of cancer.

Ovarian tissue cryopreservation typically requires surgical removal of one ovary with a laparoscopic outpatient procedure followed by separation of primordial follicle rich cortex from the stroma. The cortex is then divided into 0.5 × 0.5 cm pieces of 1–2 mm thickness to be treated with cryoprotectants and cryopreserved with a slow freezing protocol using an automated system [5].

Despite the encouraging results within the last 2 decades, ovarian cryopreservation/transplantation is still evolving and it requires further refinements to improve success rates. One of the limitations of ovarian transplantation is that the previously frozen and then thawed ovarian cortical pieces are transplanted on a vascular surface, akin to skin grafting. This results in an avascular ischemic period especially within the first 72 h after the transplantation. This ischemic period may last up to 10 days, until the revascularization of the ovarian graft is completed. This initial ischemic phase has been shown to be a major cause of ovarian follicle loss and limits the longevity and success of ovarian transplants [6]. Our laboratory has been working on improving this revascularization process within the past decade. For example, we have shown that a ceramide-induced death inhibitor Sphingosine-1-phosphate enhances and accelerates neovascularization in human ovarian xenograft models [7]. While S1P has a very short half life and not practical for human use, its synthetic analogs may find future applications in ovarian transplantation.

Our laboratory has also shown that the extracellular matrix plays important role in primordial follicle growth and survival [8, 9]. In addition, by integrating scaffold materials with various cell types and culture environments, tissue engineers have already reached significant milestones in regeneration of various organs such as kidney, heart, lung, and pancreas [10]. Encouraged by these earlier reports, we for the first time presented the use of decellularized extracellular tissue matrix in ovarian tissue transplantation to enhance outcomes [11]. Here we will describe this innovative application of decellularized scaffolds as a regenerative platform for the reconstruction of ovarian graft.

2 Materials

Prepare all media, materials, and surgical instruments in a sterilized fashion. Diligently follow the waste disposal regulations while disposing biological and chemical waste materials (*see* **Note 1**).

1. Decellularized extracellular tissue matrix scaffold (*see* **Note 2**).
2. Warm water bath.
3. Sterile Normal Saline (0.9% Sodium Chloride).
4. Cold plate.
5. Modified Human Tubal Fluid (m-HTF) media.

6. Human Serum Albumin (HSA) solution.
7. Vacuum filtration system including a filter unit with pore size of 0.22 μm.
8. Forceps, 10–12 cm in length, for handling ovarian cortical strips.
9. Glass petri dish, 150 mm.
10. Suture material: Monocryl Plus 5–0 (11 mm 3/8 undyed reverse cutting, absorption profile: 91–119 days), 12–15 cm in length.
11. Needle holder.
12. Surgical scalpel.
13. Sterile gloves.

3 Methods

3.1 Preparation of Ovarian Tissue Media

1. Prepare 50 mL transplant (wash) media by using m-HTF with a final concentration of 10% HAS (100 mg/mL).
2. Sterilize transplant media using a filtering system and store the media at 4 °C for later use (*see* **Note 3**).

3.2 Preparation of Decellularized Scaffold

1. Uncase the decellularized scaffold in sterilized fashion and place in to the water bath filled with sterile normal saline prewarmed to 37 °C (*see* **Note 4**).
2. Gently rinse the decellularized scaffold for 30 min until completely rehydrates.
3. Then place in to the glass petri dish filled with transplant media and keep at room temperature.

3.3 Preparation of Ovarian Tissue Graft

1. Fill in the glass petri dish with transplant (wash) media.
2. After thawing, transfer ovarian cortical pieces to the transplant media (*see* **Note 5**).
3. Place the glass dish with ovarian tissue pieces on to the cold plate to keep the tissue at 4 °C during the preparation of the graft (*see* **Note 5**).
4. Take decellularized tissue scaffold and place into the glass petri dish.
5. Suture ovarian cortical pieces one by one on decellularized tissue scaffold in a position which the stromal side of the cortical pieces is exposed.
6. Then trim the decellularized scaffold all around the tissue leaving approximately 5 mm tissue free zone to facilitate the

suturing process during transplantation of the constructed graft [10].

4 Notes

1. Ovarian tissue transplantation is still an experimental procedure.
2. In our experience, Alloderm Regenerative Tissue Matrix (Life-Cell Corp., Branchburg, NJ, USA) with medium thickness works the best for ovarian reconstruction [11].
3. Use MILLIPORE Stericup Filter Unit with pore size of 0.22 μm for sterilization of the transplant (wash) media. It is best to prepare the media fresh and not to store more than 48 h.
4. Thawing procedure varies based on the tissue frozen with slow freezing or vitrification. Please see ref. [12] for more detailed information on ovarian tissue cryopreservation.
5. While waiting for reconstruction ensure to keep ovarian cortical pieces in transplant (wash) media on a cold plate (4 °C) to minimize ischemic cellular damage [13].

Acknowledgement

K.O. is supported by Eunice Kennedy Shriver National Institute of Child Health and Human Development (NICHD—RO1HD053112 and R21HD061259) and National Cancer Institute (NCI).

References

1. Demeestere I, Simon P, Emiliani S, Delbaere A, Englert Y (2009) Orthotopic and heterotopic ovarian tissue transplantation. Hum Reprod Update 15:649–665
2. Oktay K, Karlikaya G (2000) Ovarian function after transplantation of frozen, banked autologous ovarian tissue. N Engl J Med 342:1919
3. Oktay K, Buyuk E, Veeck L, Zaninovic N, Xu K, Takeuchi T et al (2004) Embryo development after heterotopic transplantation of cryopreserved ovarian tissue. Lancet 363:837–840
4. Rodriguez-Wallberg KA, Oktay K (2012) Recent advances in oocyte and ovarian tissue cryopreservation and transplantation. Best Pract Res Clin Obstet Gynaecol 26:391–405
5. Sonmezer M, Oktay K (2004) Fertility preservation in female patients. Hum Reprod Update 10:251–266
6. Lee J, Kong HS, Kim EJ, Youm HW, Lee JR, Suh CS et al (2016) Ovarian injury during cryopreservation and transplantation in mice: a comparative study between cryoinjury and ischemic injury. Hum Reprod 31:1827–1837
7. Soleimani R, Heytens E, Oktay K (2011) Enhancement of neoangiogenesis and follicle survival by sphingosine-1-phosphate in human ovarian tissue xenotransplants. PLoS One 6:e19475
8. Oktay K, Karlikaya G, Akman O, Ojakian GK, Oktay M (2000) Interaction of extracellular matrix and activin-A in the initiation of follicle

growth in the mouse ovary. Biol Reprod 63:457–461
9. Oktem O, Oktay K (2007) The role of extracellular matrix and activin-A in in vitro growth and survival of murine preantral follicles. Reprod Sci 14:358–366
10. Tapias LF, Ott HC (2014) Decellularized scaffolds as a platform for bioengineered organs. Curr Opin Organ Transplant 19:145–152
11. Oktay K, Bedoschi G, Pacheco F, Turan V, Emirdar V (2016) First pregnancies, live birth, and in vitro fertilization outcomes after transplantation of frozen-banked ovarian tissue with a human extracellular matrix scaffold using robot-assisted minimally invasive surgery. Am J Obstet Gynecol 214:94.e1–94.e9
12. Suzuki N, Donnez J (eds) (2016) Gonadal tissue cryopreservation in fertility preservation. Springer, Osaka
13. Kamoshita K, Okamoto N, Nakajima M, Haino T, Sugimoto K, Okamoto A et al (2016) Investigation of in vitro parameters and fertility of mouse ovary after storage at an optimal temperature and duration for transportation. Hum Reprod 31:774–781

Methods in Molecular Biology (2018) 1577: 183–208
DOI 10.1007/7651_2017_36

Published online: 11 May 2017

Composite Bioscaffolds Incorporating Decellularized ECM as a Cell-Instructive Component Within Hydrogels as In Vitro Models and Cell Delivery Systems

Arthi Shridhar, Elizabeth Gillies, Brian G. Amsden, and Lauren E. Flynn

Abstract

Decellularized tissues represent promising biomaterials, which harness the innate capacity of the tissue-specific extracellular matrix (ECM) to direct cell functions including stem cell proliferation and lineage-specific differentiation. However, bioscaffolds derived exclusively from decellularized ECM offer limited versatility in terms of tuning biomechanical properties, as well as cell–cell and cell–ECM interactions that are important mediators of the cellular response. As an alternative approach, in the current chapter we describe methods for incorporating cryo-milled decellularized tissues as a cell-instructive component within a hydrogel carrier designed to crosslink under mild conditions. This composite strategy can enable in situ cell encapsulation with high cell viability, allowing efficient seeding with a homogeneous distribution of cells and ECM. Detailed protocols are provided for the effective decellularization of human adipose tissue and porcine auricular cartilage, as well as the cryo-milling process used to generate the ECM particles. Further, we describe methods for synthesizing methacrylated chondroitin sulphate (MCS) and for performing UV-initiated and thermally induced crosslinking to form hydrogel carriers for adipose and cartilage regeneration. The hydrogel composites offer great flexibility, and the hydrogel phase, ECM source, particle size, cell type(s) and seeding density can be tuned to promote the desired cellular response. Overall, these systems represent promising platforms for the development of tissue-specific 3-D in vitro cell culture models and in vivo cell delivery systems.

Keywords: Cell delivery vehicle, Cell therapy, Cell-instructive biomaterials, Composite scaffolds, Decellularization, Extracellular matrix (ECM), Hydrogels, In vitro, In vivo, Methacrylated chondroitin sulphate, Stem cells

1 Introduction

The tissue-specific extracellular matrix (ECM) is a key component of the complex and dynamic cellular microenvironment that mediates cellular processes including survival, adhesion, migration, proliferation and differentiation [1]. The relationship between cells and their ECM is bidirectional, with cells sensing and responding to biochemical and/or biomechanical signals from the cellular milieu, as well as being able to actively remodel the surrounding ECM [2]. Given the important role of the ECM in directing cell behaviour, there has been growing interest in the development of

3-D in vitro model and cell delivery systems that mimic the properties of the native ECM [3–6]. Ideally, these systems should be designed to allow for tuning of cell–cell and cell–ECM interactions, to be able to develop a deeper understanding of the effects of these parameters on cellular behaviour and refine the engineered microenvironments to promote the desired cellular response. In addition, they should allow for the convenient and efficient seeding of the cell populations to achieve a homogeneous cell distribution, while maintaining high cell viability.

Decellularized tissues represent an abundant source of tissue-specific ECM for the development of cell-instructive scaffolds to promote tissue regeneration [7, 8]. In general, the strategy in decellularization is to apply a series of physical, chemical and/or enzymatic treatment methods to eliminate immunogenic cellular components from the tissues and obtain tissue-specific ECM with retained structural, biochemical and biomechanical properties [9]. Decellularized matrices have been prepared from virtually every tissue in the body and investigated for use as cell culture and delivery platforms [7, 10–12]. In our lab, we have developed customized decellularization protocols for human adipose tissue [10] and porcine auricular cartilage (adapted from Woods et al. [13]), which generate scaffolds with well-preserved ECM structure and composition. Further, we are applying these scaffolds in varying formats to explore the instructive effects of the decellularized adipose tissue (DAT) and decellularized cartilage tissue (DCT) on the lineage-specific differentiation of human adipose-derived stem/stromal cells (ASCs). Supporting the rationale for a tissue-specific approach, we have demonstrated that DAT-derived scaffolds [10, 14], microcarriers [15], foams [16] and hydrogel composites [17, 18] provide a naturally inductive microenvironment that promotes the adipogenic differentiation of human ASCs in culture, as well as in vivo fat formation in a subcutaneous Wistar rat model [14, 16, 17]. More recently, in cell culture studies with 3-D cell aggregates incorporating micronized ECM, we found that the adipogenic differentiation of human ASCs was enhanced in aggregates incorporating DAT particles, while the expression of chondrogenic differentiation markers was augmented in aggregates incorporating DCT [19]. These findings are consistent with an increasing body of literature that supports that the ECM has tissue-specific instructive effects on in vitro stem cell differentiation [20]. Moreover, decellularized matrices have been shown to facilitate tissue and site-specific regeneration and remodelling of various tissues in both in vivo models and in human clinical trials [8, 21, 22].

While the inherent bioactivity of decellularized tissues is an advantage in terms of their capacity to mediate cell function, a limitation is that these biomaterials on their own offer limited flexibility in terms of tuning the scaffold properties to control both cell–cell and cell–ECM interactions [23]. To address this challenge, a strategy adopted by our group and others has been to incorporate processed decellularized ECM in the form of micronized particles within naturally derived or synthetic hydrogel carriers to generate composite bioscaffolds with tissue-specific instructive properties [17, 18, 24–27]. Hydrogels provide a more highly tunable platform that can be designed to mimic the mechanical properties of native soft tissues, while allowing for the permeability of oxygen, nutrients and other water-soluble metabolites, making them effective scaffolds for cell encapsulation [28]. Depending on the specific application and crosslinking method, the hydrogels can be crosslinked ex vivo within a mold to generate scaffolds with a defined shape and volume, or alternatively could be designed to crosslink in situ to allow for minimally invasive cell delivery at a target site for tissue repair [2, 29–31].

The composite hydrogel approach incorporating tissue-specific decellularized ECM particles represents a versatile platform. A range of different hydrogels could be utilized to enable cell encapsulation with high cell viability, including methacrylated glycol chitosan [17, 32], thiol-modified hyaluronic acid [33], alginate [34, 35] or polyethylene glycol-based materials [36, 37], offering flexibility depending on the properties required for the target application. In the protocols described below, we have selected methacrylated chondroitin sulphate (MCS) as the hydrogel carrier. In previous studies, we found that human ASCs could be encapsulated with high cell viability in MCS-based hydrogels, and that adipogenic differentiation was enhanced in MCS + DAT composites relative to both MCS and DAT controls [17]. From the perspective of cartilage regeneration, chondroitin sulphate (CS) is a major structural and bioactive component of the ECM within the native tissues, providing a rationale for its selection for this application [38]. Furthermore, CS-based biomaterials support the viability of encapsulated chondrocytes [39] and promote the chondrogenic differentiation of encapsulated stem cell populations [40, 41]. While we are demonstrating the composite approach using DAT and DCT, the strategy could be applied with other ECM sources to generate an array of tissue-specific instructive biomaterials. Offering additional layers of tunability, the ECM particle size and concentration, as well as the cell types and seeding densities, can be refined to promote the desired cellular response. For example, we previously reported that ASC adipogenesis was enhanced in MCS + DAT composites incorporating DAT particles with an average diameter of 38 ± 6 μm as compared to 278 ± 3 μm, with the effects likely mediated by greater cell–cell contact in the ASCs clustered around the smaller particles [18].

In this chapter, we describe in detail our decellularization protocols for human adipose tissue and porcine auricular cartilage, as well as the further processing of the DAT and DCT through cryomilling to generate fine particles. Our encapsulation approach allows for a high seeding efficiency and the homogeneous distribution of the cryo-milled ECM and ASCs within the MCS hydrogel phase. We also provide methods for synthesizing the MCS and for its UV-initiated free radical crosslinking using the photo-initiator Irgacure 2959 and thermally initiated free radical crosslinking using the water-soluble ammonium persulphate (APS) and tetramethylethylenediamine (TEMED) initiation system. UV-crosslinking is rapid and effective, but could cause cellular damage or arrest if the wavelength and intensity are not carefully controlled [42]. Thermally initiated crosslinking circumvents the need for exposing the hydrogel to the light source and can be designed to enable crosslinking at physiological temperatures, thereby representing a more practical approach for minimally invasive in situ cell encapsulation for many applications [43–45].

2 Materials

2.1 Human Adipose Tissue Decellularization

1. Human adipose tissue (*see* **Note 1**).
2. Solution A: Prepare 10× stock solution with 10 mM tris (hydroxymethyl)aminomethane (Tris), 5 mM ethylenediaminetetraacetic acid (EDTA). Add 900 mL of deionized water to an empty graduated cylinder. Weigh 12.1 g of Tris and 14.6 g of EDTA, transfer to the cylinder and stir until dissolved. Top up deionized water to a final volume of 1 L and stir until fully mixed. Store at room temperature. To prepare a 1× working solution, add 100 mL of the 10× stock solution to a graduated cylinder and dilute with deionized water to a final volume of 1 L. Adjust the pH to 8.0 (*see* **Note 2**).
3. Sorensen's phosphate buffer rinsing solution (SPB rinse): Prepare 10× stock SPB rinse with 0.55 M sodium phosphate dibasic heptahydrate ($Na_2HPO_4{\cdot}7H_2O$), 0.17 M potassium phosphate (KH_2PO_4). Add 900 mL of deionized water to an empty graduated cylinder. Weigh 147 g of $Na_2HPO_4{\cdot}7H_2O$ and 23 g of KH_2PO_4, transfer to the cylinder and stir until dissolved. Top up with deionized water to a final volume of 1 L. Stir until fully mixed and store at room temperature. To prepare a 1× working solution, add 100 mL of 10× stock solution to a graduated cylinder and dilute with deionized water to a final volume of 1 L. Adjust pH to 7.3.
4. Sorensen's phosphate buffer digest solution (SPB digest): Prepare 10× stock SPB digest with 0.55 M sodium phosphate dibasic heptahydrate ($Na_2HPO_4{\cdot}7H_2O$), 0.17 M potassium phosphate (KH_2PO_4), 0.049 M magnesium sulphate heptahydrate ($MgSO_4{\cdot}7H_2O$). Add 900 mL of deionized water to an empty graduated cylinder. Weigh 147 g of $Na_2HPO_4{\cdot}7H_2O$,

23 g of KH_2PO_4 and 12 g of $MgSO_4{\cdot}7H_2O$, transfer to the cylinder and stir until dissolved. Top up with deionized water to a final volume of 1 L. Stir until fully mixed and store at room temperature. To prepare a 1× working solution, add 100 mL of 10× stock solution to a graduated cylinder and dilute with deionized water to a final volume of 1 L. Adjust pH to 7.3.

5. Deoxyribonuclease (DNAse) stock solution: Weigh 0.175 g of NaCl, dissolve it in 10 mL deionized water and carefully adjust the pH to 7.3. Add 10 mL of glycerol to this solution and stir to mix well. Add 15 mL of the NaCl/glycerol solution to 200,000 Units of DNase I (Sigma-Aldrich, ≥2,000 U/mg from bovine pancreas) and mix thoroughly. Store 1 mL aliquots in cryovials at −20 °C. Thaw aliquots at room temperature immediately prior to use and do not re-freeze.
6. Ribonuclease (RNase) stock solution: Dissolve 0.06 g of Tris and 0.44 g of NaCl in 50 mL of deionized water and adjust the pH to 7.5. Add 500 mg of RNase type III-A (Sigma-Aldrich, 85–140 Kunitz U/mg from bovine pancreas) to 40 mL of the Tris/NaCl solution. Store 1 mL aliquots in cryovials at −20 °C. Thaw aliquots at room temperature immediately prior to use and do not re-freeze.
7. Lipase stock solution: Prepare 5 mM calcium chloride ($CaCl_2$) solution. Weigh out 36.75 mg of $CaCl_2{\cdot}H_2O$ and dissolve it in 50 mL of deionized water in a volumetric flask. Mix the solution well and chill to 4 °C. Add 50 mL of chilled $CaCl_2$ solution to 100,000 Units of lipase type VI-S (Sigma-Aldrich, ≥20,000 U/mg from porcine pancreas). Aliquot 2 mL of the solution into cryovials and store at −20 °C. Thaw aliquots at room temperature immediately prior to use and do not re-freeze.
8. Phenylmethylsulfonyl fluoride (PMSF) solution: Add 1 g of PMSF to a foil-wrapped bottle containing 28.7 mL of absolute isopropanol and stir until clear. Store at room temperature and protect from exposure to light and moisture (*see* **Note 3**).
9. Trypsin-EDTA: 0.25% (1×) with EDTA·4Na.
10. 100× Antibiotic/Antimycotic (ABAM) (10,000 units/mL penicillin, 10,000 μg/mL streptomycin, 25 μg/mL amphotericin B).
11. Phosphate buffered saline (PBS), pH 7.4.
12. Bovine serum albumin (BSA).
13. Ethanol: 70%.
14. Isopropanol, absolute (*see* **Note 4**).
15. Sterile deionized water.
16. Hydrochloric acid (HCl): 6 M (For pH adjustment).

17. Sodium hydroxide (NaOH): 6 M (For pH adjustment).
18. 500 mL polypropylene specimen jars (autoclavable).
19. 50 mL centrifuge tubes.
20. Set of stainless steel mesh filters (1 mm pore size): for use as a tissue press for lipid extraction.
21. Forceps.
22. Surgical scissors.
23. Pipette aid and serological pipettes.
24. Micropipettors (2–20 and 100–1,000 μL single channel).
25. Magnetic stir bar and stir plate.
26. Incubator shaker (capable of 100 rpm and maintaining 37 °C).
27. Freezers (−20 and −80 °C).
28. Electronic balance.
29. pH meter.
30. Lyophilization system.
31. Dessicator (*see* **Note 5**).

2.2 Porcine Cartilage Decellularization

1. Porcine auricular cartilage tissue.
2. Tris buffer (50 mM): Transfer 900 mL of deionized water to an empty graduated cylinder. Weigh out 6.057 g of Tris, transfer it to the cylinder and stir until completely dissolved. Top up the cylinder with deionized water to a final volume of 1 L. Stir until fully mixed and store at room temperature. Adjust the pH to 9.0.
3. Solution A (as described in Sect. 2.1).
4. Solution B: Prepare a working solution with 50 mM Tris, 1.5 M potassium chloride (KCl) and 2% Triton X-100. Transfer 900 mL of deionized water to an empty graduated cylinder. Weigh 6.057 g of Tris and 111.82 g of KCl and transfer into the cylinder. Stir until completely dissolved. Add 20 mL of Triton X-100 and top the volume up to 1 L with deionized water. Stir until completely mixed and carefully adjust the pH to 8.0. Store at room temperature.
5. Solution C: Prepare the working solution of 50 mM Tris and 1% tributyl phosphate (TBP). Weigh 6.057 g of Tris and dissolve in 900 mL of deionized water. Add 10 mL of TBP to the resulting solution and top the volume up to 1 L with deionized water. Stir until completely mixed and carefully adjust the pH to 8.0. Store at room temperature.
6. SPB Rinse (as described in Sect. 2.1).
7. SPB Digest (as described in Sect. 2.1).

8. RNase (as described in Sect. 2.1).
9. DNase (as described in Sect. 2.1).
10. PBS, pH 7.4.
11. 100× ABAM (as described in Sect. 2.1).
12. PMSF (as described in Sect. 2.1).
13. Ethanol: 70%.
14. Sterile deionized water.
15. 6 M NaOH (for pH adjustment).
16. 6 M HCl (for pH adjustment).
17. 5 mm biopsy punches.
18. Scalpel blade, sterile.
19. 500 mL polypropylene specimen jars (autoclavable).
20. 50 mL centrifuge tubes.
21. Pipette aid and serological pipettes.
22. Micropipettors (2–20 and 100–1,000 μL single channel).
23. Magnetic stir bar and stir plate.
24. Incubator shaker (capable of 250 rpm and maintaining 37 °C).
25. Freezers (−20 and −80 °C).
26. Electronic balance.
27. pH meter.
28. Lyophilization system.
29. Dessicator (*see* **Note 5**).

2.3 Cryo-Milling of DAT and DCT

1. Liquid nitrogen.
2. Surgical scissors.
3. 10 mL glass vials.
4. Stainless steel test sieves with desired mesh size for particle size separation (e.g. 200, 100 and 50 μm).
5. Ball milling system (e.g. Retsch mixer mill MM 400 ball mill grinder with 25 mL stainless steel milling chamber).
6. Stainless steel milling balls (recommended 2 × 10 mm).
7. Dessicator (*see* **Note 5**).

2.4 Fluorescent Labelling of the Cryo-Milled ECM Particles

1. Alexa Fluor® 350 *N*-hydroxysuccinimidyl (NHS) Ester (Thermo Fisher Scientific CAT# A10168): To prepare a 10 mg/mL stock solution, dilute 10 mg of the Alexa Fluor® 350 dye in 1 mL of dimethyl sulfoxide (DMSO).

2. 0.15 M Sodium bicarbonate solution ($NaHCO_3$): Prepare a stock solution by dissolving 126 mg of $NaHCO_3$ in 10 mL of deionized water. Adjust the pH to 8.3.
3. 1.5 M Hydroxylamine solution: Weigh out 420 mg of hydroxylamine hydrochloride and dissolve in 2 mL of deionized water. Carefully adjust the pH to 8.5. Add an additional 2 mL of deionized water and stir until well mixed (*see* **Note 6**).
4. Sterile PBS, pH 7.4.
5. Ethanol: 70%.
6. Cryo-milled DAT and DCT (From Sect. 2.3).
7. 50 mL centrifuge tubes.
8. Centrifuge.
9. Electronic balance.
10. pH meter.

2.5 Preparation of Methacrylated Chondroitin Sulphate (MCS)

1. Chondroitin sulphate A sodium salt (~50 kDa) (LKT Technologies CAT# 9007-28-7).
2. Methacrylic anhydride ($C_8H_{10}O_3$).
3. 0.2 M Sodium phosphate monobasic solution (NaH_2PO_4): Prepare a stock solution by dissolving 24 g of anhydrous NaH_2PO_4 in 1 L of deionized water. Stir until completely dissolved and store at room temperature.
4. 2 M NaOH (for pH adjustment).
5. 2 M HCl (for pH adjustment).
6. Dialysis membrane (3.5 kDa MWCO).
7. Sterile deionized water.
8. Metal spatula.
9. Ethanol, absolute.
10. Micropipettors (2–20 and 100–1,000 μL single channel).
11. 10 mL glass vials.
12. Magnetic stir bar and stir plate.
13. 2 L glass beaker.
14. Round bottom flask.
15. pH paper/pH meter.
16. Lyophilization system.
17. Electronic balance.
18. Argon gas.
19. Liquid nitrogen.

2.6 Preparation of Materials Prior to Crosslinking

1. Biological safety cabinet equipped with a UV lamp.
2. Vortex.
3. Electronic balance.
4. Centrifuge (10,000 × *g* minimum capacity).
5. Micropipettor (100–1,000 μL single channel).
6. Sterile PBS, pH 7.4.
7. Petri dishes, sterile (60 mm × 15 mm).
8. 2 mL microcentrifuge tubes, sterile.
9. Sterile metal spatula.

2.7 Incorporation of the Cryo-Milled ECM and Cells Within the MCS Hydrogels Using UV-Initiated Crosslinking

1. Methacrylated chondroitin sulphate (From Sect. 2.5) (*see* **Note 7**).
2. Irgacure 2959 (2-hydroxy-4′-(2-hydroxyethoxy)-2-methyl-propiophenone) (5 mg/mL). Weigh 5 mg of Irgacure 2959 and add to 1 mL of sterile PBS, vortexing until completely dissolved. Store away from light (*see* **Note 8**).
3. Cryo-milled DAT and DCT (From Sect. 2.3).
4. Sterile PBS, pH 7.4.
5. Human adipose-derived stem/stromal cells (ASCs) or other cell type of interest.
6. Proliferation medium [e.g. for human ASCs: DMEM:Ham's F12 with 10% fetal bovine serum (FBS) and 100 U/mL of penicillin with 0.1 mg/mL streptomycin (1% Pen-strep)].
7. 1 mL sterile plastic syringes or other desired mold.
8. Sterile metal spatula.
9. 18G needle, sterile.
10. UV light source (filtered to 320–390 nm, 10 mW/cm^2).
11. Retort stand for UV light source.
12. Micropipettors (2–20 and 100–1,000 μL single channel).
13. 2 mL microcentrifuge tubes, sterile.
14. 0.2 μm filter membrane.
15. Incubator shaker (capable of 100 rpm and maintaining at 37 °C).
16. Electronic balance.

2.8 Incorporation of the Cryo-Milled ECM and Cells Within the MCS Hydrogels Using Thermally Initiated Crosslinking

1. Methacrylated chondroitin sulphate (From Sect. 2.5) (*see* **Note 7**).
2. Ammonium persulphate (APS): Prepare a 400 mM working solution by dissolving 45.8 mg of APS in 500 μL of PBS. Mix thoroughly, filter to sterilize and store protected from light at 4 °C (*see* **Note 9**).

3. Tetramethylethylenediamine (TEMED): Prepare a 400 mM working solution by mixing 30 μL of TEMED in 500 μL of PBS. Mix thoroughly, sterile filter and store protected from light at 4 °C (*see* **Note 9**).
4. Cryo-milled DAT and DCT (From Sect. 2.3).
5. Sterile PBS, pH 7.4.
6. Human ASCs or other cell type of interest.
7. Proliferation medium (e.g. for human ASCs: DMEM:Ham's F12 with 10% FBS and 1% Pen-strep).
8. 1 mL sterile plastic syringe or other desired mold.
9. Sterile metal spatula.
10. 18G needle, sterile.
11. Micropipettors (2–20 and 100–1,000 μL single channel).
12. 2 mL microcentrifuge tubes, sterile.
13. Incubator shaker (capable of 100 rpm and maintaining at 37 °C).
14. Electronic balance.
15. Ice bath.

2.9 Scaffold Preparation and Culture

1. Biological safety cabinet.
2. Razor blades, sterile.
3. 12-well tissue culture plates.
4. 12-well cell culture inserts.
5. Proliferation medium (from Sects. 2.6 and 2.7).
6. Pipette aid and serological pipettes.

3 Methods

3.1 Decellularization of Adipose and Cartilage Tissue

The decellularization protocols established by our lab comprise a series of freeze-thaw cycles, mechanical agitation and chemical/enzymatic treatments that generate 3-D matrices substantially devoid of cells and cellular components, while retaining tissue-specific bioactivity. Figure 1a highlights the macroscopic appearance of the two tissue types during the various processing stages.

3.1.1 Adipose Tissue Decellularization

Day 1:

1. Remove any cauterized tissue and blood vessels from the fresh human adipose tissue samples using sterile forceps and sharp surgical scissors. Cut the tissue samples into ~2 cm × 2 cm × 2 cm (~25 g) pieces in order to maximize the surface area of tissue in contact with the reagents used for decellularization (*see* **Note 1**).

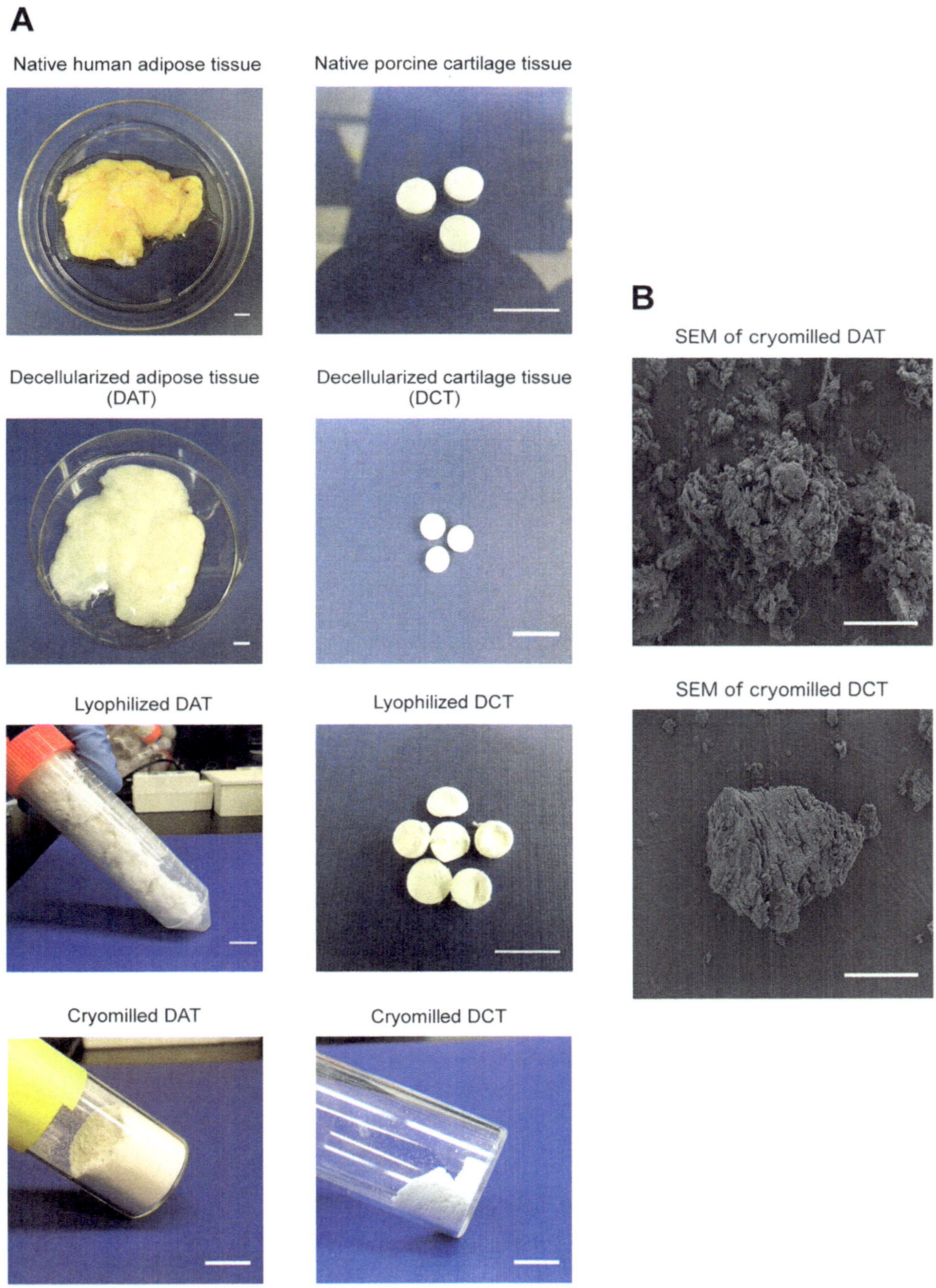

Fig. 1 Overview of human adipose tissue and porcine auricular cartilage decellularization and processing to generate the DAT and DCT particles for composite hydrogel fabrication. (**a**) Macroscopic overview of the tissues during the key processing stages including decellularization, lyophilization and cryo-milling into particles. Scale bars represent 1 cm. (**b**) Representative scanning electron micrographs of unsieved DAT and DCT particles generated through cryo-milling, showing that the DAT has a rougher surface topography. Scale bars represent 50 μm

2. Transfer four ~25 g pieces (100 g total) of adipose tissue to a new sterile specimen jar containing 99 mL of 1× Solution A, 7 μL of PMSF solution and 1 mL of 100× ABAM (*see* **Note 10**). Freeze the sample at −80 °C until completely frozen.
3. Remove the sample from the freezer and thaw in an incubator shaker at 37 °C and 100 rpm.
4. Once completely thawed, replace the solution with 99 mL of fresh 1× Solution A, 7 μL of PMSF solution and 1 mL of 100× ABAM. Freeze at −80 °C until completely frozen.
5. Repeat steps 3 and 4 two more times for a total of 3 freeze-thaw cycles.
6. At the end of the third freeze-thaw cycle, transfer the tissue to a fresh specimen jar with 99 mL of 0.25% trypsin-EDTA and 1 mL 100× ABAM. Incubate the tissue overnight (16 h) at 37 °C under agitation in the incubator shaker at 100 rpm.

Day 2:

7. Transfer the tissue to a fresh specimen jar with 99 mL of absolute isopropanol, 1 mL of 100× ABAM and 7 μL of PMSF. Incubate at 37 °C under agitation at 100 rpm for 8 h.
8. Press the tissue pieces between the two stainless steel mesh filters to aid in the extraction of lipid.
9. Replace the solution with 99 mL of fresh absolute isopropanol, 7 μL of PMSF solution and 1 mL of 100× ABAM and incubate at 37 °C under agitation at 100 rpm for 16 h (*see* **Note 11**).

Day 3:

10. Transfer the tissue to a new sterile specimen jar with 99 mL of fresh absolute isopropanol, 7 μL of PMSF solution and 1 mL of 100× ABAM and incubate at 37 °C under agitation at 100 rpm for 8 h (*see* **Note 11**).
11. Replace the solution with 99 mL of fresh absolute isopropanol, 7 μL of PMSF solution and 1 mL of 100× ABAM and incubate at 37 °C under agitation at 100 rpm for 16 h (*see* **Note 11**).

Day 4:

12. Transfer the tissue to a new sterile specimen jar and rinse in 99 mL of SPB rinse solution supplemented with 7 μL of PMSF solution and 1 mL of 100× ABAM for 30 min under agitation at 37 °C and 100 rpm (*see* **Note 12**).
13. Repeat step 12 twice more for a total of 3 rinses. Supplement the solution each time with fresh PMSF solution and ABAM.

14. After rinsing, incubate the sample in 99 mL of 0.25% trypsin-EDTA supplemented with 1 mL of 100× ABAM for a total of 6 h under agitation at 37 °C and 100 rpm.
15. Transfer the tissue to a new sterile specimen jar and rinse in 99 mL of SPB rinse solution supplemented with 7 μL of PMSF and 1 mL of 100× ABAM for 30 min under agitation at 37 °C and 100 rpm.
16. Repeat step 15 twice more for a total of 3 rinses, supplementing the solution with 7 μL of PMSF and 1 mL of 100× ABAM each time.
17. Incubate the tissue in 194 mL of SPB digest solution with 1 mL of DNase stock solution, 1 mL of RNase stock solution and 2 mL of lipase stock solution. Supplement with 2 mL of 100× ABAM. Incubate under agitation at 100 rpm at 37 °C for 16 h (*see* **Note 13**).

Day 5:

18. Transfer the tissue to a new sterile specimen jar with 99 mL of absolute isopropanol, 7 μL of PMSF solution and 1 mL of 100× ABAM. Incubate at 37 °C under agitation at 100 rpm for 8 h.
19. Rinse the tissue sample in 99 mL of SPB rinse solution supplemented with 7 μL of PMSF solution and 1 mL of 100× ABAM for 30 min under agitation at 37 °C and 100 rpm.
20. Repeat step 19 twice for a total of 3 rinses.
21. Rinse tissue in fresh 70% ethanol for 30 min under agitation at 100 rpm at room temperature for a total of 3 rinses.
22. Finally, rinse the DAT in sterile deionized water three times for 30 min under agitation at 100 rpm at room temperature to remove residual ethanol (*see* **Note 14**).
23. Place the rinsed DAT in a 50 mL centrifuge tube. Freeze overnight at −20 °C (*see* **Note 15**).
24. Keep samples frozen while transporting to the lyophilizer. Lyophilize samples for at least 48 h or until completely dry.
25. Store the lyophilized ECM in a dessicator away from moisture until further use (*see* **Note 5**).

3.1.2 Cartilage Tissue Decellularization

Day of Harvest:

1. Using a scalpel, carefully remove any dermal tissue attached to the porcine auricular cartilage. Subsequently, section the cartilage into small disks using a 5 mm biopsy punch.
2. Transfer the disks to a sterile specimen jar and rinse with 99 mL of PBS supplemented with 1 mL of 100× ABAM for 15 min to remove any excess blood.

3. Repeat step 2 two times for a total of 3 rinses.
4. Transfer 15–20 disks of cartilage tissue to a new sterile specimen jar containing 99 mL of 1× Solution A, 7 μL of PMSF solution and 1 mL of 100× ABAM (*see* **Note 10**).
5. Freeze the sample at −80 °C until completely frozen.
6. Remove the sample from the freezer and thaw in an incubator shaker at 37 °C and 250 rpm (*see* **Note 16**).
7. Once completely thawed, replace the solution with 99 mL of fresh 1× Solution A, 7 μL of PMSF solution and 1 mL of 100× ABAM.
8. Repeat steps 5–7 two more times for a total of 3 freeze-thaw cycles.
9. After the final thaw, replace the solution with 99 mL of fresh 1× Solution A, 7 μL of PMSF solution and 1 mL of 100× ABAM and agitate at 37 °C for 24 h at 250 rpm (*see* **Note 16**).

Days 2–3:

10. Transfer the tissue to a fresh specimen jar with 99 mL of Solution B, 1 mL of 100× ABAM and 7 μL of PMSF. Incubate at 37 °C under agitation at 250 rpm for 8 h.
11. Replace the solution with 99 mL of fresh Solution B, 1 mL of 100× ABAM and 7 μL of PMSF solution and incubate at 37 °C under agitation at 250 rpm for 16 h.

Day 3:

12. Transfer the tissue to a new sterile specimen jar and rinse in 99 mL of SPB rinse solution supplemented with 7 μL of PMSF solution and 1 mL of 100× ABAM for 30 min under agitation at 37 °C and 250 rpm.
13. Repeat step 12 two more times with fresh solution changes for a total of 3 rinses.
14. After rinsing, transfer the sample into a fresh sterile specimen jar with 97 mL of SPB digest solution supplemented with 1 mL of DNase stock solution, 1 mL of RNase stock solution and 1 mL of 100× ABAM. Incubate at 37 °C under agitation at 250 rpm for 5 h (*see* **Note 13**).
15. Replace the solution with 99 mL of Solution C supplemented with 7 μL of PMSF solution and 1 mL of 100× ABAM and incubate at 37 °C under agitation at 250 rpm for 16 h.

Day 5:

16. Transfer the tissue to a new sterile specimen jar and rinse in 99 mL of Solution C supplemented with 7 μL of PMSF solution and 1 mL of 100× ABAM for 8 h under agitation at 37 °C and 250 rpm.

17. Replace the solution with 99 mL of Solution C supplemented with 7 μL of PMSF solution and 1 mL of 100× ABAM and incubate at 37 °C under agitation at 250 rpm for 16 h.

Days 6–7:

18. Transfer the tissue to a new sterile specimen jar and rinse in 99 mL of sterile deionized water supplemented with 7 μL of PMSF solution and 1 mL of 100× ABAM for 30 min under agitation at 37 °C and 250 rpm.
19. Repeat step 18 two times for a total of 3 washes.
20. Transfer the tissue to a new sterile specimen jar and rinse in 99 mL of PBS supplemented with 7 μL of PMSF solution and 1 mL of 100× ABAM for 30 min under agitation at 37 °C and 250 rpm.
21. Repeat step 20 for a total of 3 rinses.
22. After the final rinse, replace the solution with 99 mL of 50 mM Tris buffer (pH 9) supplemented with 7 μL of PMSF solution and 1 mL of 100× ABAM and incubate at 37 °C under agitation at 250 rpm for 48 h, with solution changes every 12 h.

Day 8:

23. Rinse the tissue sample in 99 mL of PBS supplemented with 7 μL of PMSF solution and 1 mL of 100× ABAM for 15 min under agitation at 100 rpm and 37 °C.
24. Repeat step 23 twice for a total of 3 rinses.
25. Rinse the sample in 70% ethanol for 30 min for a total of 3 rinses.
26. Finally, rehydrate the DCT by rinsing in sterile deionized water 3 times for 30 min under agitation at room temperature to remove residual ethanol.
27. Place the rinsed DCT in a 50 mL centrifuge tube. Freeze overnight at −20 °C (*see* **Note 15**).
28. Keep samples frozen while transporting to the lyophilizer. Lyophilize samples for at least 48 h or until completely dry.
29. Store lyophilized ECM in a dessicator away from moisture until further use (*see* **Note 5**).

3.2 Cryo-Milling of DAT and DCT

Cryo-milling involves the physical processing of the decellularized tissue to generate micronized particles that can be easily dispersed within the hydrogel phase to provide a tissue-specific cell-instructive component within the hydrogel composites. Figure 1b shows representative scanning electron micrographs of the unsieved cryo-milled DAT and DCT particles.

1. Finely mince the lyophilized ECM with sharp scissors into 1–2 mm^3 pieces.
2. Transfer the minced tissue to the milling chamber along with 2 stainless steel milling balls (10 mm diameter).
3. Seal the chamber and submerge in liquid nitrogen for 3 min to snap freeze.
4. Secure the frozen chamber in the ball milling system and mill for 3 min continuously at 30 Hz or 1,800 rpm (*see* **Note 17**).
5. Store the cryo-milled ECM in a glass vial at room temperature in a dessicator away from moisture.
6. Stainless steel test sieves of different pore sizes may be used to separate the cryo-milled ECM to obtain a desired particle size range (*see* **Note 18**).

3.3 Fluorescence Labelling of the Cryo-Milled ECM Particles (Optional)

For imaging studies, the cryo-milled ECM can be pre-labelled with an amine-reactive Alexa Fluor®-conjugated succinimidyl ester to enable visualization by confocal microscopy (Fig. 3a). We frequently use the Alexa Fluor® 350 NHS Ester, which is resistant to photo-bleaching and compatible with LIVE/DEAD™ staining. However, a broad range of Alexa Fluor® dyes are commercially available, which offer flexibility for immunohistochemical characterization studies.

1. Transfer 350 mg of cryo-milled DAT or DCT into an empty foil-wrapped 50 mL centrifuge tube and add 10 mL of $NaHCO_3$ buffer.
2. Add 25 μL of the Alexa Fluor® 350 stock solution to the centrifuge tube and carefully agitate in a secured position for 1 h at room temperature protected from light.
3. After 1 h, centrifuge the tube at 1,500 × *g* for 10 min to collect the ECM particles, discarding the supernatant.
4. Add 10 mL of the prepared 1.5 M hydroxylamine solution to the ECM particles to stop the labelling process and agitate at room temperature for 1 h.
5. Centrifuge the suspension at 1,500 × *g* for 10 min to collect the labelled ECM particles. Replace the solution with 10 mL of sterile PBS and gently agitate for 5 min at room temperature.
6. Repeat step 5 for a total of 3 rinses.
7. After the final rinse, resuspend the particles in 70% ethanol for storage, protected from light to prevent photo-bleaching.

3.4 Preparation of Methacrylated Chondroitin Sulphate (MCS)

Figure 2a highlights the chemical structure of chondroitin sulphate and the MCS prepared by methacrylation with methacrylic anhydride.

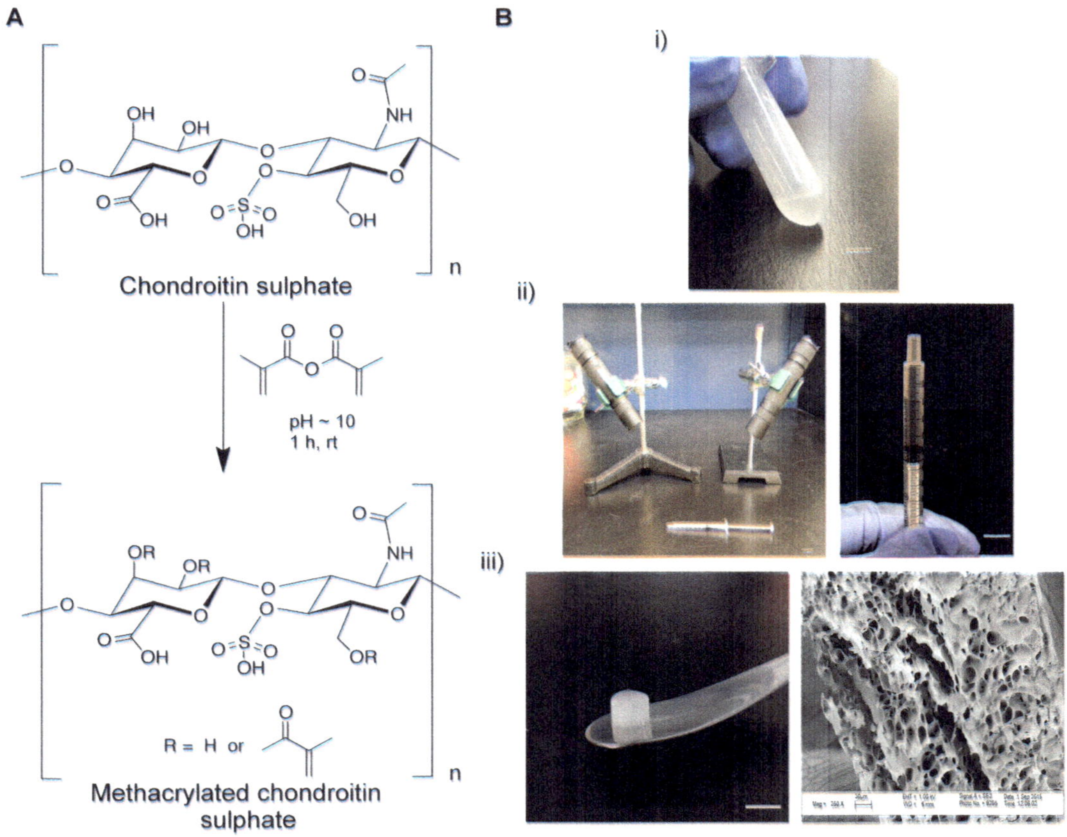

Fig. 2 Overview of the methacrylation of chondroitin sulphate and the synthesis of composite hydrogels incorporating milled DAT and DCT using UV- and thermally initiated mechanisms. (**a**) Chemical structure of chondroitin sulphate before and after methacrylation using methacrylic anhydride at pH ~ 10 for 1 h. (**b**) Flow chart highlighting the key steps involved in preparing the composite hydrogels. (i) MCS is dissolved in PBS, forming a viscous solution. The ECM particles, cells and crosslinking reagents are incorporated, and the mixture is carefully transferred into a 1 mL syringe. (ii) The syringe is then UV-crosslinked with light of wavelength in the range of 320–390 nm for UV-initiated crosslinking (*left*) or placed in an incubator at 37 °C for thermally initiated crosslinking (*right*). (iii) After crosslinking, the hydrogels are cut to form samples of the desired dimensions (3 mm × 7 mm shown). Representative SEM image of the hydrogels highlights the internal porous structure. Scale bars represent 1 cm

1. Weigh out 200 mg of chondroitin sulphate and transfer it to a glass vial wrapped in foil.
2. Add 1 mL of 0.2 M sodium phosphate monobasic buffer (NaH_2PO_4) and stir until completely dissolved. Carefully and slowly add 2 M HCl dropwise and adjust the pH to ~10 (*see* **Note 19**).
3. Add 60 μL of methacrylic anhydride dropwise with continuous stirring (*see* **Note 20**).

4. Allow the reaction to proceed for 1 h at room temperature with constant pH adjustment to maintain the pH at 10, using 2 M NaOH (*see* **Note 21**).
5. Add 10 mL of absolute ethanol to precipitate the product (*see* **Note 22**).
6. Decant the excess solvent and dissolve the precipitate in 10 mL of deionized water.
7. Transfer the solution to a 3.5 kDa MWCO dialysis membrane.
8. Dialyze against 2 L of deionized water for 48 h, with solvent changes every 8 h. The sample must be protected from light to prevent crosslinking.
9. After the dialysis, transfer the dialyzed MCS solution into a round bottom flask, neutralize the pH and snap freeze in liquid nitrogen.
10. Keep the samples frozen while transporting to the lyophilizer. Lyophilize samples for at least 48 h or until completely dry. The samples should be wrapped in foil and protected from light.
11. Following lyophilization, store the samples under argon and away from light and moisture at −20 °C (*see* **Note 23**).

3.5 Incorporation of Cryo-Milled ECM and Cells Within the MCS Hydrogels

The protocols developed in our lab are designed for a 1 mL pre-gel solution and can be scaled depending on the number of scaffolds required. While DAT or DCT particles can be stably incorporated within the hydrogel phase at concentrations ranging from 1 to 10% (w/v), the parameters described below (2% composites) were optimized to minimize the time required for gelation using UV- or thermally initiated crosslinking, while maximizing the stability of the gels over 14 days in culture. It is important to note that the concentration of ECM particles that can be incorporated within the hydrogels is highly dependent on the particle size, with larger particles generally yielding less stable gels, particularly at higher concentrations. We frequently use 1 mL sterile plastic syringes as a simple and convenient mold, but alternative molds can be used to fabricate scaffolds in a wide range of geometries. Figure 2b represents a schematic of the key steps required to synthesize composite hydrogels incorporating the cryo-milled ECM and the cell population(s) of interest. The cell density should be optimized based on the cell type and target application. We routinely encapsulate ASCs at densities between 5 and 10 million cells/mL of the pre-gel suspension, generating 50 μL hydrogels containing 2.5×10^5–5×10^5 cells/scaffold. These scaffolds can be cultured and analyzed using a variety of cell and molecular biology techniques to investigate in vitro cell function. For instance, cell viability can be assessed by confocal microscopy using the LIVE/DEAD™ assay, as demonstrated with human ASCs encapsulated in MCS + DAT and MCS + DCT composites in Fig. 3.

3.5.1 Preparation of Materials Prior to Crosslinking

1. Weigh 200 mg of MCS and transfer it to a sterile petri dish. Sterilize by exposure to the UV lamp in a biological safety cabinet for 30 min.
2. Weigh 20 mg of cryo-milled ECM and sterilize by exposure to a UV lamp in a biological safety cabinet for 30 min.

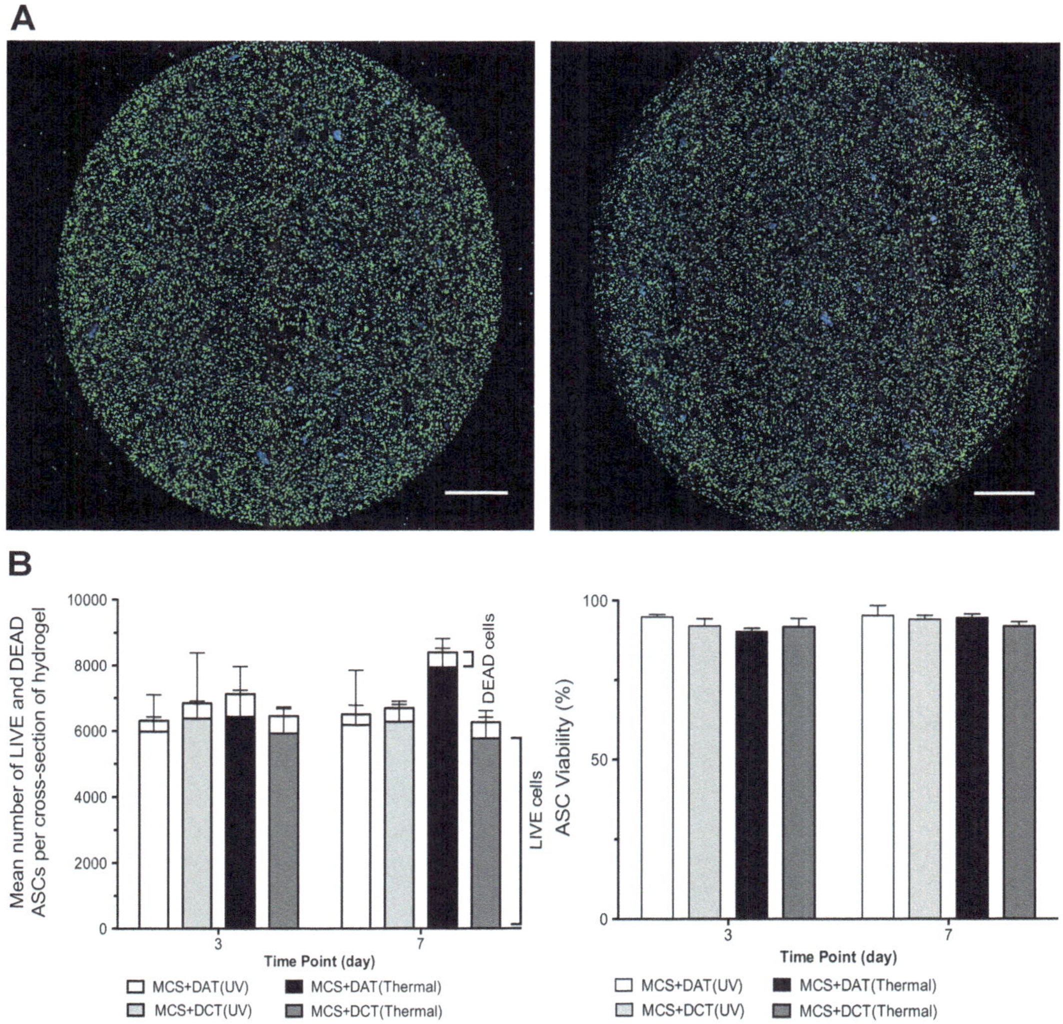

Fig. 3 Cell viability analysis within the hydrogel composites assessed using the LIVE/DEAD™ assay kit with confocal imaging. (**a**) Representative images of transverse sections of the MCS + DAT (*left*) and MCS + DCT (*right*) composite hydrogels, showing the cellular distribution of live (*green*) and dead (*red*) cells along with ECM particles labelled with Alexa Fluor® 350 NHS Ester (*blue*). The images demonstrate the distribution of the ECM particles within the hydrogel phase. Scale bars represent 1 mm. (**b**) Average number of viable and dead ASCs per x–y plane (*left*) and ASC viability over a culture period of 7 days (*right*). The graphs indicate high retention and viability of the human ASCs post-encapsulation, which is sustained throughout 7 days in culture with both crosslinking mechanisms

3. Transfer the MCS to a 2 mL sterile microcentrifuge tube using the sterile spatula. Dissolve the MCS in 700 μL of sterile PBS and vortex until completely dissolved. Centrifuge briefly at 10,000 × *g* for 10 s if necessary to remove any bubbles.

3.5.2 UV-Initiated Crosslinking

1. Prepare the UV system for crosslinking the hydrogels. For the Irgacure 2959 initiator, the light system should emit between the 320 and 390 nm range (365 nm target), with an intensity of 10 mW/cm^2 to enable crosslinking with high cell viability (*see* **Note 24**).
2. Dissolve 5 mg of Irgacure 2959 in 1 mL of sterile PBS and sterile filter into a fresh sterile microcentrifuge tube.
3. Using a sterile spatula, carefully add the ECM particles into the tube containing the MCS pre-polymer dissolved in PBS (From Sect. 3.5.1). Add 100 μL of the Irgacure stock solution and carefully mix through gentle stirring with an 18G needle. Ensure that the samples are protected from light at all steps.
4. Resuspend the desired number of cells of interest in 200 μL of proliferation medium containing serum and pipette the cells into the pre-gel/ECM suspension from step 3. Mix thoroughly with an 18G needle until evenly dispersed.
5. Working quickly, carefully transfer the pre-gel mixture into a sterile 1 mL syringe, ensuring not to create any bubbles (*see* **Note 25**).
6. Place the syringe under the light source and expose the sample to the UV light for 2 min to allow for crosslinking, flip the syringe and allow for an additional 2 min of crosslinking for a total of 4 min.

3.5.3 Thermally Initiated Crosslinking

1. Add the ECM particles carefully into the tube containing the MCS pre-polymer dissolved in PBS (from Sect. 3.5.1) and thoroughly mix with an 18G needle.
2. Resuspend the desired number of cells of interest in 200 μL of proliferation media containing serum and pipette into the pre-gel mixture from step 1. Mix thoroughly with an 18G needle until evenly dispersed (*see* **Note 25**).
3. Add 25 μL of APS to the microcentrifuge tube from step 2 and mix thoroughly with an 18G needle for at least 30 s. During this step, keep the sample and APS stock solution on ice to prevent crosslinking.
4. Add 25 μL of TEMED to the microcentrifuge tube from step 3 and mix thoroughly with an 18G needle for at least 30 s. During this step, keep the sample on ice.
5. Working quickly, transfer the pre-gel mixture into a 1 mL sterile syringe, being careful to prevent bubble formation.

6. Place the syringe in an upright position in an incubator at 37 °C for 15 min to allow for crosslinking. The crosslinking time must be adjusted depending on the size of the mold, as well as the volume and concentration of the polymer mixture.

3.5.4 Scaffold Preparation and Culture

1. Working in a sterile biological safety cabinet, use a sterile razor blade to remove the bore of the syringe. Gently extrude the crosslinked gel from the syringe, carefully cutting the hydrogel perpendicular to the cross-section using the razor blade to generate 50 μL hydrogel samples. The resulting gels should have a diameter of ~7 mm and height of ~3 mm.
2. Transfer the gels to cell culture inserts within 12-well tissue culture plates filled with ~3 mL of culture medium.
3. Replace the medium in the wells with fresh medium after 15 min for a total of two rinses (*see* **Note 26**).
4. Culture the samples under the desired media conditions to perform in vitro cell function studies or use as implants for in vivo cell delivery applications.

4 Notes

1. Transport freshly isolated adipose tissue to the lab on ice in a sterile tub with sterile PBS supplemented with 2% BSA. Although this protocol has been developed for use with excised tissue blocks, lipoaspirates can also be processed.
2. Adjust the pH of solutions using HCl and NaOH. Ensure that the pH of Solution A is at 8 to minimize any protease activity [46].
3. PMSF is a serine protease inhibitor that renders proteases inactive during cell lysis. The PMSF stock solution must be prepared with dry solvents (i.e. absolute isopropanol), foil wrapped for protection from light and stored away from moisture.
4. Isopropanol is miscible in both water and oil, making it an effective solvent for lipid extraction.
5. For long-term storage, it is recommended that the DAT and DCT be stored in a lyophilized state at room temperature in a desiccator, as described in Sect. 3.2. Storing in a dry state enhances the stability of the proteinaceous ECM. Storage within a desiccator is required as the ECM is highly hygroscopic and will readily absorb moisture from the atmosphere.
6. The hydroxylamine hydrochloride solution is used to stop the labelling reaction with the Alexa Fluor® dye. This solution is

unstable and must be freshly prepared immediately before use. The pH is highly unstable and must be adjusted very carefully.

7. The methacrylate groups in MCS are highly sensitive to light, oxygen and moisture. Unintentional exposure can cause cross-linking of the methacylate groups by free oxygen radicals or photo-sensitive crosslinking. For long-term storage of MCS, store in an inert atmosphere under argon, wrapped in foil at −20 °C.
8. Irgacure 2959 is a photo-initiator that must be stored away from light under dry conditions. As it is sparingly soluble in water, the stock solution should be prepared and stirred for at least 20 min. The stock solution has poor stability and a fresh sample should be prepared on the day of encapsulation.
9. APS is the initiator while TEMED is the accelerator for this thermally initiated free radical polymerization system. Both reagents are highly unstable in aqueous solutions and should be freshly prepared on the day of encapsulation. The reagents must be protected from light and stored in an ice bath or at 4 °C prior to use.
10. Scale the reagents as required, based on the size and volume of tissue prior to decellularization. For larger tissue volumes, ensure that the samples are well mixed during mechanical agitation.
11. Prior to the polar solvent extraction, either manually massage or press the tissue between two stainless steel mesh filters to effectively remove lipids. This step is critical in the decellularization process.
12. Day 4 requires extended processing with many solvent changes and rinses. It is recommended to start the protocol as early as possible in the day to complete the decellularization procedure.
13. SPB digest solution contains a high salt concentration, which may precipitate during storage. Ensure that the solution is well mixed prior to use.
14. At the end of the decellularization, the DAT appears off-white. DAT that appears yellow is indicative of ineffective lipid removal, which may be resolved with additional rinses in absolute isopropanol.
15. When freezing the samples, place the 50 mL centrifuge tube on its size to maximize surface area during lyophilization and reduce drying time. For the DAT and DCT, do not exceed more than 30 mL of tissue within the centrifuge tube.
16. As cartilage tissue has a very dense ECM structure, it is important that the decellularization process includes high-speed agitation (e.g. 250 rpm minimum) during all indicated processing steps to ensure effective decellularization.

17. The particle size of the ECM can be tailored by varying the cryo-milling time and speed. A smaller particle size has been shown to enhance cell–cell contact and promote adipogenesis [18].
18. To obtain ECM particles of desired size ranges, separate the cryo-milled particles using stainless steel mesh sieves with sequentially decreasing pore sizes. Gently tap and agitate the sieves to promote transfer.
19. For all further steps, ensure that the chondroitin sulphate solution is foil-wrapped and protected from light.
20. The reaction conditions have been adjusted to obtain MCS with a 17% degree of methacrylation, which enables crosslinking while maintaining high cell viability. In order to tune methacrylation, the molar equivalents of methacrylic anhydride can either be increased or decreased compared to the CS. A higher degree of methacrylation generally results in stiffer gels, but may also impair cell viability [32].
21. pH adjustment is critical throughout the reaction. If the pH becomes acidic, this can cause cleavage of the glycosidic linkages in the chondroitin sulphate backbone, thereby leading to product degradation and ineffective crosslinking. Test the pH using indicator strips every 5 min to maintain it at pH 10.
22. The absolute ethanol must be added in excess to ensure a high yield of product during precipitation.
23. Ensure that the sample is stored under argon and wrapped in foil as any oxygen radicals from the environment can cause crosslinking of the material. The photo-sensitive MCS should also be stored away from light at all times (i.e. wrapped in foil).
24. The apparatus for UV crosslinking can be selected according to individual requirements. The UV intensity for crosslinking can be manipulated by adjusting the distance between the sources and the height from the sample. A higher UV intensity can be used to crosslink the gels more rapidly, but may also lead to cell damage by the formation of pyrimidine dimers. Thus, care must be taken to optimize the UV intensity to enable crosslinking while maximizing cell viability. The wavelength of the UV source should also be monitored carefully, as any UV light in the UV-B or UV-C regions (<300 nm) can be damaging to the cells.
25. Avoiding bubbles is very crucial to this step. Any trapped air in the pre-gel mixture can interfere with the crosslinking and form unstable gels. Additionally, bubbles may also affect the cell response following encapsulation and culture.
26. Excess radicals not involved in the crosslinking must be extracted to minimize toxicity to the cells.

Acknowledgements

Scholarship funding to support this work was provided by the NSERC CREATE CONNECT! Program and Western's Collaborative Graduate Training Program in Musculoskeletal Health Research (CMHR). Operational grant funding support was provided by NSERC. The authors would like to thank Cody Brown for technical assistance with the SEM imaging and the Western Nanofabrication facility for access to their imaging equipment. Finally, we would like to acknowledge Drs. Aaron Grant, Brian Evans and Robert Richards for their clinical collaborations in providing human adipose tissue samples, as well as the Mount Brydges abattoir for providing the porcine auricular cartilage tissue.

References

1. Frantz C, Stewart KM, Weaver VM (2010) The extracellular matrix at a glance. J Cell Sci 123:4195–4200. doi:10.1242/jcs.023820
2. Jabbari E, Leijten J, Xu Q, Khademhosseini A (2015) The matrix reloaded: the evolution of regenerative hydrogels. Mater Today 19:190–196. doi:10.1016/j.mattod.2015.10.005
3. Knight E, Przyborski S (2015) Advances in 3D cell culture technologies enabling tissue-like structures to be created in vitro. J Anat 227:746–756. doi:10.1111/joa.12257
4. Hinderer S, Layland SL, Schenke-Layland K (2016) ECM and ECM-like materials—biomaterials for applications in regenerative medicine and cancer therapy. Adv Drug Deliv Rev 97:260–269. doi:10.1016/j.addr.2015.11.019
5. Sart S, Tsai A-C, Li Y, Ma T (2014) Three-dimensional aggregates of mesenchymal stem cells: cellular mechanisms, biological properties, and applications. Tissue Eng Part B Rev 20:365–380. doi:10.1089/ten.TEB.2013.0537
6. Geckil H, Xu F, Zhang X et al (2010) Engineering hydrogels as extracellular matrix mimics. Nanomedicine 5:469–484. doi:10.2217/nnm.10.12
7. Benders KEM, van Weeren PR, Badylak SF et al (2013) Extracellular matrix scaffolds for cartilage and bone regeneration. Trends Biotechnol 31:169–176. doi:10.1016/j.tibtech.2012.12.004
8. Hoshiba T, Lu H, Kawazoe N, Chen G (2010) Decellularized matrices for tissue engineering. Expert Opin Biol Ther 10:1717–1728. doi:10.1517/14712598.2010.534079
9. Crapo PM, Gilbert TW, Badylak SF (2011) An overview of tissue and whole organ decellularization processes. Biomaterials 32:3233–3243. doi:10.1016/j.biomaterials.2011.01.057
10. Flynn LE (2010) The use of decellularized adipose tissue to provide an inductive microenvironment for the adipogenic differentiation of human adipose-derived stem cells. Biomaterials 31:4715–4724. doi:10.1016/j.biomaterials.2010.02.046
11. Huang S-P, Hsu C-C, Chang S-C et al (2012) Adipose-derived stem cells seeded on acellular dermal matrix grafts enhance wound healing in a murine model of a full-thickness defect. Ann Plast Surg 69:656–662. doi:10.1097/SAP.0b013e318273f909
12. Dainese L, Guarino A, Burba I et al (2012) Heart valve engineering: decellularized aortic homograft seeded with human cardiac stromal cells. J Heart Valve Dis 21:125–134
13. Woods T, Gratzer PF (2005) Effectiveness of three extraction techniques in the development of a decellularized bone–anterior cruciate ligament–bone graft. Biomaterials 26:7339–7349. doi:10.1016/j.biomaterials.2005.05.066
14. Han TTY, Toutounji S, Amsden BG, Flynn LE (2015) Adipose-derived stromal cells mediate in vivo adipogenesis, angiogenesis and inflammation in decellularized adipose tissue bioscaffolds. Biomaterials 72:125–137. doi:10.1016/j.biomaterials.2015.08.053
15. Turner AEB, Yu C, Bianco J et al (2012) The performance of decellularized adipose tissue microcarriers as an inductive substrate for human adipose-derived stem cells. Biomaterials 33:4490–4499. doi:10.1016/j.biomaterials.2012.03.026

16. Yu C, Bianco J, Brown C et al (2013) Porous decellularized adipose tissue foams for soft tissue regeneration. Biomaterials 34:3290–3302. doi:10.1016/j.biomaterials.2013.01.056
17. Ki H, Tian T, Han Y et al (2014) Biomaterials composite hydrogel scaffolds incorporating decellularized adipose tissue for soft tissue engineering with adipose-derived stem cells. Biomaterials 35:1914–1923. doi:10.1016/j.biomaterials.2013.11.067
18. Brown CFC, Yan J, Han TTY et al (2015) Effect of decellularized adipose tissue particle size and cell density on adipose-derived stem cell proliferation and adipogenic differentiation in composite methacrylated chondroitin sulphate hydrogels. Biomed Mater 10:45010. doi:10.1088/1748-6041/10/4/045010
19. Heinbuch D (2016) Adipose-derived stem cell differentiation in cell aggregates supplemented with micronized, decellularized extracellular matrix. University of Western Ontario, London
20. Agmon G, Christman KL (2016) Controlling stem cell behavior with decellularized extracellular matrix scaffolds. Curr Opin Solid State Mater Sci 20:193–201. doi:10.1016/j.cossms.2016.02.001
21. Kang H, Peng J, Lu S et al (2014) In vivo cartilage repair using adipose-derived stem cell-loaded decellularized cartilage ECM scaffolds. J Tissue Eng Regen Med 8:442–453. doi:10.1002/term.1538
22. Quint C, Kondo Y, Manson RJ et al (2011) Decellularized tissue-engineered blood vessel as an arterial conduit. Proc Natl Acad Sci U S A 108:9214–9219. doi:10.1073/pnas.1019506108
23. Badylak SF, Freytes DO, Gilbert TW (2009) Extracellular matrix as a biological scaffold material: structure and function. Acta Biomater 5:1–13. doi:10.1016/j.actbio.2008.09.013
24. Tsai C-H, Chou M-Y, Jonas M et al (2002) A composite graft material containing bone particles and collagen in osteoinduction in mouse. J Biomed Mater Res 63:65–70
25. Mazzitelli S, Luca G, Mancuso F et al (2011) Production and characterization of engineered alginate-based microparticles containing ECM powder for cell/tissue engineering applications. Acta Biomater 7:1050–1062. doi:10.1016/j.actbio.2010.10.005
26. Chang C-H, Chen C-C, Liao C-H et al (2014) Human acellular cartilage matrix powders as a biological scaffold for cartilage tissue engineering with synovium-derived mesenchymal stem cells. J Biomed Mater Res Part A 102:2248–2257. doi:10.1002/jbm.a.34897
27. Beck EC, Barragan M, Libeer TB et al (2016) Chondroinduction from naturally derived cartilage matrix: a comparison between devitalized and decellularized cartilage encapsulated in hydrogel pastes. Tissue Eng Part A 22 (7–8):665–679. doi:10.1089/ten.tea.2015.0546
28. Drury JL, Mooney DJ (2003) Hydrogels for tissue engineering: scaffold design variables and applications. Biomaterials 24:4337–4351. doi:10.1016/S0142-9612(03)00340-5
29. Cai S, Liu Y, Zheng Shu X, Prestwich GD (2005) Injectable glycosaminoglycan hydrogels for controlled release of human basic fibroblast growth factor. Biomaterials 26:6054–6067. doi:10.1016/j.biomaterials.2005.03.012
30. Lim HL, Hwang Y, Kar M, Varghese S (2014) Smart hydrogels as functional biomimetic systems. Biomater Sci 2:603. doi:10.1039/c3bm60288e
31. Spiller KL, Maher SA, Lowman AM (2011) Hydrogels for the repair of articular cartilage defects. Tissue Eng Part B Rev 17:281–299. doi:10.1089/ten.teb.2011.0077
32. Hayami JWS, Waldman SD, Amsden BG (2015) Photo-cross-linked methacrylated polysaccharide solution blends with high chondrocyte viability, minimal swelling, and moduli similar to load bearing soft tissues. Eur Polym J 72:1–11. doi:10.1016/j.eurpolymj.2015.01.038
33. Flynn LE, Prestwich GD, Semple JL, Woodhouse KA (2008) Proliferation and differentiation of adipose-derived stem cells on naturally derived scaffolds. Biomaterials 29:1862–1871. doi:10.1016/j.biomaterials.2007.12.028
34. Kong HJ, Smith MK, Mooney DJ (2003) Designing alginate hydrogels to maintain viability of immobilized cells. Biomaterials 24:4023–4029. doi:10.1016/S0142-9612(03)00295-3
35. Rouillard AD, Berglund CM, Lee JY et al (2011) Methods for photocrosslinking alginate hydrogel scaffolds with high cell viability. Tissue Eng Part C Methods 17:173–179. doi:10.1089/ten.tec.2009.0582
36. Koh W, Revzin A, Pishko MV (2002) Poly (ethylene glycol) hydrogel microstructures encapsulating living cells. Langmuir 18:2459–2462. doi:10.1021/la0115740
37. Thanh T, Thi H, Lee JS et al (2016) Enhanced cellular activity in gelatin-poly (ethylene glycol) hydrogels without compromising gel stiffness 16(3): 334–340
38. Wang D-A, Varghese S, Sharma B et al (2007) Multifunctional chondroitin sulphate for

cartilage tissue-biomaterial integration. Nat Mater 6:385–392. doi:10.1038/nmat1890
39. Li Q, Williams CG, Sun DDN et al (2004) Photocrosslinkable polysaccharides based on chondroitin sulfate. J Biomed Mater Res 68A:28–33. doi:10.1002/jbm.a.20007
40. Goude MC, McDevitt TC, Temenoff JS (2014) Chondroitin sulfate microparticles modulate transforming growth factor-β1-induced chondrogenesis of human mesenchymal stem cell spheroids. Cells Tissues Organs 199:117–130. doi:10.1159/000365966
41. Varghese S, Hwang NS, Canver AC et al (2008) Chondroitin sulfate based niches for chondrogenic differentiation of mesenchymal stem cells. Matrix Biol 27:12–21. doi:10.1016/j.matbio.2007.07.002
42. Ravanat J-L, Douki T, Cadet J (2001) Direct and indirect effects of UV radiation on DNA and its components. J Photochem Photobiol B 63:88–102. doi:10.1016/S1011-1344(01)00206-8
43. Park H, Temenoff JS, Holland TA et al (2005) Delivery of TGF-β1 and chondrocytes via injectable, biodegradable hydrogels for cartilage tissue engineering applications. Biomaterials 26:7095–7103. doi:10.1016/j.biomaterials.2005.05.083
44. Park H, Temenoff JS, Tabata Y et al (2007) Injectable biodegradable hydrogel composites for rabbit marrow mesenchymal stem cell and growth factor delivery for cartilage tissue engineering. Biomaterials 28:3217–3227. doi:10.1016/j.biomaterials.2007.03.030
45. Temenoff JS, Park H, Jabbari E et al (2004) Thermally cross-linked oligo(poly(ethylene glycol) fumarate) hydrogels support osteogenic differentiation of encapsulated marrow stromal cells in vitro. Biomacromolecules 5:5–10. doi:10.1021/bm030067p
46. Gilbert TW, Sellaro TL, Badylak SF (2006) Decellularization of tissues and organs. Biomaterials 27:3675–3683. doi:10.1016/j.biomaterials.2006.02.014

Methods in Molecular Biology (2018) 1577: 209–226
DOI 10.1007/7651_2017_37

Published online: 11 May 2017

3D-Printing Composite Polycaprolactone-Decellularized Bone Matrix Scaffolds for Bone Tissue Engineering Applications

Alexandra N. Rindone, Ethan Nyberg, and Warren L. Grayson

Abstract

Millions of patients worldwide require bone grafts for treatment of large, critically sized bone defects from conditions such as trauma, cancer, and congenital defects. Tissue engineered (TE) bone grafts have the potential to provide a more effective treatment than current bone grafts since they would restore fully functional bone tissue in large defects. Most bone TE approaches involve a combination of stem cells with porous, biodegradable scaffolds that provide mechanical support and degrade gradually as bone tissue is regenerated by stem cells. 3D-printing is a key technique in bone TE that can be used to fabricate functionalized scaffolds with patient-specific geometry. Using 3D-printing, composite polycaprolactone (PCL) and decellularized bone matrix (DCB) scaffolds can be produced to have the desired mechanical properties, geometry, and osteoinductivity needed for a TE bone graft. This book chapter will describe the protocols for fabricating and characterizing 3D-printed PCL:DCB scaffolds. Moreover, procedures for culturing adipose-derived stem cells (ASCs) in these scaffolds in vitro will be described to demonstrate the osteoinductivity of the scaffolds.

Keywords Bone tissue engineering, 3D-printing, Decellularized bone matrix, Polycaprolactone, Adipose-derived stem cells

1 Introduction

Over two million people are treated annually with bone grafts for conditions such as trauma, back pain, congenital defects, and cancer [1]. Due to high demand, bone grafts had a market size of $4.4 billion in 2014 and is projected to account for $5.3 billion in sales by 2019 [1]. Current bone grafts used in the clinic include autografts, allografts, and synthetic grafts [2]. Autografts, which are harvested from the patient's own tissue, are the current gold standard for treatment since they are biologically active and pose no risk of immune rejection or disease transmission. However, use of autografts in the clinic is limited due to lack of tissue availability and donor site morbidity. Allografts and synthetic grafts are more customizable and readily available

Alexandra N. Rindone and Ethan Nyberg contributed equally to this work.

than autografts, but issues such as immune rejection and inferior bioactivity prevent their widespread use in the clinic.

To combat the issues associated with current bone grafts, researchers have focused on developing tissue engineered (TE) grafts that regenerate functional bone tissue and provide a permanent treatment for bone defects. Typical bone TE approaches involve the combination of multipotent stem cells with porous, biodegradable biomaterial scaffolds to create a regenerative bone graft. The scaffold serves to promote osteogenesis, provide mechanical support, allow for nutrient diffusion and cell infiltration through its pores, and degrade at a rate comparable to tissue regeneration within the graft [3]. Techniques that have been used to create porous, biodegradable TE scaffolds include solvent casting and particulate leaching, thermally induced phase separation, gas foaming, electrospinning, microsphere sintering, emulsion freeze drying, and 3D-printing [4, 5]. Of these technologies, 3D-printing is the most widely studied due to its ability to produce scaffolds that have precise and customizable geometries [3]. The 3D-printing fabrication process involves the generation of a CAD file that defines the overall geometry and pore design of the scaffold (Fig. 1). This CAD file is then converted into a G-code that is uploaded into the 3D-printer. The scaffold material is printed layer-by-layer onto a bed according to the instructions in the G-code. Once the scaffold is printed, post-processing steps such as removal of support material, sintering, or surface modification may be performed to attain desired scaffold geometry and properties.

Materials used for 3D-printing bone TE scaffolds include ceramics, such as tricalcium phosphate and hydroxyapatite, and synthetic thermoplastic polymers. One of the most widely studied materials is

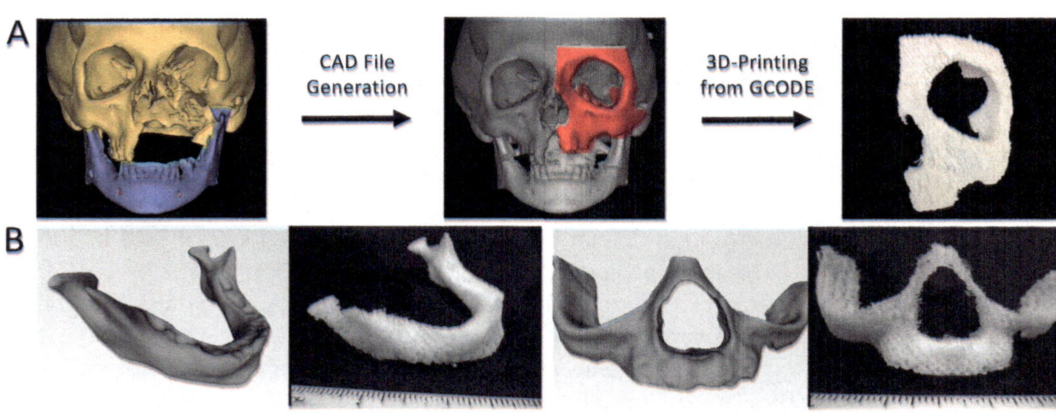

Fig. 1 3D-printing of patient-specific bone TE scaffolds. (**a**) Workflow of 3D-printing process. The patient's CT scan is used to create an STL file for the graft that precisely fits into the defect site [2]. This STL file is then converted into GCODE and 3D-printed into a customized TE scaffold [3]. (**b**) Examples of porous, biodegradable polycaprolactone (PCL) scaffolds that were 3D-printed based upon the specific geometry of the mandible (*left*) and maxilla (*right*). **b** was adapted from Ref. [6] with permission

polycaprolactone (PCL), a thermoplastic, aliphatic polyester. PCL is a desirable material for bone TE because it is biocompatible, has mechanical properties similar to trabecular bone, and has a slower biodegradation rate than other thermoplastics [7, 8]. Moreover, its material properties, such as its low melting temperature and good melt viscosity, allow PCL to be readily fabricated with 3D-printing [6, 9]. However, PCL alone is not osteoinductive, so ceramic or biological additives are needed to promote bone regeneration on PCL scaffolds [10].

Decellularized bone matrix (DCB) is a natural biomaterial that enhances the osteoinductivity of PCL scaffolds [11]. DCB consists of inorganic hydroxyapatite mineral and organic matrix components, including collagen, glycosaminoglycans, and growth factors [12, 13]. The biochemical composition and micro- and nanostructure of DCB promotes stem cell attachment, proliferation, and differentiation into bone cells [13, 14]. When added to 3D-printed PCL scaffolds, DCB significantly enhances in vitro adipose-derived stem cell (ASC) osteogenesis, even in the absence of other osteogenic factors [11, 15]. This chapter will describe the protocols for processing DCB and printing composite PCL:DCB scaffolds. Furthermore, the procedures for assessing in vitro bone regeneration by ASCs will be described to demonstrate the osteoinductivity of composite PCL:DCB scaffolds.

2 Materials

2.1 Bone Processing

1. Bovine calf knees, fresh or frozen at −20 or −80 °C
2. Scalpel handle and blade
3. Toothed forceps
4. Band saw suitable for cutting bone (i.e., Mar-Med Inc. Bone Band Saw; Cleveland, OH)
5. Faucet with high-water pressure (small diameter nozzle attachment may be required)
6. Lyophilizer
7. 6875 Freezer/Mill with mid or large polyvial set and large vial opener (SPEX SamplePrep; Metuchen, NJ)
8. Liquid nitrogen
9. Sieve shaker with a mesh size of ≤40 μm
10. Mass balance and weighing trays

2.2 Decellularization Reagents

1. Wash Solution 1: 0.1 % EDTA in 1 × PBS without Ca^{2+}. Adjust the pH to 8.0 with NaOH to allow EDTA to completely dissolve. Store at room temperature.

2. Wash Solution 2: 0.1 % EDTA, 10 mM Tris buffer in 1× PBS without Ca^{2+}. Adjust the pH to 8.0 with NaOH to allow EDTA to completely dissolve. Store at room temperature.
3. Wash Solution 3: 0.5 % SDS, 10 mM Tris buffer in 1× PBS without Ca^{2+}. Store at room temperature.
4. Wash Solution 4: 50 units/mL DNAse I, Grade II (Sigma-Aldrich) and 1 unit/mL RNAse in Wash Solution 2. Prepare solution just before using.

2.3 3D-Printing Software

1. CAD Software (TinkerCAD, Autodesk, www.tinkercad.com)
2. Slicing Software (Slic3r, www.slic3r.org)
3. Communication with the Printer (Mach3, ArtSoft, USA)
4. ImageJ (NIH, www.imagej.nih.gov)
5. DiameterJ (www.imagej.net/DiameterJ)
6. PowerPoint (Microsoft)

2.4 3D-Printing Manufacturing Tools and Materials

1. Scaffold material: PCL (powdered, 43,000–50,000 MW, #25090 Polysciences Inc., Warrington, PA) with <40 μm bone particles (*see* **Note 1**).
2. XYZ Positioning Arm, such as a Syil X4 CNC Mill (Syil America, Coos Bay, OR)
3. Stainless Steel extrusion and thermal chamber. Base should have 3/8″ cylindrical stock and standard male screw threads, while the top should have an adaptor for tubing (Nordson EFD 5225K711)
4. Pressure regulator (Ultimus V, Nordson EFD, Providence, RI)
5. Heating Element (Type K thermocouple, Watlow)
6. PID Controller (Series CV Temperature Controller, Watlow)
7. Heated bed, for example, a Flexible Heater Rectangular (115V 270W, Rapid Industrial Supply) connected to a Variable Voltage Regulator (4098 Variac Transformer TDGC-0.5KM, Circuit Specialists)
8. Extrusion nozzle (460 μm diameter, Nordson EFD #322018)
9. Air compressor
10. Stainless steel sifter (400 μm mesh)
11. 4 mm circular punch (McMaster #3424A14)
12. Hammer
13. Cutting board
14. Ruler

2.5 Cell Culture

1. ASCs isolated from lipoaspirate (*see* **Note 2**)
2. Expansion medium: 10 % fetal bovine serum (Atlanta Biologicals), 1 % penicillin–streptomycin, and 1 ng/mL FGF-2 (Peprotech) in high-glucose DMEM with sodium pyruvate (Gibco). Add FGF-2 to growth medium just before feeding. Expansion medium without FGF-2 may be stored at 4 °C for up to 2 weeks
3. Osteogenic medium: 10 % fetal bovine serum, 1 % penicillin–streptomycin, 10 mM β-glycerophosphate (Sigma-Aldrich), and 50 μm ascorbic acid (Sigma-Aldrich) in low-glucose DMEM with sodium pyruvate (Gibco). Osteogenic medium may be stored at 4 °C for up to 1 week
4. Fibrinogen solution: 10 mg/mL of fibrinogen (Sigma-Aldrich) in PBS without Ca^{2+}. Sterile filter prior to use. Solution may be stored at 4 °C for up to 1 week.
5. Thrombin solution: 10 units/mL of thrombin (Sigma-Aldrich) in PBS with Ca^{2+}. Sterile filter prior to use. Prepare solution just before seeding scaffolds.
6. Phosphate buffered saline (PBS) with and without Ca^{2+} (Mediatech)
7. 0.25 % Trypsin-EDTA (Gibco)

2.6 Osteogenic Assays

1. Alizarin Red stain: 40 mM Alizarin Red S (Sigma-Aldrich) in dH_2O. Vortex solution to dissolve and adjust the pH to 4.1–4.3 with NaOH. Prepare the solution on the same day as staining
2. Quant-iT PicoGreen dsDNA Assay Kit (Invitrogen)
3. Calcium (CPC) LiquiColor Test (Stanbio Laboratory; Boerne, TX)
4. DNA lysis buffer: 10 mM Tris buffer, 1 mM EDTA, 0.1 % Triton X-100 (Sigma-Aldrich) in dH_2O. Add 0.1 mg/mL of proteinase K (Quality Biologicals) just before use. Solution without proteinase K may be stored at RT for 6 months
5. 0.5 M HCl
6. 96-well plate reader

3 Methods

3.1 Bone Harvesting

1. Thaw calf knee(s) overnight at 4 °C, and then bring knee(s) to room temperature 1 h before harvesting.
2. Separate the femur from the tibia by cutting the tissue around the knee socket. Use a scalpel blade and toothed forceps to cut away the tissue.
3. Once the joint is separated, cut away all muscle, ligaments, and other connective tissues from the knee. Connective tissue will interfere with the bone sawing process, so it is essential to remove as much of it as possible.

4. Cut bone into approximately 1 cm^3 pieces using a band saw. Use a couple of plastic or wooden blocks to guide the bone through the saw to avoid hand contact with the saw blade. Discard any cortical bone (present in the diaphysis) or cartilage (present on joint surface and in growth plates). You may need to use a scalpel and forceps to remove cartilage from the trabecular bone pieces. A representative image of a sawed bone piece is shown in Fig. 2.
5. Proceed to the decellularization procedure or store the bone pieces at −20 °C.

3.2 Decellularization

1. If the sawed bone is frozen, thaw bone pieces overnight at 4 °C.
2. Wash the bone pieces with a high-pressure water faucet to clear out the bone marrow. The bone pieces should have an off-white color after washing is complete (Fig. 2). Bone pieces may be stored at −20 °C after washing, but avoid multiple freeze-thaw cycles if possible.
3. Place bone in a large container (i.e., 50–500 mL centrifuge tubes), and fill the container with Wash Solution 1. The volume of the solution should be 2–3 times the bone volume so that all bone pieces are immersed in the solution. Incubate the bone particles on an orbital shaker (250 rpm) at room temperature for 1 h. Use the same solution volume and shaker RPM for all remaining washes.
4. Discard Wash Solution 1 from the container. Fill the container with Wash Solution 2, and incubate the bone pieces at 4 °C for at least 12 h.
5. Wash the bone pieces with PBS for 1 h at room temperature.
6. Incubate the bone pieces in Wash Solution 3 at room temperature for at least 24 h.
7. Wash the bone pieces successively with PBS until most bubbles in the container disappear.
8. Incubate the bone pieces in Wash Solution 4 for 5 h at room temperature.

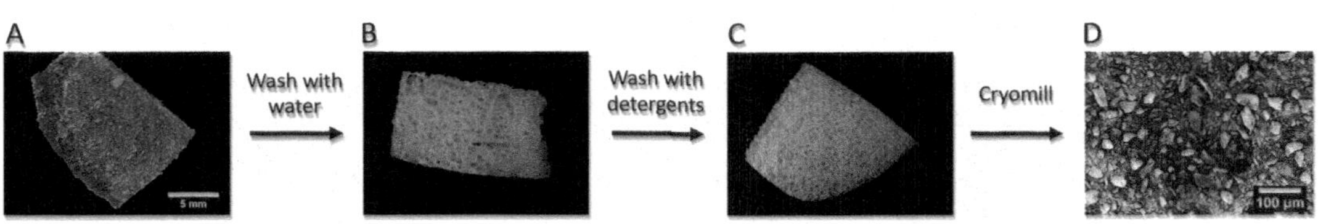

Fig. 2 Obtaining bone particles for 3D-printing. Stereoscope images of bone after: (**a**) sawing, (**b**) washing with water, (**c**) and decellularizing/lyophilizing. (**d**) Scanning electron micrograph of <40 μm bone particles after cryomilling. Scale bars for stereoscope and scanning electron microscope images are 5 mm and 100 μm, respectively

9. Wash the bone pieces with PBS 2× at room temperature for at least 1 h each. One of the washes may be performed overnight.
10. Discard PBS, and remove the lids from the containers. Fold a KimWipe in half and secure it to the top of the container with a rubber band. Freeze the bone pieces at −20 °C for at least 4 h to prepare for lyophilization
11. Lyophilize the bone pieces in the KimWipe-sealed container overnight.
12. Store the dried bone pieces in a sealed container at room temperature.

3.3 Cryomilling

1. Load bone pieces and impactor into medium- or large-sized grinding vials. Shake the vial to make sure that the impactor is mobile. If not, remove bone pieces until the impactor freely moves in the grinding vial.
2. Load the grinding vial into the SPEX cryomill according the manufacturer's instructions. Make sure that the grinding vial fits tightly into the cryomill. A loose vial may cause bone to leak out of the vial during milling.
3. Fill the cryomill with liquid N_2 until it is ~75–80 % full. Do not overfill the cryomill (*see* **Note 3**).
4. Slowly close the cryomill, and turn the cryomill power on.
5. Adjust the settings of the cryomill to the following:
   a. Cycles: 15
   b. Precool: 3 min
   c. Run: 1 min
   d. Cool: 3 min
   e. Rate: 10 CPS
6. Start the cryomill, and monitor every 15 min to make sure that the liquid N_2 levels do not get too low. If the liquid N_2 levels get too low, pause the run and add more liquid N_2 before resuming the run.
7. Remove the cryovial once the run is complete. Keep the lid open to allow the liquid N_2 to evaporate.
8. Immediately remove one of the cryovial plugs (or lids) with a vial extractor to avoid pressure buildup in the vial.
9. Let bone particles sit in the vial for 30 min, or until the vial reaches room temperature.

3.4 Particle Size Separation and Characterization

1. Load bone particles into the top sieve of a sieve shaker.
2. Run the sieve shaker according to manufacturer's instructions.
3. Carefully transfer the bone particles from each sieve to a plastic weighing tray. Use a separate weight tray for each sieve.

4. Measure the mass of bone particles from each sieve.
5. Determine the percentage of bone particles by weight for each sieve. These percentages represent bone particle size range and weight distribution for a given batch.
6. Set aside <40 μm bone particles for 3D-printing. Particles >40 μm may be saved for further analysis or other applications.
7. If bone particles are not being used immediately, store at −20 °C.

3.5 Evaluation of Decellularization Efficiency

1. Prepare 1.5 or 2 mL tubes with 10 mg of bone ($n \geq 3$ tubes).
2. Add 1 mL of DNA lysis buffer to each tube, mixing up and down a few times with a pipet.
3. Incubate the samples on a shaker at room temperature for 5–10 min.
4. Store samples at −20 °C or proceed to overnight incubation step.
5. Incubate samples at 50 °C for 15–16 h. If samples were stored at −20 °C, thaw samples at room temperature prior to incubation.
6. Vigorously mix or vortex the samples for a few seconds, and let them come to RT prior to the assay. Samples may also be stored at 4 °C up to a week before analysis.
7. Centrifuge the samples for 1 min at >1,000 g to concentrate the bone particles at the bottom of the tubes.
8. Perform the PicoGreen DNA Assay according to manufacturer's instructions. Make sure to protect all solutions containing PicoGreen from light (*see* **Note 4**).
9. DNA quantity should be <50 ng/mg of bone [16]. If the DNA quantity is higher, do not use this bone batch for 3D-printing scaffolds, as the DNA could elicit an immune response if the scaffolds are implanted in vivo.

3.6 3D Design of Scaffolds

1. Open TinkerCAD and create a new design.
2. Add a box object by dragging it from the sidebar onto the print bed.
3. Click on the ruler tool, and then dimension the object as desired:
   a. For two-layer sheets: 3 cm × 3 cm and 0.782 mm in height.
   b. For compression cubes, 1 cm on edge.
   c. Design a sheet to cut into scaffolds. For two-layer scaffolds with a 460 μm nozzle, use 3 cm in *x* and *y*, and 782 μm in height (0.7 × 460 μm first layer + 460 μm second layer).
4. Download file for 3D-printing: File > Export > .STL.
5. Open Slic3r.
6. Import the .STL file downloaded in step 4.

7. Apply the following settings:
   a. Printer Settings (Extruder 1):
      - Nozzle diameter: 0.46 mm (or the diameter of installed nozzle).
   b. Print Settings:
      - Layers and Perimeters:
        - Layer height: 0.46 mm (matching in 7.a.i).
        - First layer height: 70 %.
        - Perimeters: 0.
        - Horizontal shells. Top: 0; bottom: 0.
      - Infill:
        - Fill density: 40 %.
        - Fill pattern: Rectilinear.
        - Fill angle: 0°.
      - Skirt and Brim:
        - Loops: 0 for no priming loops.
        - Brim Width: 0.
      - Speed:
        - Infill: 1.682 mm/s. This value will be manually changed once the extrusion speed of the material is measured.
        - Travel: 50 mm/s.
8. Export .GCODE file and transfer to the printing computer.

3.7 Scaffold Material Preparation

1. Weigh DCB particles and PCL pellets. Mix these to obtain 30 % DCB and 70 % PCL by mass. Use between 3 and 10 g of total mixture. Prepare enough to fill at least 1/2 of the extrusion chamber (*see* **Note 5**).
2. Mix the bone and PCL using the sifter. Repeat 4 times.
3. Pack the mixture into the chamber.
4. Heat to melt temperature (80 °C) for 30 min using the PID controller and the heating element.
5. Set the heated bed to 50–65 °C.
6. Load the prepared GCODE into the Mach 3 software.
7. Zero the *z*-axis so that print nozzle is just touching the print surface using the jog function. Set the zero point just close enough that a folded sheet of paper can barely be pulled from between the nozzle and the bed.
8. Retract the nozzle upwards 50 mm and perform a test extrusion by activating the pressure regulator for 1 min.

9. Measure the length of material extruded during the previous step and calculate the mm/min extrusion rate.
10. Modify the GCODE to be set at the extrusion rate:
    a. Edit GCODE.
    b. Find and replace any "F###" in the GCODE with the calculated rate "F###" (Edit > Replace. . .). For instance, if the rate is 100 mm/min, type in "F100" wherever an *F* value appears in the GCODE.
11. Launch the print from the software and activate the pressure.
12. Monitor the print for any inconsistencies.
13. At the end of the print, quickly turn off the pressure and move the extrusion head up 50 mm to prevent extra deposition across the printed scaffold.
14. Allow the scaffold to cool for 20 min before removing from the print surface.
15. Store in desiccating conditions at room temperature. Representative images of scaffolds with various PCL and DCB compositions are shown in Fig. 3.

3.8 Print Quality Assessment

1. Take a high-resolution picture of a two-layer sheet on a stereoscope (5–8×).
2. In PowerPoint or another graphics software, design an ideal rectilinear porous geometry at the same resolution (pixels/μm) and dimensions (Fig. 4). This is most easily done by drawing a series of rectangles with exact dimensions over one of the images from the stereoscope, and saving the drawings as a separate image at the same resolution.
3. Use ImageJ to threshold both images to B&W:
    a. First, threshold image using Image > Adjust > Threshold.
    b. Then, transform to a binary image using Process > Binary > Make Binary.
    c. If needed, invert the images so that the scaffolds are white and the background is black using Edit > Invert.
    d. Save image files.
4. Multiply the two images together using Process > Image Calculator:
    a. The result shown is a measure of the resulting overlap signal.
    b. Record the percent overlap using Analyze > Histogram > List, value of 0 (black pixels) for both the perfect image and the multiplied image.
    c. Record the print quality using this formula: $1-\frac{\text{Area Multiplied Image}}{\text{Area Perfect Image}}$.

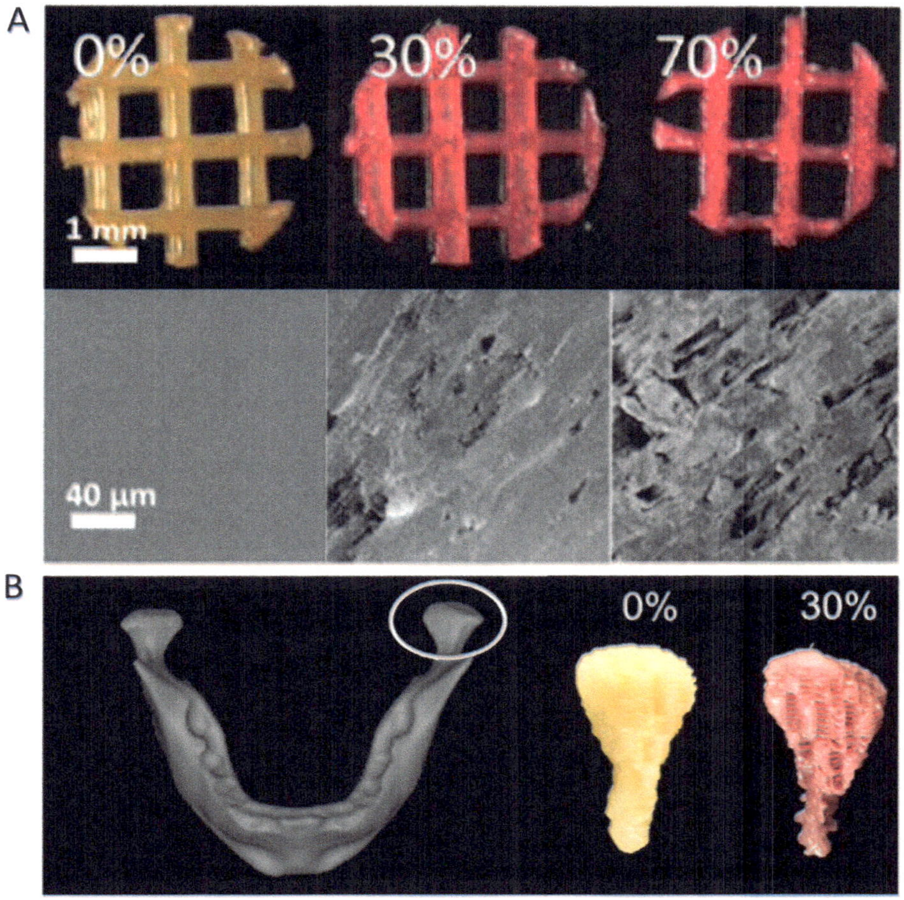

Fig. 3 3D-printed, porous decellularized bone matrix (DCB):PCL scaffolds. (**a**) Two-layer PCL scaffolds with various weight percentages of DCB. The Alizarin Red stain (*top row*) increases with the amount of DCB in the scaffold, demonstrating that the scaffolds have higher mineral content. Surface roughness also increases with the amount of DCB in the scaffold, as shown by the SEM images (*bottom row*). (**b**) 3D-printed PCL and PCL: DCB scaffolds for the human temporomandibular joint condyle. Alizarin Red staining demonstrates that DCB particles are dispersed throughout the DCB:PCL scaffold

5. Additionally, use the ImageJ plugin DiameterJ to calculate the pore areas and average strut width, and compare to the original design values.

3.9 Mechanical Analysis

1. Print a minimum of six 1 cm^3 scaffolds for testing in the print axis ($n = 3$) and in the shear axis ($n = 3$).
2. Measure the exact dimensions along the axis of compression and cross-sectional area for each scaffold using a handheld caliper.
3. Prepare a compressive testing machine with a 50 kN load cell.

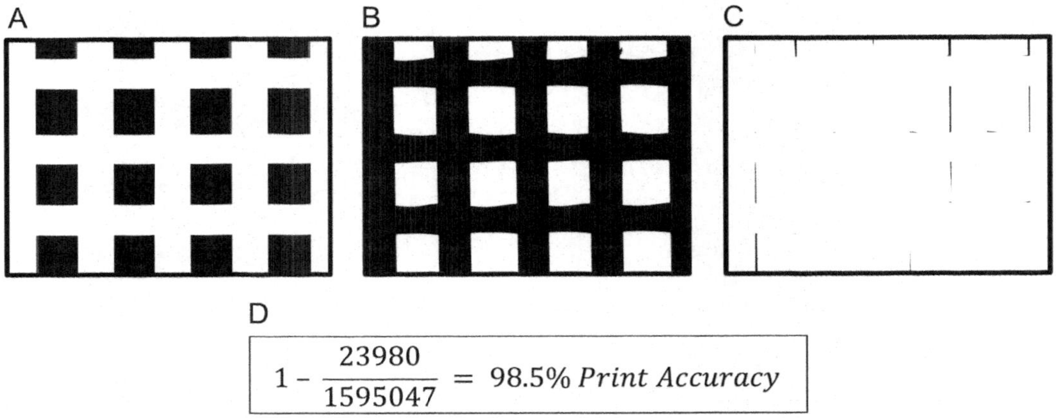

Fig. 4 Print quality analysis. Images of the ideal rectilinear lattice (**a**) and actual example scaffold (**b**) are multiplied together to obtain an overlap of the two images. The ratio of *black pixel area* in the overlapped versus ideal image is used to determine print accuracy (**c**) Representative calculation for determining the print accuracy (**d**)

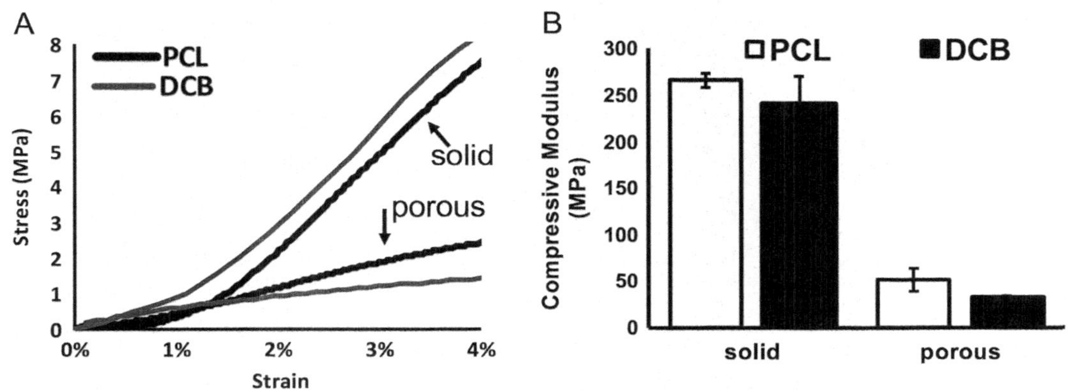

Fig. 5 Mechanical testing of 3D-printed PCL and PCL:DCB scaffolds. (**a**) Stress–strain curves for PCL and PCL: DCB in porous or solid states. (**b**) Compressive moduli for solid and porous scaffolds

4. For each sample:
   a. Place the sample in between the platens.
   b. Lower the upper platen until it just touches the sample and the load cell registers to approximately 1 N.
   c. Begin an automated compression for 10–50 % strain at 4 mm/min.
   d. Save the time (seconds), load (Newtons), and position (millimeters) data.
   e. Create a curve of stress vs. strain (Fig. 5), and measure the slope in the linear region between 2 and 10 % strain. This is compressive modulus.

5. The compressive modulus of solid PCL is approximately 120 MPa [7, 15]. Porous scaffolds typically have a lower modulus that is proportional to the reduction in the load-bearing cross-sectional area (Fig. 5).

3.10 Scaffold Preparation and Sterilization for Cell Culture or Implantation

1. Punch scaffolds out of a printed sheet using a 4 mm punch and hammer on a cutting board.
2. Place scaffolds in a 1.5 or 2 mL centrifuge tubes (maximum of ten scaffolds per tube).
3. Add enough 3 M NaOH to cover the scaffolds.
4. Degas in a vacuum chamber (with tube open) for 10 min.
5. Close the tube, lay it sideways on an orbital shaker, and incubate the samples for 1 h at room temperature. Excess incubation will result in scaffold degradation and weakening of its mechanical properties.
6. Wash scaffolds with PBS 3 × 10 min on the orbital shaker.
7. Sterilize scaffolds by filling 75 % of the tube with 70 % EtOH. Spray all portions of the open tube before closing. Incubate scaffolds on the orbital shaker for 1 h at room temperature.
8. In a cell culture hood, aspirate the EtOH, and wash 3 × 20 min with sterile PBS. Store at room temperature in sterile conditions until the day before seeding.
9. One day before seeding, incubate the scaffolds in expansion medium overnight at 37 °C.

3.11 Scaffold Seeding with Adipose-Derived Stem Cells

1. Determine the number of cells needed for scaffold seeding. Cells should be seeded in a fibrin gel at a concentration of 20×10^6 cells/mL. Seeding volume should be determined by calculating the amount of void space within the scaffold:

$$\text{Seeding volume} = \text{scaffold volume} \times (1 - \text{infill density})$$

2. Plate passage 1 ASCs at a density of 5,000 cells/cm^3 in T-175 flasks (*see* **Note 2**). Culture the ASCs in expansion medium until they reach about 80 % confluency. Do not let the flasks become over-confluent, as this may compromise the bone regeneration capacity of ASCs. A typical ASC yield is 4–6 million cells per flask.
3. Just before scaffold seeding, prepare scaffolds by drying them on a sterile KimWipe and placing them in a well plate (*see* **Note 6**).
4. Trypsinize the ASCs and centrifuge at 300 **g** for 5 min. Set aside some cells to count before centrifuging.
5. Count the number of cells in each flask using a hemocytometer. Determine the volume of fibrinogen needed to resuspend the cells at a concentration of 25×10^6 cells/mL.

6. Aspirate as much media from the cell pellet as possible, and resuspend the cells in fibrinogen.
7. Prepare aliquots of the cell-fibrinogen suspension in sterile 500 μL tubes. The volume of each aliquot should be calculated beforehand using the equation below, where 0.8 and 1.1 account for the volume fraction of fibrinogen present within fibrin and for volume loss while pipetting, respectively. Seed no more than four scaffolds at a time since the fibrin will clot quickly once thrombin is added.

$$\text{Aliquot volume} = \text{no.scaffolds} \times 0.8\,(\text{seeding volume}) \times 1.1$$

8. Set two pipets to the seeding volume and thrombin volume (*see* **Note 7**). The thrombin volume can be calculated by the equation below. The final ratio of fibrinogen to thrombin should be 4:1.

$$\text{Thrombin volume} = \frac{(\text{aliquot volume})}{4}$$

9. Pipet thrombin into an aliquot, and briefly mix in order to disperse the thrombin throughout the cell-fibrinogen suspension.
10. Briefly mix the solution with the "seeding" pipet. Pipet the fibrin-cell suspension directly onto the scaffolds. This process should be performed quickly to avoid premature clotting of the fibrin.
11. Perform steps 9–10 until all scaffolds are seeded.
12. Incubate scaffolds in a cell culture incubator for 30 min to allow the fibrin to completely crosslink. Meanwhile, warm up osteogenic medium to 37 °C.
13. Add osteogenic medium to each well, and place the scaffolds back into the incubator.
14. Feed scaffolds 3 times per week with osteogenic medium for 3–4 weeks to allow for bone regeneration to occur.
15. Harvest scaffolds for analysis according to the instructions for each assay explained below.

3.12 Alizarin Red Staining

1. Fix scaffolds with 4 % paraformaldehyde at 4 °C overnight (*see* **Note 8**).
2. Wash cells once with dH_2O. If not proceeding immediately to staining, store samples in dH_2O at 4 °C for up to 1 week.
3. Discard dH_2O and add enough Alizarin Red stain to immerse the entire scaffold. Incubate the scaffolds at room temperature for 10–20 min.

4. Discard the stain and wash samples with dH_2O at room temperature until solution runs clear. For most effective results, wash samples overnight with gentle agitation for 2–4 days to allow excess stain to diffuse out of the scaffolds.
5. Image the scaffolds using a camera or stereoscope. Example images of noncultured scaffolds are shown in Fig. 3.

3.13 DNA Quantification

1. Wash cells once with PBS in a sterile environment.
2. Transfer scaffolds into individual tubes or containers with DNA lysis buffer. For scaffolds that fit into 2 mL centrifuge tubes, 250–1,000 μL of lysis buffer should be sufficient to immerse the entire scaffold.
3. Incubate the scaffolds in DNA lysis buffer at room temperature for 5–10 min.
4. Store samples at −20 °C if not proceeding immediately to overnight incubation step.
5. Incubate samples at 50 °C for 15–16 h. If samples were stored at −20 °C, thaw samples at room temperature prior to incubation.
6. Vigorously mix or vortex the samples for a few seconds, and let them come to RT prior to the assay. Samples may also be stored at 4 °C for a week before analysis.
7. Perform the PicoGreen DNA Assay according to manufacturer's instructions to determine the mass of DNA in each scaffold. Make sure to protect all solutions containing PicoGreen from light.

3.14 Ca^{2+} Quantification

1. Wash cells once with PBS in a sterile environment.
2. Transfer scaffolds into individual tubes or containers with 0.5 M HCl. For scaffolds that fit into 2 mL centrifuge tubes, 250–1,000 μL of 0.5 M HCl should be sufficient to immerse the entire scaffold.
3. Incubate the samples on a shaker at 4 °C overnight.
4. Store samples at −20 °C or proceed to calcium quantification assay.
5. Perform calcium quantification assay according to manufacturer's instructions. If samples were stored at −20 °C, thaw samples at room temperature prior to performing the assay.
6. Obtain a normalized Ca^{2+}/DNA ratio for scaffolds by dividing the mass of Ca^{2+} by the mass of DNA, as determined during DNA quantification. Representative Ca^{2+}/DNA data is shown in Fig. 6. Further in vitro and in vivo data is shown in Figs. 6 and 7 to demonstrate the osteoinductivity of DCB:PCL scaffolds.

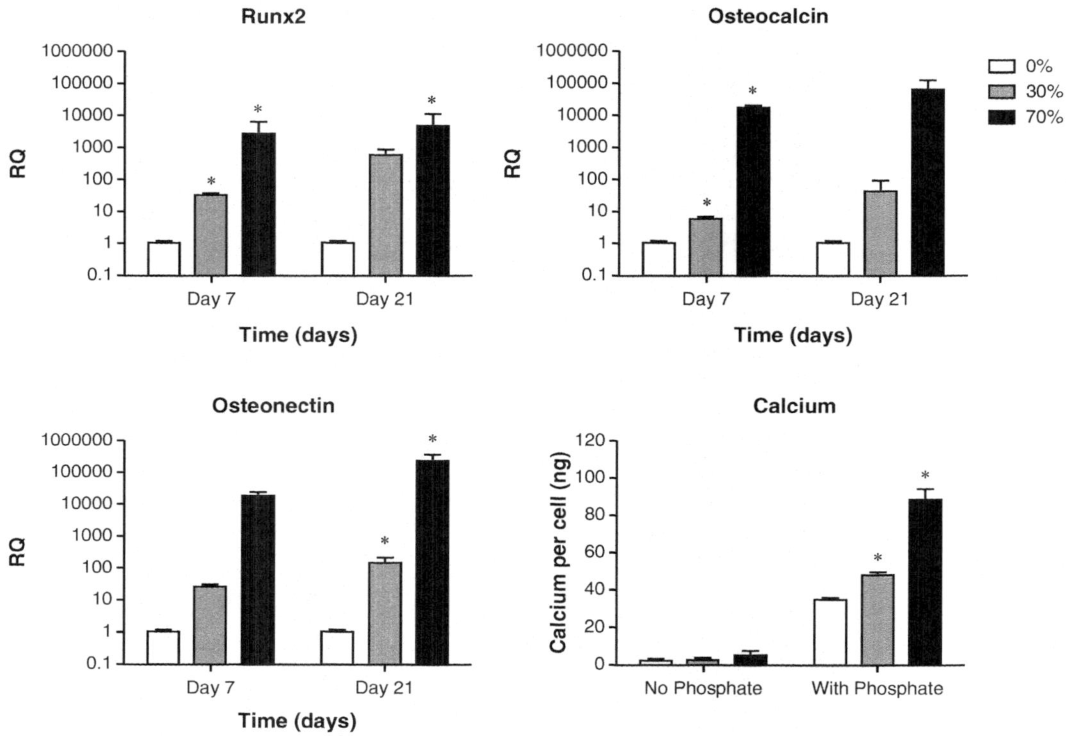

Fig. 6 In vitro osteogenesis of adipose-derived stem cells (ASCs) on 3D-printed DCB:PCL scaffolds. Expression of osteogenic genes Runx2, osteocalcin, and osteonectin is upregulated in DCB:PCL versus PCL only scaffolds. Calcium per cell (or DNA) also increases when ASCs are cultured on DCB:PCL scaffolds (*bottom right*). *$p < 0.05$ compared to cells cultured in pure polycaprolactone scaffolds by one-way ANOVA with posthoc Tukey test

4 Notes

1. Other thermoplastic materials may be substituted for PCL. 3D-printing parameters, such as extrusion pressure and melt temperature, will need to be optimized for a given thermoplastic material.
2. Other multipotent stem cells or osteogenic cell lines may be used instead of ASCs. However, in vitro bone formation results may vary between different cell types.
3. Wear cryoprotective gloves whenever handling liquid N_2 or items cooled with liquid N_2.

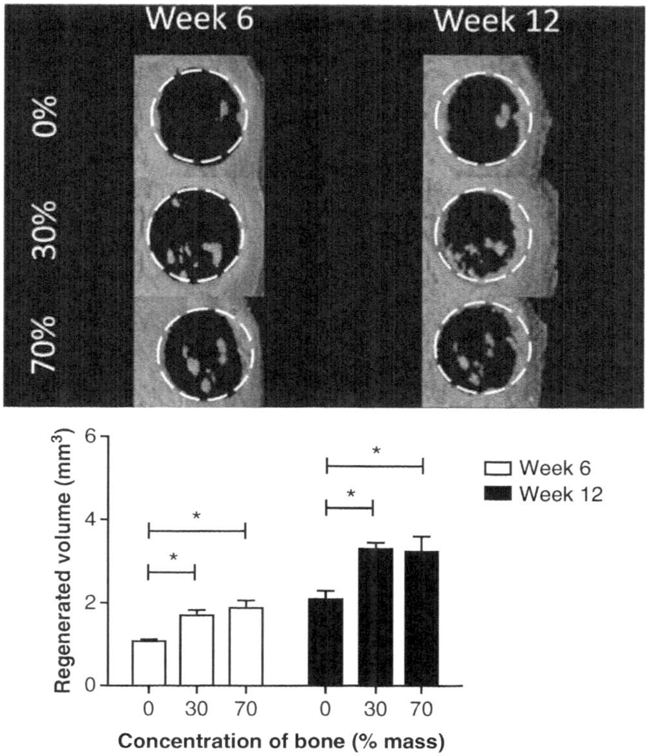

Fig. 7 In vivo bone regeneration within ASC-seeded, 3D-printed DCB:PCL scaffolds. DCB:PCL scaffolds result in greater bone regeneration in comparison to PCL scaffolds when implanted into critically sized, murine calvarial defects. All data was obtained using computed tomography (CT)

4. Be careful when pipetting solution in and out of the sample tubes. If the bone pellet becomes disrupted during the assay, centrifuge again before continuing the assay.
5. Other additives may be substituted for comparison, such as hydroxyapatite or tricalcium phosphate [15].
6. Coating the well plates with a thin layer of 2 % sterile agarose (~150 μL in 24-well plate) may help prevent outgrowth of ASCs onto the well plate as the scaffolds are cultured. Non-adherent well plates may also be used to prevent ASC outgrowth from the scaffolds.
7. Pre-wetting the tip for the "seeding pipet" may prevent premature clotting of fibrin. For best results, pre-wet the tip before seeding scaffolds from each aliquot.
8. Harvesting samples is best performed by transferring the scaffolds into individual 2 mL centrifuge tubes.

References

1. Elder M (2014) Advanced orthopedic technologies, implants and regenerative products. BCC Research, Wellesley, MA
2. Gainer K (2015) Regenerative medicines: bone and joint applications. BCC Research, Wellesley, MA
3. Bose S, Roy M, Bandyopadhyay A (2012) Recent advances in bone tissue engineering scaffolds. Trends Biotechnol 30:546–554
4. Stevens MM (2008) Biomaterials for bone tissue engineering. Mater Today 11:18–25
5. Liu X, Ma PX (2004) Polymeric scaffolds for bone tissue engineering. Ann Biomed Eng 32:477–486
6. Temple JP, Hutton DL, Hung BP et al (2014) Engineering anatomically shaped vascularized bone grafts with hASCs and 3D-printed PCL scaffolds. J Biomed Mater Res A 102:4317–4325
7. Williams JM, Adewunmi A, Schek RM et al (2005) Bone tissue engineering using polycaprolactone scaffolds fabricated via selective laser sintering. Biomaterials 26:4817–4827
8. Shor L, Güçeri S, Wen X et al (2007) Fabrication of three-dimensional polycaprolactone/hydroxyapatite tissue scaffolds and osteoblast-scaffold interactions in vitro. Biomaterials 28:5291–5297
9. Woodruff MA, Hutmacher DW (2010) The return of a forgotten polymer—polycaprolactone in the 21st century. Prog Polym Sci 35:1217–1256
10. García-Gareta E, Coathup MJ, Blunn GW (2015) Osteoinduction of bone grafting materials for bone repair and regeneration. Bone 81:112–121
11. Hung BP, Naved BA, Nyberg EL et al (2016) Three-dimensional printing of bone extracellular matrix for craniofacial regeneration. ACS Biomater Sci Eng 2(10):1806
12. Penel G, Delfosse C, Descamps M et al (2005) Composition of bone and apatitic biomaterials as revealed by intravital Raman microspectroscopy. Bone 36:893–901
13. Benders KEM, van Weeren PR, Badylak SF et al (2013) Extracellular matrix scaffolds for cartilage and bone regeneration. Trends Biotechnol 31:169–176
14. Datta N, Holtorf HL, Sikavitsas VI et al (2005) Effect of bone extracellular matrix synthesized in vitro on the osteoblastic differentiation of marrow stromal cells. Biomaterials 26:971–977
15. Nyberg E, Rindone A, Dorafshar A et al (2016) Comparison of 3D-printed poly-ε-caprolactone scaffolds functionalized with tricalcium phosphate, hydroxyapatite, bio-Oss, or decellularized bone matrix. Tissue Eng Part A. doi:10.1089/ten.TEA.2016.0418
16. Crapo PM, Gilbert TW, Badylak SF (2011) An overview of tissue and whole organ decellularization processes. Biomaterials 32:3233–3243

Methods in Molecular Biology (2018) 1577: 227–237
DOI 10.1007/7651_2017_49

Published online: 13 July 2017

Decellularization of Large Tendon Specimens: Combination of Manually Performed Freeze-Thaw Cycles and Detergent Treatment

Susanne Pauline Roth, Ina Erbe, and Janina Burk

Abstract

Reliable decellularization techniques applicable to tendon tissue play a critical role in the field of current tissue engineering. Particularly, an application as three-dimensional culture model for in vitro research and translational approaches to establish graft-based tendon repair as a routine clinical tool represent two main application fields for decellularized tendon scaffolds. Considering methodological issues of tendon decellularization, one of the major challenges lies in the preservation of the tendon-specific extracellular matrix (ECM) architecture to reflect natural tissue characteristic as best as possible. Concurrently, further requirements for high-quality decellularized biological tendon scaffolds include not only the reduction of resident cells, but also an ensured cytocompatibility.

To date, a large number and a wide variety of decellularization protocols for natural tendon tissue have already been investigated and usually, physical as well as chemical and/or enzyme-based treatments are used for the purpose of decellularization. However, to the best of our knowledge, there is a lack of evidence-based protocols for the processing of full-thickness large tendon samples, such as the equine flexor tendons.

Therefore, the here presented protocol describes a reliable procedure to decellularize equine superficial digital flexor tendons by using a combined treatment of physical decellularization in the form of repetitive freeze-thaw cycles, and of chemical decellularization with the non-ionic detergent Triton X-100. The decellularization effectiveness evaluated by reduction of cell and DNA content, the influence of decellularization on the morphology of the tendon extracellular matrix (ECM) as well as the cytocompatibility of the decellularized tendon scaffolds obtained have been investigated previously. Based on this previous study, the here present protocol is an effective procedure, particularly applicable for large tendon specimens.

Keywords: Regenerative medicine, Tissue engineering, Decellularization, Scaffold, Freeze-thaw cycles, Triton X-100, Detergent treatment, Tendon, Horse

1 Introduction

Decellularization of tendon tissue is a key procedure in the toolbox of orthopaedic tissue engineering related to connective tissues. Not only the in vitro application in various research areas as a three-dimensional culture model, but also the translation of graft-based tendon restoration into clinics represents a crucial field of application for decellularized tendon scaffolds [1, 2]. Regardless of a specific purpose for use, the ultimate methodological objective is the removal of cellular elements as effective as possible while preserving the natural, residue-free, tendon-specific extracellular matrix (ECM) composition. Thereby it should be emphasized that a decellularized tendon scaffold with a naturally structured ECM works as a best possible reflection of in vivo conditions for cell attachment, proliferation, and differentiation and further maintains tissue-specific biomechanical functionality and biocompatibility. Additionally, in terms of clinically applicable tendon grafts for musculoskeletal reconstructive surgery, best possible removal of cellular remnants facilitates re-seeding procedures as well as minimizes the risk of rejection after implantation due to cell associated immune response [3, 4].

Although a great number of decellularization protocols for tendon tissue have already been investigated, to date there is no methodological gold standard for the effective decellularization of large tendon specimens [5–7]. Among favoured tendon tissue donors for decellularization, the horse is considered as the fourth most frequently used species [3]. Regarding translational research strategies, especially the large equine superficial digital flexor tendon has been subject of numerous studies due to its functional, structural, and pathological characteristics, which are similar to those of the human Achilles tendon [8–10]. Further, decellularized equine tendon tissue offers certain advantages regarding the increasing relevance of commercially available xenografts for reconstructive tendon repair in human medicine. Those benefits include a comfortable access related to breeding and slaughter, an adequate size of received xenogeneic graft material, and a manageable number of zoonotic agents [3].

The here presented method for decellularization of tendon tissue was successfully developed on full-thickness equine superficial digital flexor tendon samples, thus being particularly suitable for large tendon specimens. Burk et al. have shown the high process reliability leading to a reduction in resident cells to 1% and to a residual DNA-content of 20% while maintaining cytocompatibility and a natural structure of the tendon ECM [5]. Furthermore, a partial automation of this decellularization procedure by using a customized, liquid nitrogen-based controlled rate freezer has been published in a pilot study [11].

Here we present the previously developed protocol for decellularization of tendon tissue including a 7-day procedure applicable especially for large tendon specimens. After sample collection performed with a particular focus on sterile working conditions and an overnight storage at 4 °C, the dissected tendon specimens were washed in phosphate buffered saline (PBS) and 70% ethanol two times each to prevent bacterial contamination. The subsequently applied decellularization protocol includes a physical treatment consisting of five freeze-thaw cycles combined with a chemical detergent-based (Triton X-100) treatment. In respect of a reliable reproducibility, it is appreciable that freeze-thawing is considered as the key procedural step within the here presented protocol. Generally, physical treatments, such as freeze-thawing, are used to disrupt membranes of resident cells, detach the cells within their ECM network, and facilitate further rinsing to remove cellular remnants [3, 12]. Rapid freeze-thaw procedures are commonly used for physical decellularization of natural tendon tissue [13–15]. The herein intended effect of applied fast cooling rates is a direct cell injury with minimal alterations of the ECM. More precisely, the fast drop in temperature initiates the formation of intracellular ice crystals that results almost always in cell death due to disintegrated organelles and cell membranes [16]. Further steps of decellularization following the freeze-thawing are carried out at room temperature as well as under continuous agitation and include firstly a 48 h incubation in hypotonic solution. This step does not only serve the purpose of rinsing, but also induce cell lysis by osmotic effects. Secondly, tendon specimens undergo a chemical decellularization by 1 % Triton X-100 in 1M TRIS solution for 48 h. The non-ionic detergent Triton X-100 has the function to solubilize cell membranes and dissociate DNA from proteins. Due to a lack of ionic charge in its molecular structure, Triton X-100 has a weak impact on protein structures and therefore it is among the most frequently used reagents for tissue decellularization [17, 18]. As third step, the tendon scaffolds are subjected to hypotonic solution twice. These intermediate washing steps mainly aim to remove residual chemicals as well as cellular leftovers. Flushing action and therefore decellularization effectiveness is increased through the renewal of the flushing solution. A following incubation in cell culture medium for 24 h provides further rinsing to remove cellular remnants as well as residual chemicals, and prepares the tendon scaffolds for cell culture optimally. Finally, the tendon samples undergo a last 24 h incubation in PBS. The storage of decellularized scaffolds in customized size takes place in sterile containers at −80 °C.

2 Materials

The following listing gives an overview of required materials for decellularization of tendon tissue on a daily basis. Starting with collection and storage of native tendon specimens, the subsequent processing for decellularization includes a 7-day procedure. It should be noted that on day two as well as on day four of the decellularization procedure, there is no processing step needed to be performed due to an incubation time of 48 h from the previous day.

2.1 Sample Collection

- Clippers
- Cleansing and antiseptic skin preparation:
  - Povidone iodine scrub solution (Jodosept®; Vétoquinol)

  - 70% isopropyl alcohol (2-Propanol 70%; Braun)
  - Povidone iodine solution (Braunoderm®; Braun)
- Sterile gloves
- Sterile surgical tools:
  - Scalpel (blade no. 22)
  - Forceps
  - Scissors

2.2 Storage of Native Tendon Specimens Until Further Treatment

- Specimen transport medium:
  - Sterile phosphate buffered saline (PBS) (10×) (PBS pH 7.4 (1×); Thermo Fisher Scientific) or sterile 0.9% sodium chloride (NaCl 0.9% B. Braun®; Braun) each supplemented with 2% penicillin-streptomycin (Penicillin-Streptomycin 10.000 U/ml; Thermo Fisher Scientific) and 0.1% gentamycin (Gentamycin-Lösung 50 mg/ml, CELLPURE®; Carl Roth).
- Sterile surgical tools:
  - Forceps
  - Scissors
- Ethanol 70% (Ethanol 70% denatured; Carl Roth)
- Phosphate buffered saline (PBS) (10×) (PBS pH 7.4 (1×); Thermo Fisher Scientific)
- Sterile plastic sampling bags (Whirl Pak®; Carl Roth)

2.3 Decellularization Procedure

2.3.1 Day 1: Freeze-Thaw Cycles and Subsequent Incubation in Hypotonic Solution

- Liquid nitrogen
- Protective clothing required for handling of liquid nitrogen
- Cryogenic dewar transporting container (volume of 2–4 l)
- Phosphate buffered saline (PBS) (10×) (PBS pH 7.4 (1×); Thermo Fisher Scientific) heated to 37 °C
- Sterile distilled water
- Sterile surgical tools:
  - Long forceps (blunt, length 300 mm)
- Shaker

2.3.2 Day 3: Detergent Treatment

- 1 M TRIS solution:
  - 121.14 g TRIS (TRIS PUFFERAN®; Carl Roth) dissolved in 800 ml distilled water
  - Set pH value to 7.6 by titration with HCl 37% (hydrochloric acid 37% pure; Carl Roth)
  - Filled up to a volume of 1000 ml with distilled water

- 1% Triton X-100 in TRIS (sterile-filtered): 10 ml Triton X-100 (Triton® X-100 pure; Carl Roth) in 990 ml 1 M TRIS solution
- Shaker

2.3.3 Day 5: Washing Step and Incubation in Cell Culture Medium

- Distilled water
- DMEM low glucose (Dulbecco's Modified Eagle Medium (1×) low glucose, pyruvate; Thermo Fisher Scientific) supplemented with 10% fetal bovine serum (FBS) (Fetal Bovine Serum; Thermo Fisher Scientific), 1% penicillin-streptomycin (Penicillin-Streptomycin 10.000 U/ml; Thermo Fisher Scientific), and 0.1% gentamycin (Gentamycin-Solution 50 mg/ml, CELLPURE®; Carl Roth)
- Shaker

2.3.4 Day 6: Incubation in Phosphate Buffered Saline (PBS)

- Phosphate buffered saline (PBS) (10×) (PBS pH 7.4 (1×); Thermo Fisher Scientific)
- Shaker

2.3.5 Day 7: Customizing and Storage of Decellularized Tendon Scaffolds

- Surgical tools:
  - Blades
  - Forceps
  - Scissors
- Sterile plastic sampling bags (Whirl-Pak®; Carl Roth)

3 Methods

3.1 Sample Collection (See Note 1)

1. Use fresh equine cadaver limbs as source of tendon tissue and place them in a dissecting room for processing.
2. Once clipped, the metatarsal/metacarpal region of the limbs undergoes antiseptic skin preparation by scrubbing with scrub brushes and an antibacterial liquid soap containing polyvinylpyrrolidon (PVP) iodine and 96% ethanol.
3. Remove antibacterial scrub by using gauze squares soaked with 70% ethanol.
4. Perform further antiseptic skin preparation by using 70% isopropyl alcohol and povidone iodine solution, following the appropriate contact times.
5. Lay out ready sterile gloves and sterile surgical equipment for tendon dissection.
6. Shortly recap that the designation "superficial digital flexor tendon" stands for the final chord of the superficial digital flexor muscle. Along the entire palmar/plantar aspect of the

third metacarpal/metatarsal bone the crescent-shaped tendon is the most superficial part of the digital flexor tendon bundle and lies directly adjacent dorsally to the skin and plamary/plantary to the deep digital flexor tendon. In order to further trouble-free processing, it should be noted that an accurate as well as a quick dissection of the tendon must be ensured and that surrounding connecting tissues need to be removed properly.

7. Perform a window-like incision of the skin (no. 22 scalpel blade) and dissect the superficial digital flexor tendon along its entire length.

3.2 Storage of Native Tendon Specimens Until Further Treatment (See Note 2)

1. Put the dissected tendon specimens in sterile phosphate buffered saline (PBS) or sterile 0.9% sodium chloride supplemented with 2% penicillin-streptomycin and 0.1% gentamycin (storage solution should be prepared in advance) immediately after their harvest.
2. Store them overnight at 4 °C. It should be noted that all recovered tendon specimens need to be covered with fluid during storage.
3. Any further procedural steps take place at the laminar flow in a cell culture laboratory in accordance with general principles of working under sterile conditions.
4. Place harvested tendon specimens in cell culture dishes of adequate size.
5. Remove remaining connective tissue carefully by using sterile surgical tools.
6. Meanwhile, rinse tendon pieces with PBS from time to time to avoid drying-out of the tissue.
7. Wash tendon samples 5 min each in PBS and 70% ethanol two times.
8. From here onwards, tendon specimens can either be stored at −80 °C (which is then said to be the first freezing regarding the decellularization process) and processed on a later time point, or subjected directly to freeze-thawing by the use of liquid nitrogen.

3.3 Decellularization Procedure (See Note 3)

3.3.1 Day 1: Freeze-Thaw Cycles and Subsequent Incubation in Hypotonic Solution

1. In the here presented protocol, collected tendon samples undergo in total a number of five freeze-thaw cycles: Specimens which were freshly harvested on the day before are freeze-thawed manually five times. However, specimens which were stored for a variable period of time at −80 °C are freeze-thawed manually four times, since storage at −80 °C followed by complete thawing is said to be the first out of five freeze-thaw cycles.
2. Thawing procedure:

(a) Heat PBS to 37 °C by using a water bath.

(b) Transfer tendon pieces from their −80 °C—storage into preheated PBS and place them in a water bath (37 °C) for approximately 10–20 min.

(c) Successfully thawed tendon samples show a more flexible, softened texture in comparison to still frozen specimens.

3. Freezing procedure:

   (a) Wipe out the utilized cryogenic dewar container using 70% ethanol.

   (b) Fill the cryogenic dewar container with liquid nitrogen.

   (c) Use long forceps and put tendon samples inside the prepared liquid nitrogen until they are frozen completely (this takes usually 2 min), clearly discernible with a more rigid tissue texture.

   (d) Hereafter frozen tendons undergo thawing in PBS heated to 37 °C as described earlier in this section.

   (e) To ensure a reliable freezing procedure, renew liquid nitrogen for each single cycle of freeze-thawing.

4. After completion of five freeze-thaw cycles, place tendon samples in groups of ten or less specimens in a sterile laboratory bottle, made of borosilicate glass and covering a volume of 1000 ml.

5. Incubation in hypotonic solution (48 h):

   (a) Fill every bottle with sterile distilled water so that the surface of the tendon pieces is just covered.

   (b) Measure the required volume of distilled water for each bottle separately, in order to calculate the amount of fluids required for further decellularization steps.

   (c) Position all bottles safely on a shaker, to allow continuous agitation during the entire incubation.

3.3.2 Day 3: Detergent Treatment

1. Thoroughly replace distilled water by using serological pipettes.

2. The use of a suction device at the very beginning helps effectively to remove lather, but care needs to be taken not to suck in tiny pieces of remaining connective tissue and thereby block the suction unit.

3. Incubation in 1% Triton X-100 in 1 M tris solution (48 h):

   (a) Add the already calculated amount of the prepared 1% Triton X-100 in 1 M TRIS solution (sterile filtrated) to the tendon pieces.

   (b) Place all bottles on a shaker again, to allow continuous agitation during the entire incubation.

3.3.3 Day 5: Washing Step and Incubation in Cell Culture Medium

1. Exchange fluids in the same way as described above in the previous section.
2. Perform two consecutive 15 min washing steps: For these the appropriate amount of sterile distilled water is filled into the bottles and the latter are placed on the shaker to provide thorough agitation.
3. Finally, remove all washing fluids.
4. Incubation in cell culture medium (24 h):
   (a) Add an adequate volume of cell culture medium supplemented with 10% fetal bovine serum (FBS), 1% penicillin-streptomycin, and 0.1% gentamycin to the tendon samples.
   (b) Position all bottles again on a shaker, to allow continuous agitation during the entire incubation.

3.3.4 Day 6: Incubation in Phosphate Buffered Saline (PBS)

1. Exchange fluids in the same way as described earlier in the previous section.
2. Incubation in PBS (24 h):
   (a) Add an appropriate amount of PBS to the tendon pieces.
   (b) Place all bottles on a shaker again, to allow continuous agitation during the entire incubation.

3.3.5 Day 7: Customizing and Storage of Decellularized Tendon Scaffolds

1. Place all processed tendon samples in a sterile dish.
2. During further optional customizing by using sterile surgical tools, the tendon pieces need to be covered with PBS in order to avoid drying-out of the tissue.

For a safe storage of the decellularized scaffolds, we recommend the use of sterile plastic sampling bags, which allow the visual assessment of the bags' content without their opening and further make sample storage in a flat position possible.

4 Notes

1. Sample collection.

   The effectiveness of the here presented protocol as well as the final quality of decellularized scaffolds depends among others on the quality of harvested tendon samples. We recommend to order cadaver limbs on request at a local abattoir to perform the decellularization procedure on a high quantity of samples. This allows to exclude limbs with obvious injuries or any signs of inflammation, while ensuring a sufficient amount of tendon specimens. However, there is still a remaining intraspecific variation and there are aspects like breed, gender, age, and previous use, which influence tissue quality to a varying extent. To state precisely, it must be said that tendon specimen is not all the same.

For maintaining a high quality of harvested tissue, following processing of the selected limbs should avoid any bacterial contamination. Therefore, the whole process of sample collection and handling needs to be done with a particular focus on sterile working conditions since it includes the tissue transfer between its naturally sterile environment and the artificially sterile laboratory environment. Before starting the dissection of tendon specimens, spacious clipping and carefully performed antiseptic skin preparation needs to be done. During antibacterial scrubbing, an overzealous scrubbing should be avoided since it transports bacteria within hair follicles to the skin surface and can lead to irritations or abrasions that are extremely rapid colonized by bacteria [19].

2. Storage of native tendon specimens until further treatment.
   To further prevent bacterial contamination during overnight storage of freshly harvested tendon pieces, we recommend to utilize PBS, respectively, sodium chloride-containing containers of a volume of 1000 ml for storing of 12 up to 16 tendon specimens. By using a volume of 800 ml fluid supplemented with 16 ml penicillin-streptomycin and 0.8 ml gentamycin, all contained tendons were fully covered with fluid.

3. Decellularization procedure.
   Besides an ensured high quality of the starting tendon samples that shall be subjected to further decellularization, the damaging and detaching of resident cells (by freeze-thawing, incubation in hypotonic solution, detergent treatment) represent core issues regarding decellularization effectiveness. Additionally, the removal of cellular remnants and of residual chemicals (intermediate washing steps at day 3, incubation in cell culture medium at day 5, and in PBS at day 6, continuous agitation) is strongly required in terms of an effective decellularization.

   Therefore, alterations of these steps within the here presented protocol lead to variability in the results obtained. Particular care has to be applied to keep the time frames described above, to carefully prepare the solutions to be used, to avoid an overloading of laboratory bottles with tendon samples, and to ensure a full operability of the shaker.

   Aiming at more standardized conditions and a potential processing of even higher sample quantities, the key procedure of freeze-thawing already has been automated by using a liquid nitrogen-controlled rate freezer within a pilot study [11]. The study compared the here presented manual regime for decellularization with an automated processing including two different cooling and heating rates. Finally, the automated procedure was as effective as the manually performed freeze-thaw cycles. Due to the requirement of a customized controlled rate freezer and the so far limited characterization of the tendon matrices that underwent automated freeze-thaw cycles, an automation of freeze-thawing is practicable but needs further endeavours.

Regarding manually performed freeze-thawing, a very brief review of safe handling and use of liquid nitrogen is given. However, we highly recommend to recap appropriate safety precautions in detail, before starting manual freeze-thaw cycles for tendon decellularization by using liquid nitrogen. The potential hazards are mainly based on two certain chemical characteristics of liquid nitrogen: (1) At atmospheric pressure, liquid nitrogen boils at −196 °C and therefore it is extremely cold. (2) Vaporization of 1 l liquid nitrogen releases 0.7 m^3 gas, accordingly very small volumes of liquid nitrogen vaporize into large amounts of gas. Resulting precautions can be summarized as follows: (a) handle liquid nitrogen carefully, (b) wear protective clothing (face shield or safety goggles, cryo gloves, high-top shoes, trousers should be worn outside the shoes), (c) use only cryogenic liquid containers (do not ever cover or plug the container opening, do not overfill the container, use only a fully functional container), (d) store and use liquid nitrogen only in well-ventilated places.

To perform the herein presented manually freeze-thawing, it is reasonable to work in a team of two competent persons. This is not only recommended for safety reasons, but also to ensure proper positioning for all processed tendon samples inside the liquid nitrogen. Usually, 10 up to 12 samples can be placed in a 2 l cryogenic dewar container and all samples need to be flipped several times during freezing. Moreover, the tendon pieces tend to stick together once they are in contact with liquid nitrogen and therefore care must be taken to keep them separate while freezing.

The disposal of small residual quantities of liquid nitrogen should be done outdoors in a safe place. Cryogenic dewar containers should be emptied slowly on gravel or bare earth, so that remaining liquid nitrogen can evaporate safely. Liquid nitrogen should not be poured on the pavement.

After freeze-thawing, tendon samples were placed in laboratory bottles to undergo several further steps of decellularization. In the sections above we describe an exchange of utilized fluids by pipetting and by suction. Alternatively, it is also possible to change incubation fluids by placing the tendon pieces in new laboratory bottles and add the required amount of the respective incubation solution.

References

1. Cheng CW, Solorio LD, Alsberg E (2014) Decellularized tissue and cell-derived extracellular matrices as scaffolds for orthopaedic tissue engineering. Biotechnol Adv 32(2):462–484
2. Youngstrom DW, Barrett JG (2016) Engineering tendon: scaffolds, bioreactors, and models of regeneration. Stem Cells Int 2016:3919030
3. Lovati AB, Bottagisio M, Moretti M (2016) Decellularized and engineered tendons as biological substitutes: a critical review. Stem Cells Int 2016:7276150
4. Schulze-Tanzil G, Al-Sadi O, Ertel W, Lohan A (2012) Decellularized tendon extracellular matrix—a valuable approach for tendon reconstruction? Cell 1(4):1010–1028
5. Burk J, Erbe I, Berner D, Kacza J, Kasper C, Pfeiffer B et al (2014) Freeze-thaw cycles enhance decellularization of large tendons. Tissue Eng Part C Methods 20(4):276–284
6. Youngstrom DW, Barrett JG, Jose RR, Kaplan DL (2013) Functional characterization of detergent-decellularized equine tendon extracellular matrix for tissue engineering applications. PLoS One 8(5):e64151
7. Bottagisio M, Pellegata AF, Boschetti F, Ferroni M, Moretti M, Lovati AB (2016) A new strategy for the decellularisation of large equine tendons as biocompatible tendon substitutes. Eur Cell Mater 32:58–73
8. Dowling BA, Dart AJ (2005) Mechanical and functional properties of the equine superficial digital flexor tendon. Vet J 170(2):184–192
9. Thorpe CT, Clegg PD, Birch HL (2010) A review of tendon injury: why is the equine superficial digital flexor tendon most at risk? Equine Vet J 42(2):174–180
10. Patterson-Kane JC, Rich T (2014) Achilles tendon injuries in elite athletes: lessons in pathophysiology from their equine counterparts. ILAR J 55(1):86–99
11. Roth SP, Glauche SM, Plenge A, Erbe I, Heller S, Burk J (2017) Automated freeze-thaw cycles for decellularization of tendon tissue—a pilot study. BMC Biotechnol 17(1):13
12. Gilbert TW, Sellaro TL, Badylak SF (2006) Decellularization of tissues and organs. Biomaterials 27(19):3675–3683
13. Azuma C, Tohyama H, Nakamura H, Kanaya F, Yasuda K (2007) Antibody neutralization of TGF-β enhances the deterioration of collagen fascicles in a tissue-cultured tendon matrix with ex vivo fibroblast infiltration. J Biomech 40 (10):2184–2190
14. Omae H, Zhao C, Sun YL, An K-N, Amadio PC (2009) Multilayer tendon slices seeded with bone marrow stromal cells: a novel composite for tendon engineering. J Orthop Res 27(7):937–942
15. Stewart AA, Barrett JG, Byron CR, Yates AC, Durgam SS, Evans RB et al (2009) Comparison of equine tendon-, muscle-, and bone marrow-derived cells cultured on tendon matrix. Am J Vet Res 70(6):750–757
16. Gage AA, Baust J (1998) Mechanisms of tissue injury in cryosurgery. Cryobiology 37 (3):171–186
17. Crapo PM, Gilbert TW, Badylak SF (2011) An overview of tissue and whole organ decellularization processes. Biomaterials 32 (12):3233–3243
18. Gilbert TW (2012) Strategies for tissue and organ decellularization. J Cell Biochem 113 (7):2217–2222
19. Smeak DD, Olmstead ML (1984) Infections in clean wounds: the roles of the surgeon, environment, and host. The compendium on continuing education for the practicing veterinarian. Comp Cont Educ 6:626

Methods in Molecular Biology (2018) 1577: 239–254
DOI 10.1007/7651_2017_50

Published online: 03 August 2017

Decellularized Bone Matrix Scaffold for Bone Regeneration

Guobao Chen and Yonggang Lv

Abstract

Decellularized bone matrix is gaining a lot of attention as implantable biomaterials and/or biological scaffolds for bone tissue repair, and shows good clinical performance. This chapter describes the processing techniques and characterization protocols of decellularized bone. For the applications of the decellularized bone scaffold in promoting bone repair and regeneration, we discuss some of the current advances, and highlight the advantages and disadvantages of these scaffolds. Fabrication and application of the hydrogel derived from decellularized bone for bone tissue engineering are also presented.

Keywords: Decellularization, Bone, Matrix, Hydrogel, Osteogenic differentiation, Bioreactor

1 Introduction

Bone has a complex internal and external structure and is a heterogeneous composite material [1]. A lot of bone grafts are needed to repair defects in bone caused by trauma, tumor, infections, and so on. A major hurdle in design of tissue engineering scaffold is that most materials cannot fulfill the microstructure, composite, mechanical characterization, and bioresorbable of nature bone tissue simultaneously. Hence, it is urgent to seek the bone scaffold with the requirements of bioactivity, degradability, and mechanical competence.

In native bone tissue, the extracellular matrix (ECM) plays an important role in maintenance and renewal of tissues. The ECM not only provides a platform for cell adhesion, but also regulates the cell behaviors. It is difficult to mimic the native bone ECM in vitro. Decellularized tissue has become popular in tissue engineering applications as the natural ECM can provide an appropriate cellular microenvironment for some kinds of stem cells [2]. Among them, decellularized bone also got a lot of attention in bone tissue engineering due to its excellent mechanical and structural profile, comparable to natural bone, and lack of immunogenicity in host tissue [3].

Decellularized bone is a naturally derived scaffold formed by removing cellular components from the native bone matrix and preserving the specific microarchitecture and composites of the bone ECM. In addition, the bioactive molecules such as growth factors and cytokines in nature bone matrix still preserve in the decellularized bone matrix. The specific microarchitecture of decellularized bone provides the mechanical support to aid the development of bone over time and facilitates the cell infiltration to the inner of the scaffold in vitro and in vivo. Discarding the cells and removing DNA from the nature bone matrix are critical for preventing the immune reactions of these scaffolds post-implantation. Lower immunogenicity enables a wide variety of decellularized bone scaffolds to become commercially available for bone tissue engineering applications [3, 4]. These preserving bioactive molecules in decellularized bone matrix endow them with high osteoinductive ability. By understanding decellularized tissues' effects on cell differentiation more thoroughly, this technology could provide a useful platform for controlling cell fate and generating regenerative therapies.

The aim of the present chapter is to describe the decellularization methods, characterization protocols of decellularized bone scaffold, and the effects of these scaffolds on bone tissue engineering. Additionally, the advances of hydrogels derived from demineralized and decellularized bone matrix are also described.

2 Decellularization Protocols of Bone

The goal of decellularizing bone tissue is to maintain all of the structural and biochemical cues in the bone ECM. Various decellularization methods of bone have been developed. Commonly used decellularization methods include chemical treatment, biological treatment, and physical treatment. All the above methods of cell removal from nature bone tissue will affect the ECM composition and cause a certain degree of ultrastructure disruption. In addition, many factors affect the cell removal efficiency, such as the tissue thickness and the density of the matrix. Therefore, it is very important to determine which method or combination is the most suitable for particular tissue types. For the decellularization protocols of bone, the chemical treatment or physical treatment always combines with the biological treatment to removal the cells from the nature bone tissue. We will introduce the details of the two combined methods in the following.

2.1 Chemical Treatment Combines with Biological Treatment

The combination of chemical and biological treatments is the most commonly used decellularization process of native bone. As described in the previous papers [5, 6], fresh cancellous bone is obtained from the femoral head of a white swine from a slaughterhouse. The femoral head is soaked with 0.9% saline and immediately transferred to the clean conditions. The femoral head is cleaned by removing the residual tissues, sectioned into desired discs, rinsed and soaked in distilled water for 12 h at 4 °C. Then, the discs are treated by 1% Triton X-100 for 2 days, degreased with methane/methyl alcohol for 1 day, incubated with DNase I at 37 °C for 2 h, and then washed in PBS at room temperature by continuous gradual shaking. The specimens are soaked in dehydrated alcohol for 4 h to remove all cellular remnants. Then, the specimens are rinsed with deionized water for 2 h. Finally, the specimens are stored at −80 °C before further application.

2.2 Physical Treatment Combines with Biological Treatment

Physical method that can be used to facilitate decellularization of different tissues includes freezing, direct pressure, sonication, agitation [7], and vacuum-assisted decellularization [8]. For decellularization of bone, high-hydrostatic pressurization (HHP) method is a unique physical method. It has been demonstrated that the HHP method can completely decellularize the porcine cortical bone and cortical bone including bone marrow (Fig. 1) [9]. Kishida's group proposed that the HHP method can destroy cell membranes at over 2000 atm, microorganism membranes at over 6000 atm, and virus capsules at over 9000 atm [10]. In addition, the HHP method avoids the use of any chemical agents to reduce the ECM damages and ultrastructure disruption [11].

As described in Hashimoto's article [9], fresh porcine femur and costa are harvested from a local slaughterhouse. The femora and costae are sectioned into small discs and washed in PBS containing penicillin (100 units/ml) and streptomycin (100 mg/ml). The specimens are completely filled with PBS and sealed in a plastic pack to prevent implosion and leakage during pressure application. The specimens are hydrostatically pressurized at 980 MPa at 30 °C for 10 min using a cold isostatic pressurization machine to dismantle the cells. Pressurization and decompression are performed at 65.3 MPa/min, and propylene glycol is used as a transmission fluid. Then, the specimens are rinsed in an EGM-2 containing DNase I (0.2 mg/ml) antibiotics under continuous shaking at 37 °C under 5% CO_2 for 21 days. Then, the specimens are washed with an 80% v/v ethanol at 37 °C for 3 days. After being washed and shaken with PBS again, the specimens are stored at 4 °C.

3 Characterization Protocols of Decellularized Bone

The preparation of decellularized bone matrix requires to removal of all cells from the source bone tissue. There is no strict standard by which to determine whether a bioscaffold is decellularized. The term decellularization has not been defined by quantitative metrics. Based upon the findings of studies in which an in vivo constructive remodeling response has been observed and adverse cell and host responses have been avoided, the following minimal criteria suffice to satisfy the intent of decellularization. Three criteria have been suggested as metrics of decellularization: (1) lack of visible nuclear

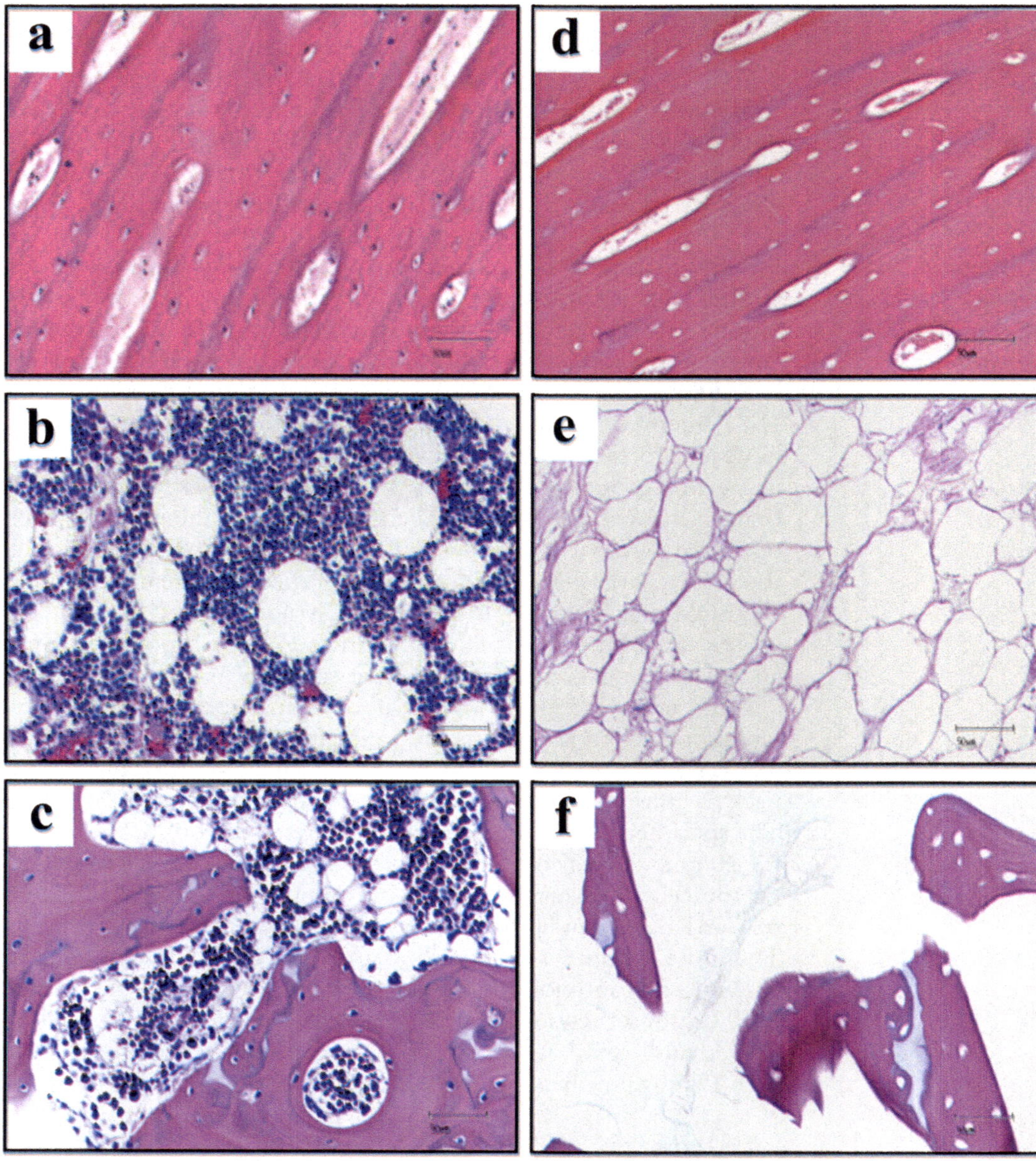

Fig. 1 H-E staining of native bone/bone marrows (**a–c**) and bone/bone marrows decellularized by high-hydrostatic pressurization (**d–f**). Images are cortical bone (**a**, **d**), bone marrow (**b**, **e**) from femora, and cancellous bone/bone marrow (**c**, **f**) from costae. Bone/bone marrows were pressurized for 10 min at 980 MPa, 30 °C. The scale bar indicates 50 μm (cited from [9])

material in tissue sections stained with 4′, 6-diamino-phenylindole (DAPI) and hematoxylin and eosin (H-E) staining; (2) less than 50 ng of double-strand DNA per mg of ECM dry weight; and (3) less than 200 base pairs fragment length of remnant DNA [12, 13].

3.1 Verification of Cell Removal

There are a number of methods available to determine the efficiency of the removal of cellular material from tissues. In general, H-E staining and DNA content can be used to evaluate the decellularization.

3.1.1 H-E Staining

Standard histological staining with H-E can serve as the first line of inspection to determine if nuclear structure can be observed. The native bone and decellularized bone are fixed in 4% paraformaldehyde at room temperature for 2 days and then decalcified with 12% EDTA-2Na for 1 month. The decalcified bone scaffold is dehydrated in ethanol and embedded by paraffin to cut 7 μm longitudinal sections using a microtome. The paraffin sections are further stained with H-E staining. The samples are analyzed under a light microscope. Figure 2 shows the images of H-E staining of native cancellous bone (Fig. 2a) and decellularized cancellous bone (Fig. 2b) [6]. Staining of cell nuclei with hematoxylin shows that the nuclei were almost completely removed from the cells in bone scaffolds. The natural porous structure and the components of bone were well maintained in decellularized bone. Eosin staining shows that natural bone matrix in the decellularized bone was also reserved.

3.1.2 DNA Quantification

To determine the amount of DNA in the decellularized bone, they were freeze-dried and weighted ($n \geq 3$). The specimens of 100 mg are pulverized and digested with proteinase-K (50 mg/ml) in 50 mM Tris–HCl, 25 mM EDTA-2Na, 100 mM NaCl, and 1% SDS solution at 55 °C for 12 h. DNA is isolated by a phenol/chloroform extraction and collected by an ethanol precipitation. The obtained DNA solution is assessed by an ultraviolet (UV) measurement at 260 nm using a UV/visual spectrophotometer. The ratios of absorbance at 260 and 280 nm are within the range of 1.6–1.9 for all samples. The amount of DNA is divided by the dry

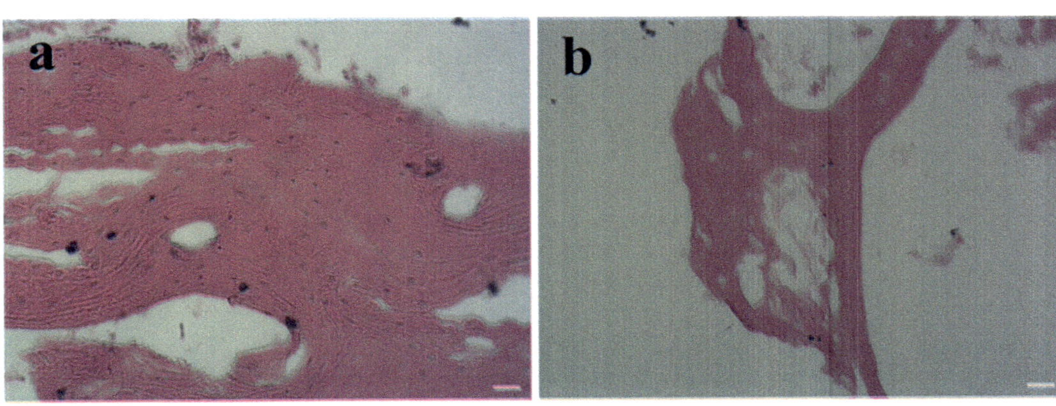

Fig. 2 H-E staining of native cancellous bone (**a**) and decellularized cancellous bone (**b**). The scale bar indicates 100 μm (cited from [6])

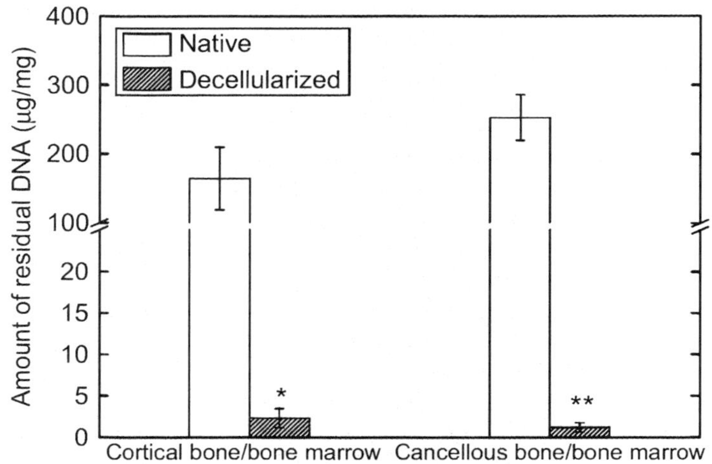

Fig. 3 Quantification analysis of amount of residual DNA in cortical and cancellous bone/bone marrow. Statistically significant differences are indicated as $*p < 0.05$ and $**p < 0.01$ and compared with native (cited from [9])

weight of the tissue samples to normalize it. Figure 3 shows the amount of DNA in the bone marrow decellularized by pressurization at 980 MPa at 30 °C [9]. The residual DNA in the decellularized bone was sufficiently decreased during the HHP process, showing the complete removal of cell debris [9].

3.2 Mechanical Testing

With verification of complete cellular removal, the effects of decellularization on the mechanical and material properties of the remaining ECM scaffold are of interest. The compressive modulus of the bone scaffold can be measured by uniaxial compression testing [6, 14]. The decellularized bone can be prepared as a cylinder for this test. The specimens are compressed at a strain rate of 0.03 mm/min with a mechanical testing machine. Results are shown as stress versus strain for strain rate. The compressive modulus of each treatment is taken from the slope of the stress-–strain curve. In addition, the local stiffness of decellularized bone can also be obtained by the atomic force microscope (AFM) or nanoindentation. For instance, the bone samples are cut into pieces with 1 mm thickness and probed with a soft silicon nitride quadratic pyramid tip. The AFM tip is applied a force at different positions along the sample and measured the deflection. The force–distance curves are collected and analyzed using the special software to obtain the Young's modulus of elasticity of different samples.

3.3 Structure Properties

Scanning electron microscopy (SEM) images and micro-computed tomography (μ-CT) scanning techniques can be used to show the surface and pore architecture of decellularized bone scaffold.

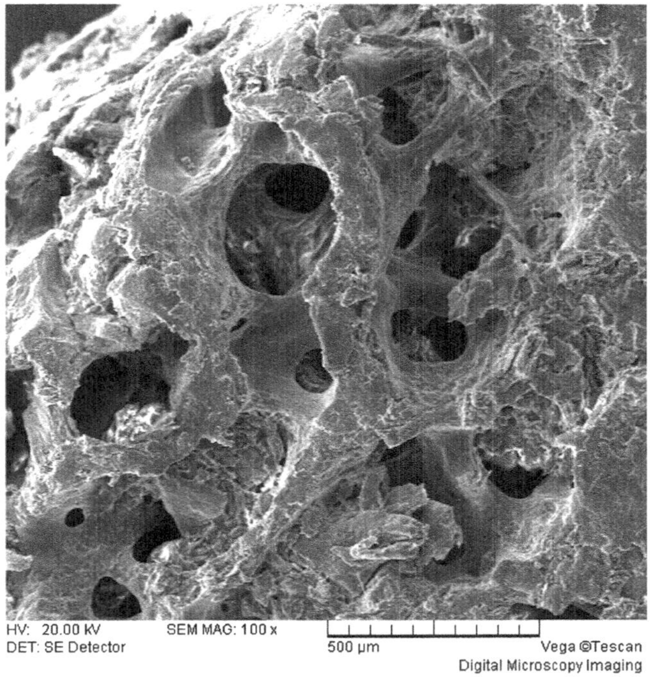

Fig. 4 SEM micrographs of decellularized cancellous bone. The SEM image shows the porous structure of decellularized cancellous bone. The scale bar indicates 500 μm (cited from [6])

3.3.1 SEM Analysis

The scaffolds are fixed in 2.5% glutaraldehyde at room temperature for 1 day. They are dehydrated in a graded series of ethanol (50, 70, 85, 90, 95, and 100%), dried at room temperature, sputter-coated with platinum for 40 s, and observed by SEM. As shown in Fig. 4, the decellularized bone scaffold had good natural porous microstructure and three-dimensional (3D) interconnectivity [6].

3.3.2 Micro-CT

The qualitative information on the decellularized bone scaffold architecture is obtained by μ-CT imaging using a microfocus X-ray CT system. Special software μ-CT v6.1 of Scanco medical AG is used to visualize the 2D X-ray section images of the layered scaffolds. The slice increment is 10.5 μm. Isotropic slice data are obtained by the system and reconstructed into 2D images. These slice images are compiled and analyzed to render 3D images and used to investigate the interconnectivity of the pores, porosity, and distribution of pore size.

4 Decellularized Bone Matrix for Bone Tissue Engineering

Decellularized bone was used to generate scaffolds for bone tissue engineering applications because of their ideal molecular composition, structural properties, and mechanical properties, enabling bone development in vitro and immediate mechanical support following graft implantation into load-bearing areas [15]. Previous studies have demonstrated the positive influences of decellularized bone scaffolds on cell adhesion, proliferation, osteogenic differentiation of osteoprogenitor cells [9, 16–20]. Among them, the scaffolds made by decellularizing native bone have found to support the osteogenic differentiation and formation of engineered bone by human mesenchymal stem cells (MSCs) [9, 16], human embryonic stem cells (ESCs) [17], induced pluripotent stem cells (iPSCs) [18], and adipose derived stem cells (ASCs) [15, 19]. Among them, Hashimoto et al. [9] found that the alkaline phosphatase (ALP) activity of MSCs on the 3D decellularized bone was significantly elevated than that on a 2D tissue culture polystyrene (TCPS) dish. Even in the absence of dexamethasone, a common osteogenic factor, MSCs in decellularized bone had more ALP activity as determined biochemically and via staining than cells grown on TCPS, suggesting that decellularized bone matrix favorably promotes early osteogenic differentiation of MSCs. Additionally, subcutaneous implantation of decellularized bone matrix into rats demonstrated cell infiltration with neovascularization after 6 months [9].

The harvesting sites of decellularized bone from the native bone will affect the bone density and matrix architecture. A study evaluated and compared the osteogenesis of mesenchymal progenitors derived from human ESCs in decellularized bone with different densities [20]. The decellularized bone scaffolds were classified into three experimental groups, namely low density group (0.281 ± 0.018 mg/mm^3), medium density group (0.434 ± 0.015 mg/mm^3), and high density group (0.618 ± 0.027 mg/mm^3) (Fig. 5a). Pore size and porosity of the three different groups decreased with increasing bone density while compressive elastic modulus increased with increasing bone density. After 5 weeks of culture, the medium density group showed the highest densities of cells and exhibited significantly higher staining of osteopontin (OPN), osteocalcin (OCN), and bone sialoprotein (BSP) than the other two groups (Fig. 5b). The results indicated that the medium density bone scaffolds have the best balance of the transport rates of nutrients and metabolites, space for cell infiltration, surface for cell attachment, and the mechanical strength [20].

In addition to the direct study the performance of the decellularized bone, Chen et al. [6] coated the decellularized bone with

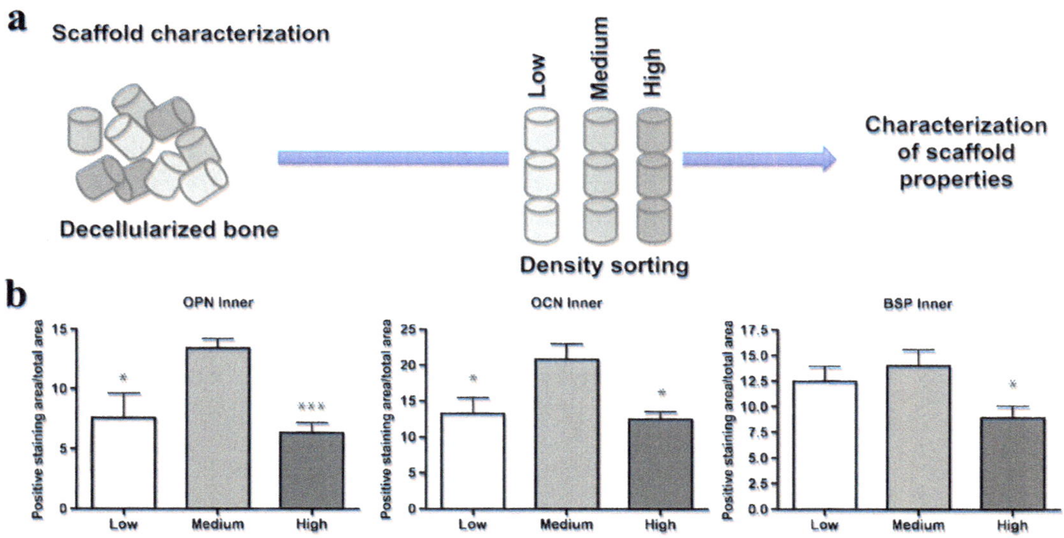

Fig. 5 Scaffolds were sorted according to the mass density into three experimental groups (low, medium, and high density) (**a**). Construct areas staining positively for OPN, OCN, and BSP were quantified by using Image J (**b**). Data represent average ± standard deviation (SD) for $n = 4$ (*$p < 0.05$; **$p < 0.01$; ***$p < 0.001$) (revised from [20])

increasing concentrations of collagen containing equal masses of hydroxyapatite (HA) to form the scaffolds with different stiffness but the same microstructure. Compared to decellularized bone alone, all decellularized bone scaffolds coated with collagen/HA exhibited increased local and bulk stiffness and enhanced osteogenic differentiation of rat MSCs in vitro. Among them, the constructs with the highest bulk modulus yielded the greatest expression of osteogenic markers in vivo when implanted in an ectopic site. Furthermore, the bone repair capacity of the scaffolds in the rabbit large bone defect model was also evaluated. Results demonstrated that the cell-free collagen/HA scaffolds of proper stiffness combined with endogenous osteoprogenitor cells could increase the bone regeneration significantly [21]. However, the number of recruiting stems cells was still low when the decellularized bone scaffold is lack of chemotaxis effect on endogenous stem cells in this study [21]. Therefore, chemokine stromal cell-derived factor-1α (SDF-1α) was loaded into the decellularized bone scaffold by coating of collagen/HA mixture with optimal ratio to increase the recruitment of endogenous MSCs and promote osteogenic differentiation in situ [22]. This study shows that MSCs osteogenesis in vitro was enhanced by matrix elasticity and SDF-1α, and endogenous MSCs recruitment in subcutaneous implantation of rat was increased by the release of SDF-1α from the scaffold, and bone regeneration in rabbit large bone defect model was significantly improved by matrix elasticity and SDF-1α [22].

One limitation in the development of tissue engineered bone grafts is the insufficient transport of nutrient delivery, oxygen supply and exchange, and metabolic-waste discharge in the inner regions of the scaffolds [23]. These limitations eventually decrease the survival rate of cells inside the scaffold, form nonuniform cellular distribution within scaffold, and limit the usable size of 3D scaffold. To overcome these drawbacks associated with static culturing system, the most common approach is to use a bioreactor system to generate a dynamic environment for cell loaded scaffold culture.

Perfusion bioreactors can allow for medium perfusion through the whole constructs that can increase new bone formation by providing more uniform nutrient supply and applying different shear stress stimulation to cells [15, 24–27]. In order to determine how medium flow velocity influences the fate of MSCs and mineral deposits in 3D scaffolds, to establish predictive correlations between perfusion rates and osteogenesis of MSCs, Grayson and colleagues [24] cultured human MSCs on cylindrical-shaped decellularized bone scaffolds using a wide range of superficial flow velocities (80, 400, 800, 1200, and 1800 μm/s) in custom-designed perfusion bioreactors for 5 weeks. It is found that the flow velocities ranging from 400 to 800 μm/s had the best overall osteogenic responses, including histological analyses, connexin expression, and bone protein contents [24]. Considering the different patterns of fluid shear stress in human bone, a recent work used a homemade oscillatory flow device to investigate the synergistic action of fluid shear stress and matrix stiffness on MSCs survival, distribution, and osteogenic differentiation in cylindrical-shaped decellularized bone scaffolds [28]. The study has proved that the oscillatory shear stress not only helps to maintain high MSCs survival rate and uniform cell distribution throughout the scaffold after a relatively long period culture in vitro but also facilitates osteogenic differentiation of MSCs. Additionally, the findings also suggest that the oscillatory perfusion and matrix stiffness synergistically affect the osteogenesis of MSCs within an appropriate range of parameters [28]. In future studies, perfusion bioreactors will play a more important role in fabricating large-sized bone-like scaffolds with uniform distribution of viable cells and more bone matrix deposition.

In addition to many in vitro study of decellularized bone in regulating and controlling the fate of osteoprogenitor cells, several researches have also focused on how to improve the regenerative ability of decellularized bone scaffold in different bone defect model. In a recent study, Bhumiratana et al. [29] have engineered one of the most geometrically complex facial bones by using autologous stem cells, native decellularized bovine bone matrix, and a perfusion bioreactor for the growth and transport of living grafts, without bone morphogenetic proteins. First, the decellularized bovine trabecular bone was crafted into an anatomically correct shape using image-guided micromilling to fit the defect. The autologous ASCs were seeded into the scaffold and cultured in a specialized bioreactor for 3 weeks and then implanted into the defects of minipigs. At 6 months after surgery, the engineered grafts still maintained their structure, enhanced graft-host bone integration, and formed larger volume of new bone and better vascular infiltration than either pure decellularized bovine trabecular bone scaffolds or untreated defects. Therefore, the osteoinductive capacity of the decellularized bovine, the osteogenic capacity of various adult stem cells, and the physiologic benefits of perfusion flow contribute to a comprehensive paradigm of bone tissue engineering.

5 Hydrogels Derived from Decellularized Bone for Bone Tissue Engineering

Although decellularized bone matrix has been widely used in bone defects as a commercial allograft biomaterial, the differences in the preparation and processing methods and donor age all have an impact on its properties and clinical performance. Additionally, some demineralized bone matrix products have shown to cause adverse responses in vivo, including nephrotoxicity and inflammation with various carriers [30]. To address this issue, Sawkins et al. [31] fabricated hydrogel scaffold derived from the decellularized and demineralized bone that have potential to be used clinically without carriers (Fig. 6). This new soluble form of decellularized and demineralized bone matrix had distinct mechanical and biological properties, including osteogenic functionality and the potential for clinical use.

The hydrogel scaffolds from decellularized bone were able to stimulate odontogenic differentiation of human dental pulp stem cells (DPSCs) by upregulating the expression levels of the odonto/

Fig. 6 Production of ECM hydrogel from bone. Bovine tibiae were processed to form (**a**) fragments and then subjected to mineral, lipid, and cell removal procedures to produce (**b**) decellularized bone (bone ECM) prior to pepsin digestion and solubilization to form an ECM hydrogel (**c**) (cited from [31])

osteogenic-specific genes and mineral deposition [32]. The same research group also found that the mRNA expression levels of the odonto/osteogenic-specific genes of DPSCs on the bone ECM hydrogel scaffolds were significantly upregulated with respect to those cultured on collagen I scaffolds or TCPS in the absence of extra odontogenic inducers. Furthermore, the expression levels of the odonto/osteogenic-specific genes of DPSCs cultured on bone ECM hydrogels were further upregulated by the addition of growth factors or osteo/odontogenic medium compared to collagen I treated cells in the same culture conditions [33]. Additional studies with bone ECM hydrogel scaffolds have shown the efficacy of bone-derived ECM hydrogels with incorporated growth factors to augment bone formation within the ex vivo chick femoral defects [34, 35]. Moreover, Gothard et al. [36] showed that alginate/bone ECM hydrogel scaffolds in combination with adult human bone marrow stromal cells were able to provide in vivo tissue mineralization and bone formation. In a recent study [37], the osteogenic potential of ECM hydrogels prepared from demineralized and decellularized bovine bone in the presence and absence of osteogenic medium was evaluated. The results in this work found that mouse primary calvarial cells expressed higher ECM proteins OPN and OCN on bone ECM and decellularized bone matrix in osteogenic and basal medium than those on collagen I hydrogel and TCPS. The findings of this study demonstrated the ability of bone ECM and decellularized bone matrix hydrogels to influence osteogenic differentiation of osteoprogenitor cells without the addition of biological factors. Therefore, hydrogels are best for creating 3D environments and can be used as injectable therapies in bone tissue engineering and clinical applications.

Despite recapitulating the appropriate anatomic geometry, obtaining clinically relevant and continuous volumes of decellularized bone for scaffold production is challenging. 3D-printing is a key technique in bone tissue engineering that can be used to fabricate functionalized scaffolds with patient-specific geometry. Recently, Huang and colleagues [38] tried using 3D-printing to fabricate novel hybrid decellularized bone matrix particles/polycaprolactone (PCL) scaffolds for craniofacial regeneration (Fig. 7). Hybrid scaffolds were readily printable at compositions of up to 70% bone by mass and displayed robust mechanical properties. This material is 3D printable while displaying osteoinductive properties in vitro and in vivo as evidenced by gene expression, calcium per cell, and in vivo bone regeneration of seeded human ASCs. Taken together, these results demonstrate that 3D printed decellularized bone matrix particles/PCL scaffolds might be effective for stimulating bone regeneration [38]. Furthermore, the same research group also compared the print quality of 3D-printed PCL scaffolds functionalized with tricalcium phosphate (TCP), HA, bio-oss (BO), or decellularized bone matrix [39]. In this 3D-

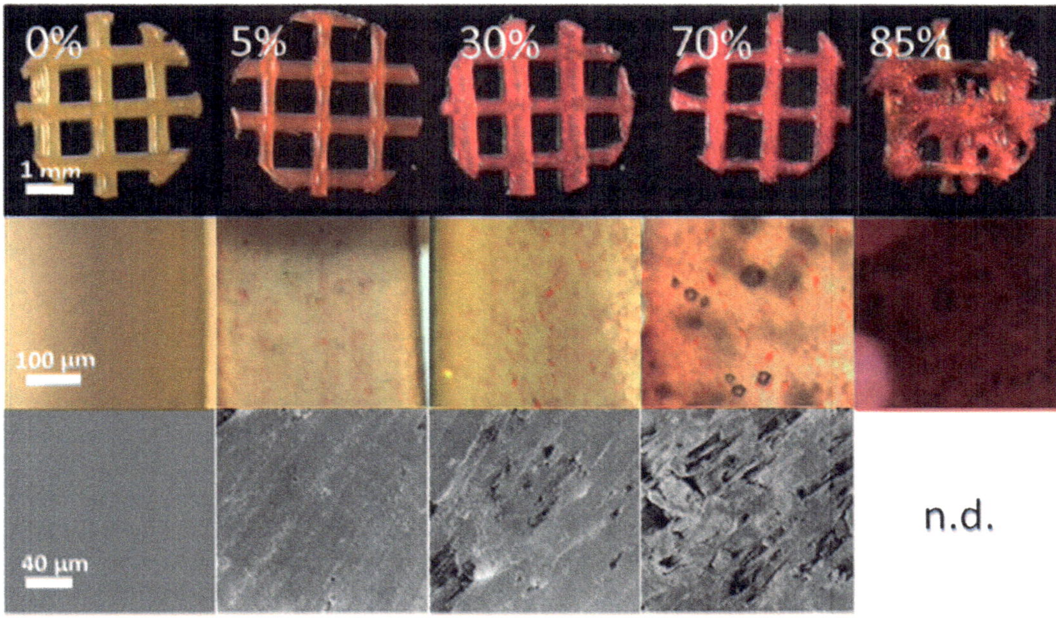

Fig. 7 Imaging of 3D-printed hybrid scaffolds. *Top*: scaffolds stained positively for Alizarin Red S in all cases except for the pure PCL case. *Middle*: magnified images of stained scaffold struts delineating the punctate stain of the mineralized particles within the PCL. *Bottom*: SEM of strut surfaces revealing rougher surface topographies in the more concentrated hybrid scaffolds (cited from [38])

printing system, PCL-BO and PCL-decellularized bone matrix exhibited greater ability for osteoinduction than synthetic materials such as PCL-HA or PCL-TCP. PCL-DCB and PCL-BO blends induced significant increases in mineral deposition and upregulation of collagen and osteonectin relative to PCL only scaffolds. This greater induction might be due to the presence of a collagen phase, the structure of the apatite, or greater presence of the BO and decellularized bone matrix particles on the surface of the struts following the printing process. These results indicate that doping 3D-printed PCL scaffolds with decellularized bone matrix or BO might better support bone healing in vivo in comparison to TCP- or HA-doped grafts.

6 Conclusions and Future Considerations

Overall, decellularized bone scaffolds are a legitimate alternative to conventional therapies of bone defects and have the huge potential to enhance the repair outcomes or provide solutions that may not be available with synthetic scaffolds. As shown in this chapter, a range of studies have demonstrated the advantages of decellularized bone matrix in bone tissue engineering. Moreover, an increasing number of studies have also paid attention to the modification of decellularized bone matrix to improve its functionality.

However, everything has its merits and demerits. In spite of the significant progress in using decellularized bone matrix as scaffolds for bone tissue engineering, a lot of issues still need to be improved and eventually solved. Firstly, a much simpler and faster approach of removing cells from the native bone tissue need to develop in the future. Less time spent on the manufacturing of decellularized bone matrix is critical for large-scale clinical applications. Secondly, researchers should pay more attention to the role of mechanical cues of the decellularized bone matrix on stem cell differentiation, not just its osteoinductive properties. Thirdly, the integration of the implanted decellularized bone with native bone was not good enough [29]. How to effectively reduce the soft fibrous tissues between the host bone and the implanted decellularized bone and enhance their connection is still one of the research priorities for some time to come. In addition, the use of cytokines or growth factors that have specific functions to further improve the function of decellularized bone matrix remains an important area of research. Furthermore, it is necessary to increase the development and applications of the hydrogels derived from decellularized bone, and endow the specific functions through targeted modification. It is important to ensure the cell viability and osteogenic potential before implanting to the bone defects. For this reason, different bioreactor systems should be invented to apply different physical stimuli, such as electrical stimulation, shear force stimulation, and pressure stimulation. Additionally, the ECM itself is a complex combination of biochemical cues and therefore it can be difficult to determine which specific components or combinations thereof are causing certain cell behaviors [40]. To characterize the decellularized bone matrix proteomic composition is critical to better comprehend the role of the ECM in directing cellular processes.

In conclusion, the decellularized bone scaffold is not just a scaffold for cells, but the control center of cell function or a niche for more complex biological functions. The wider range of clinical applications of decellularized bone scaffold will likely continue to expand as new indications are studied and a more complete understanding of the underlying mechanisms of in vivo remodeling is developed.

Acknowledgements

This work was supported in part by grants from the National Natural Science Foundation of China (11672051), the China Postdoctoral Science Foundation (2015M582521), and the Fundamental Research Funds for the Central Universities (106112017CDJQJ468823).

References

1. Boskey AL (2013) Bone composition: relationship to bone fragility and antiosteoporotic drug effects. Bonekey Rep 2:447
2. Tapias LF, Ott HC (2014) Decellularized scaffolds as a platform for bioengineered organs. Curr Opin Organ Transplant 19:145–152
3. Lee DJ, Diachina S, Lee YT et al (2016) Decellularized bone matrix grafts for calvaria regeneration. J Tissue Eng 7:2041731416680306
4. Song JJ, Ott HC (2011) Organ engineering based on decellularized matrix scaffolds. Trends Mol Med 17:424–432
5. Sun XJ, Peng W, Yang ZL et al (2011) Heparin-chitosan-coated acellular bone matrix enhances perfusion of blood and vascularization in bone tissue engineering scaffolds. Tissue Eng Part A 17:2369–2378
6. Chen G, Dong C, Yang L et al (2015) 3D scaffolds with different stiffness but the same microstructure for bone tissue engineering. ACS Appl Mater Interfaces 7:15790–15802
7. Gilbert TW, Sellaro TL, Badylak SF (2006) Decellularization of tissues and organs. Biomaterials 27:3675–3683
8. Butler CR, Hynds RE, Crowley C et al (2017) Vacuum-assisted decellularization: an accelerated protocol to generate tissue-engineered human tracheal scaffolds. Biomaterials 124:95–105
9. Hashimoto Y, Funamoto S, Kimura T et al (2011) The effect of decellularized bone/bone marrow produced by high-hydrostatic pressurization on the osteogenic differentiation of mesenchymal stem cells. Biomaterials 32:7060–7067
10. Nakamura N, Kimura T, Kishida A (2016) Overview of the development, applications, and future perspectives of decellularized tissues and organs. ACS Biomater Sci Eng. doi:10.1021/acsbiomaterials.6b00506
11. Funamoto S, Nam K, Kimura T et al (2010) The use of high-hydrostatic pressure treatment to decellularize blood vessels. Biomaterials 31:3590–3595
12. Crapo PM, Gilbert TW, Badylak SF (2011) An overview of tissue and whole organ decellularization processes. Biomaterials 32:3233–3243
13. Costa A, Naranjo JD, Londono R et al (2017) Biologic scaffolds. Cold Spring Harb Perspect Med. pii:a025676
14. Shahabipour F, Mahdavi-Shahri N, Matin MM et al (2013) Scaffolds derived from cancellous bovine bone support mesenchymal stem cells' maintenance and growth. In Vitro Cell Dev Biol Anim 49:440–448
15. Frohlich M, Grayson WL, Marolt D et al (2010) Bone grafts engineered from human adipose-derived stem cells in perfusion bioreactor culture. Tissue Eng Part A 16:179–189
16. Grayson WL, Bhumiratana S, Cannizzaro C et al (2008) Effects of initial seeding density and fluid perfusion rate on formation of tissue-engineered bone. Tissue Eng Part A 14:1809–1820
17. Marolt D, Campos IM, Bhumiratana S et al (2012) Engineering bone tissue from human embryonic stem cells. Proc Natl Acad Sci U S A 109:8705–8709
18. de Peppo GM, Marcos-Campos I, Kahler DJ et al (2013) Engineering bone tissue substitutes from human induced pluripotent stem cells. Proc Natl Acad Sci U S A 110:8680–8685
19. Chen Q, Yang Z, Sun S et al (2010) Adipose-derived stem cells modified genetically in vivo promote reconstruction of bone defects. Cytotherapy 12:831–840
20. Marcos-Campos I, Marolt D, Petridis P et al (2012) Bone scaffold architecture modulates the development of mineralized bone matrix by human embryonic stem cells. Biomaterials 33:8329–8342
21. Chen G, Yang L, Lv Y (2016) Cell-free scaffolds with different stiffness but same microstructure promote bone regeneration in rabbit large bone defect model. J Biomed Mater Res A 104:833–841
22. Chen G, Lv Y (2017) Matrix elasticity-modified scaffold loaded with SDF-1α improves the in situ regeneration of segmental bone defect in rabbit radius. Sci Rep 7:1672
23. Du D, Asaoka T, Ushida T et al (2014) Fabrication and perfusion culture of anatomically shaped artificial bone using stereolithography. Biofabrication 6:045002

24. Grayson WL, Marolt D, Bhumiratana S et al (2011) Optimizing the medium perfusion rate in bone tissue engineering bioreactors. Biotechnol Bioeng 108:1159–1170
25. Li D, Tang T, Lu J et al (2009) Effects of flow shear stress and mass transport on the construction of a large-scale tissue-engineered bone in a perfusion bioreactor. Tissue Eng Part A 15:2773–2783
26. Yu X, Botchwey EA, Levine EM et al (2004) Bioreactor-based bone tissue engineering: the influence of dynamic flow on osteoblast phenotypic expression and matrix mineralization. Proc Natl Acad Sci U S A 101:11203–11208
27. Kim J, Ma T (2012) Perfusion regulation of hMSC microenvironment and osteogenic differentiation in 3D scaffold. Biotechnol Bioeng 109:252–261
28. Chen G, Xu R, Zhang C et al (2017) Responses of MSCs to 3D scaffold matrix mechanical properties under oscillatory perfusion culture. ACS Appl Mater Interfaces 9:1207–1218
29. Bhumiratana S, Bernhard JC, Alfi DM et al (2016) Tissue-engineered autologous grafts for facial bone reconstruction. Sci Transl Med 8:343ra83
30. Markel DC, Guthrie ST, Wu B et al (2012) Characterization of the inflammatory response to four commercial bone graft substitutes using a murine biocompatibility model. J Inflamm Res 5:13–18
31. Sawkins MJ, Bowen W, Dhadda P et al (2013) Hydrogels derived from demineralized and decellularized bone extracellular matrix. Acta Biomater 9:7865–7873
32. Paduano F, Marrelli M, White LJ et al (2016) Odontogenic differentiation of human dental pulp stem cells on hydrogel scaffolds derived from decellularized bone extracellular matrix and collagen type I. PLoS One 11:e0148225
33. Paduano F, Marrelli M, Alom N et al (2017) Decellularized bone extracellular matrix and human dental pulp stem cells as a construct for bone regeneration. J Biomater Sci Polym Ed 28:730–748
34. Smith EL, Kanczler JM, Gothard D et al (2014) Evaluation of skeletal tissue repair, part 1: assessment of novel growth-factor-releasing hydrogels in an ex vivo chick femur defect model. Acta Biomater 10:4186–4196
35. Smith EL, Kanczler JM, Gothard D et al (2014) Evaluation of skeletal tissue repair, part 2: enhancement of skeletal tissue repair through dual-growth-factor-releasing hydrogels within an ex vivo chick femur defect model. Acta Biomater 10:4197–4205
36. Gothard D, Smith EL, Kanczler JM et al (2015) In vivo assessment of bone regeneration in alginate/bone ECM hydrogels with incorporated skeletal stem cells and single growth factors. PLoS One 10:e0145080
37. Alom N, Peto H, Kirkham GR et al (2017) Bone extracellular matrix hydrogel enhances osteogenic differentiation of C2C12 myoblasts and mouse primary calvarial cells. J Biomed Mater Res B Appl Biomater. doi:10.1002/jbm.b.33894
38. Hung BP, Naved BA, Nyberg EL et al (2016) Three-dimensional printing of bone extracellular matrix for craniofacial regeneration. ACS Biomater Sci Eng 2:1806–1816
39. Nyberg E, Rindone A, Dorafshar A et al (2017) Comparison of 3D-printed poly-ε-caprolactone scaffolds functionalized with tricalcium phosphate, hydroxyapatite, bio-oss, or decellularized bone matrix. Tissue Eng Part A. doi:10.1089/ten.TEA.2016.0418
40. Agmon G, Christman KL (2016) Controlling stem cell behavior with decellularized extracellular matrix scaffolds. Curr Opin Solid State Mater Sci 20:193–201

Methods in Molecular Biology (2018) 1577: 255–270
DOI 10.1007/7651_2017_56

Published online: 23 July 2017

Preparation of Decellularized Liver Scaffolds and Recellularized Liver Grafts

Yibin Chen, Sharon Geerts, Maria Jaramillo, and Basak E. Uygun

Abstract

Severe liver disease is the 12th leading cause of death in the USA, with organ transplantation often being the only viable option for treatment. However, due to the shortage of viable donor livers, it is estimated that over 1200 patients died in 2015 while waiting for liver transplantation. This highlights the need for alternative sources of viable organs. In this study, we describe a method that provides the groundwork for the development of functional liver grafts. The approach described here is for removal of cells from intact livers and subsequently repopulating them with functional liver cells. Briefly, rat livers are harvested and subjected to a series of perfusion decellularization steps using an anionic detergent such that an intact decellularized liver matrix (DLM) scaffold with preserved vascular architecture is obtained. Further, we describe methods to recellularize DLM scaffolds with adult primary hepatocytes, creating a liver graft that exhibits hepatic functions in vitro.

Keywords Decellularization, Liver tissue engineering, Recellularization, Whole liver scaffold

1 Introduction

Whole liver engineering is a new approach to tissue engineering that has gained attention over the past few years [1]. It utilizes whole organs that are devoid of cells, thus creating scaffolds that retain structural, mechanical, and chemical attributes of the native tissue, and serving as an ideal material for development of engineered tissues. One major advantage of this approach is that it takes advantage of the natural tissue matrix by preserving the intact microvascular structure, which facilitates connection of the tissue directly with patient vasculature upon transplantation, allowing immediate circulation and instant delivery of nutrients and oxygen. This approach is ideal for engineering transplantable liver grafts where constant perfusion is required for nutrient and oxygen delivery and waste removal to maintain highly metabolically active liver cells. In addition, the well-preserved extracellular matrix (ECM) enhances the cell attachment and reorganization [2].

Perfusion decellularization is the first step in whole-organ engineering that aims to eliminate resident cells and cellular components from the target organ while retaining the native composition and structure of extracellular matrix. The next step is recellularization which is repopulating the decellularized organ with the cells of the native organ. This approach has been successfully applied for tissues and organs such as trachea [3], heart [4], lung [5, 6], kidney [7], urinary bladder [8], and liver [9, 10]. Here, we describe decellularization, recellularization, and characterization techniques for liver engineering using rat livers as the model system. We also describe a nondestructive monitoring method during the decellularization that ensures the consistency of the scaffold's quality, including the extent of cell removal and the preservation of extracellular matrix components, without disrupting the scaffold's integrity using computed tomography and perfusate analysis. We recently reported that there is a strong correlation between DNA removal and Hounsfield unit (HU) where HU is a quantitative scale unit used in computed tomography for describing radiodensity. The methods described here are the first steps of creating transplantable liver grafts and can be scaled up to human size liver grafts. The recellularized liver grafts can be used as alternatives to donor livers for transplantation or can serve as a drug testing platform superior to in vitro hepatocyte cultures.

2 Materials

Solutions are prepared with ultrapure water which has a resistivity of 18 MΩ-cm at 25 °C. Reagents are subsequently stored at room temperature (unless indicated otherwise).

2.1 Liver Harvest

2.1.1 Animals

- 150–200 g Female Lewis rats (Charles River Laboratories, Wilmington, MA).

2.1.2 Chemicals and Surgical Tools

- Phosphate buffered saline (PBS).
- 70% (v/v) Isopropanol.
- Gas tank 100% O_2.
- Cotton tips (2×).
- Gauze (2×) (Kendall Versalon 4 ply 2″ × 2″).
- 3.0 and 5.0 polypropylene sutures (Ethicon).
- 18G catheter (Terumo Surflash IV catheters, Somerset, NJ).
- Cell culture dish (100 mm × 20 mm, Corning, NY).
- Tape (0.5″ wide Transpore 3 M).
- Heparin (Hospira, Inc. Lake Forest, IL).
- Large scissors (1×) (Aesculap, Roboz).
- Micro dissecting rat tooth forceps (1×) (Aesculap, Roboz).

- Large hemostats (1×) (Aesculap, Roboz).
- Fine forceps, straight and curved (Aesculap, Roboz).
- Curved micro forceps, stainless steel (Aesculap, Roboz).
- Small blunt nosed dissecting scissors (1×) (Aesculap, Roboz).
- 30 ml syringe.
- Isoflurane vaporizer (Colonial Medical Supply, Windham, NH).

2.2 Perfusion Decellularization

2.2.1 Reagents

- Phosphate buffered saline (PBS).
- Sodium dodecyl sulfate (SDS) (Sigma-Aldrich), 0.01% (w/v) in dH_2O.
- SDS, 0.1% (w/v) in dH_2O.
- SDS, 0.2% (w/v) in dH_2O.
- SDS, 0.5% (w/v) in dH_2O.
- Triton X-100 (Sigma-Aldrich), 1% (w/v) in dH_2O.
- Peracetic acid (Sigma-Aldrich).
- Gentamicin (Sigma-Aldrich).
- Amphotericin B (Sigma-Aldrich).
- Distilled water.
- Ethanol (200 proof).

2.2.2 Perfusion System Set-Up

- L/S® standard digital drive (Cole-Parmer Cat # EW-07523-80).
- Masterflex® standard pump head for L/S 15 (Cole-Parmer Cat # EW-07015-21).
- Masterflex® platinum-cured silicone tubing, L/S® 14 (Cole-Parmer Cat # EW-96410-14).
- 8-l containers.
- Bubble trap (Radnoti part# 130149).
- 10-cm petri dishes.
- 30 cc syringes.

2.3 Characterization of Decellularized Liver Matrix

2.3.1 Computed Tomography (CT) Scanning of Decellularized Livers

- Clear ziplock bag.
- CereTom CT scanner (Neurologica, Danvers, MA) with CereTom CT Scanner software.
- AMIDE open source software (http://amide.sourceforge.net).

2.3.2 Immunoassay and Biochemical Assays

Total Protein Assay

- Scienceware® Liquid Nitrogen Cooled Mini Mortar & Pestle Set (Thomas Scientific, Swedesboro, NJ).
- RIPA Lysis and Extraction Buffer (Thermo Fisher Scientific, Waltham, MA).
- Pierce BCA Assay Kit (Life Technologies, Grand Island, NY).

Western Blot

- PBS.
- JLA-20 antibody (Developmental Studies Hybridoma Bank, Iowa City, IA).
- Laemmli buffer (BIO-RAD, Hercules, CA).
- β-mercaptoethanol (Sigma-Aldrich, St. Louis, MO).
- Tris-buffered saline, 0.1% Tween 20 (TBST).
- Nitrocellulose membrane (BIO-RAD, Hercules, CA).
- Protein ladder (Li-Cor, Lincoln, NE).
- Blocking buffer (Li-Cor, Lincoln, NE).
- Transfer buffer: 20 mM Tris, 190 mM glycine, 20% methanol.
- Running buffer: 4% SDS, 10% 2-mercaptoethanol, 20% glycerol, 0.004% bromophenol blue, 0.125 M Tris–HCl.
- Odyssey CLx (Li-Cor, Lincoln, NE).

DNA Assay

- Purelink Genomic DNA kit (Life Technologies, Grand Island, NY).
- Picogreen Assay kit (Life Technologies, Grand Island, NY).

Glycosaminoglycan (GAG) Assay

- Lyophilizer (FTS SYSTEMS, Stone Ridge, NY).
- 6 N hydrochloric acid.
- 0.5 mg/ml chondroitin sulfate A (Sigma-Aldrich, St. Louis, MO).

- Dimethylene blue (Sigma-Aldrich, St. Louis, MO).
- Microplate spectrophotometer: Benchmark Plus (BIO-RAD, Hercules, CA).

Collagen Assay

- Quickzyme Total Collagen assay kit (Cedarlane Labs, Burlington, NC).

2.3.3 Corrosion Casting

- Batson's 17 anatomic corrosion kit (Polysciences Inc. Cat # 07349-1).
- 1 N potassium hydroxide.

2.3.4 Histological Analyses

- 10% formalin.
- 70% ethanol.

2.3.5 Perfusate Analysis

- Beakers (600 ml or 1000 ml).
- Conical tubes (50 ml).

2.4 Perfusion Recellularization

2.4.1 Cells

Adult primary rat hepatocytes isolated from 150 to 200 g female Lewis rats using the two-step perfusion protocol as described by Dunn et al. [11] and used at a density of 10 million cells per ml.

2.4.2 Reagents

- Seeding medium: High glucose Dulbecco's modified Eagle's medium supplemented with 10% fetal bovine serum (Peak Serum), 100 U/ml penicillin, and 100 μg/ml streptomycin (Invitrogen).
- Culture medium: High glucose Dulbecco's modified Eagle's medium with 10% fetal bovine serum (Peak Serum), 7 ng/ml glucagon (Bedford Laboratories), 7.5 μg/ml hydrocortisone (Pfizer), 0.5 U/ml insulin (Novolin), 20 ng/ml Epidermal Growth Factor (Invitrogen), 200 U/ml penicillin, and 200 μg/ml streptomycin (Invitrogen).

2.4.3 Engineering Parts

Cell Seeding Perfusion Setup

- Masterflex® standard pump head for L/S® 15 (Cole-Parmer Cat # EW-07015-21).
- Masterflex® platinum-cured silicone tubing, L/S® 14 (Cole-Parmer Cat # EW-96410-14).
- Bubble trap (Radnoti part# 130149).

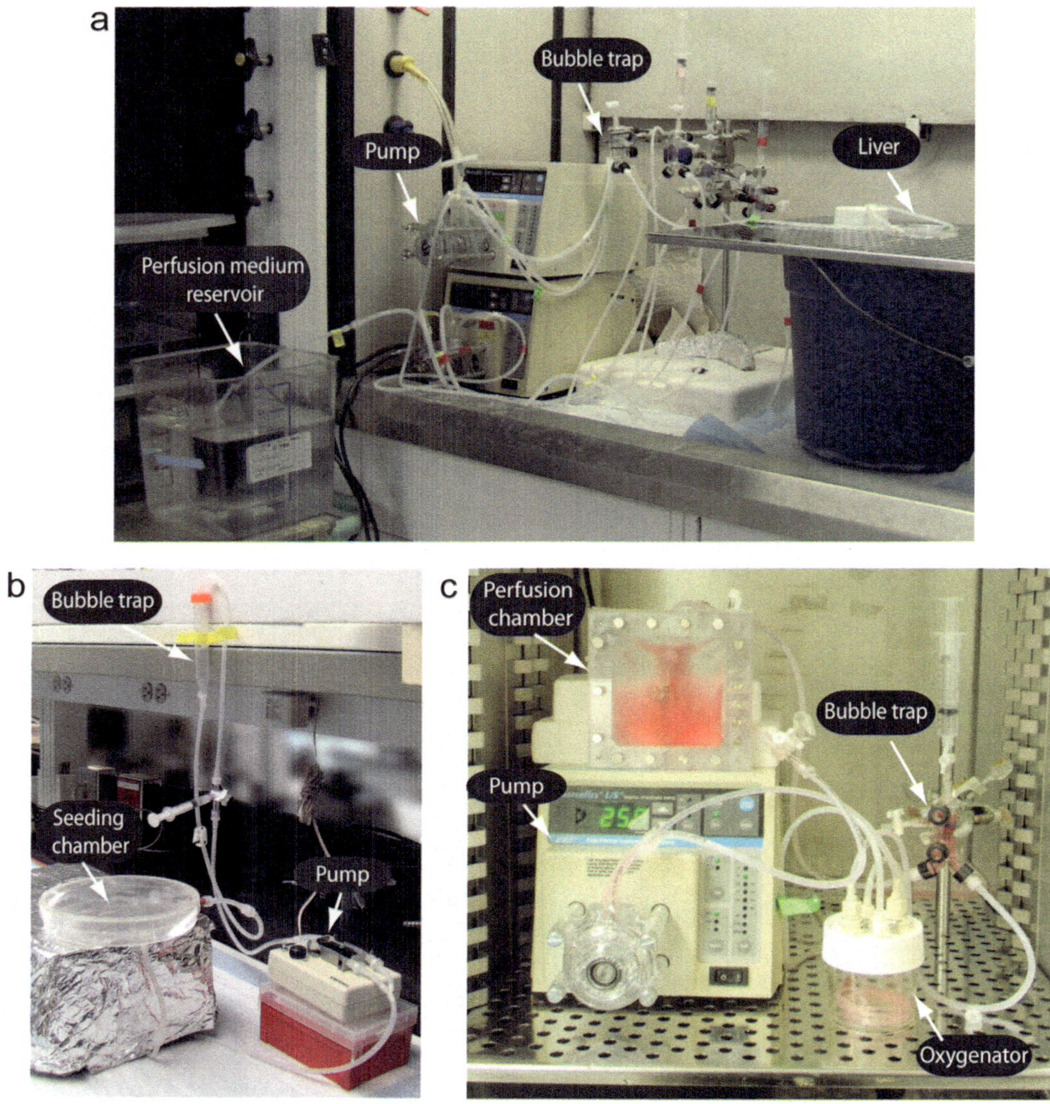

Fig. 1 (**a**) Perfusion decellularization system. The perfusion decellularization system is composed of a perfusion medium reservoir, a pump, and a bubble trap. Decellularization medium is pumped from the reservoir through the bubble trap and perfused into the liver graft which is placed in a 10-cm petri dish. (**b**) Cell seeding perfusion set-up. A 15-cm petri dish is served as the seeding chamber. The decellularized liver graft (not shown in the figure) is placed in the petri dish and connected to the system for seeding. A peristaltic pump facilitates seeding media flow from the chamber through a bubble trap into the graft. Cells are injected through a three-way valve prior to the bubble trap and thus perfused in the graft. (**c**) Perfusion culture system. The recellularized liver graft is placed in a perfusion chamber which is composed of two silicon sheets sealed by a frame and metallic screws. A three-way valve is added respectively at the inlet and the outlet of the perfusion chamber. The culture medium is circulated in the system from the chamber to an oxygenator and a bubble trap back to the liver. A pump is employed between the chamber and the oxygenator to facilitate the flow

- Three-way stopcocks with luer connections (Cole-Parmer cat # EW-30600-02).
- 15-cm petri dish equipped with inflow and an outflow tubing (Fig. 1b).

In-Vitro Culture Perfusion Setup

- L/S® standard digital drive (Cole-Parmer Cat # EW-07523-80).
- Masterflex® standard pump head for L/S® 15 (Cole-Parmer Cat # EW-07015-21).
- Masterflex® platinum-cured silicone tubing, L/S® 14 (Cole-Parmer Cat # EW-96410-14).
- Oxygenator (Radnoti part# 130144).
- Bubble trap (Radnoti part# 130149).
- Perfusion chamber (*see* **Note 1**):
  - Two square polycarbonate frames (made to order).
  - Two silicone sheets (McMaster-Carr Cat # 87315K62).
  - Masterflex® platinum-cured silicone tubing, L/S® 14 (Cole-Parmer cat # EW-96410-14).
  - 20G catheter.
  - 1-way, 3-way/4-way stopcocks with luer connections (Cole-Parmer cat # EW-30600-01, # EW-30600-02, # EW-30600-03).

Syringes (5-cc, 30-cc)
95% Oxygen, 5% Carbon Dioxide Gas Tank

3 Methods

3.1 Hepatectomy

1. Ensure animals are maintained in accordance with National Research Council guidelines. Experimental protocols must be approved by relevant Committees on Research Animal Care.
2. Induce anesthesia in an induction chamber with isoflurane (*see* **Note 2**).
3. Put the animal on surgical table in supine position with continuous administration of isoflurane through nose cone (*see* **Note 3**).
4. Shave the animal's thorax and abdomen ventrally.
5. Place animal in supine position on surgical table, and secure four limbs firmly on the table with tape.

6. Disinfect the shaved area with 70% ethanol. Check the leg reflex by pinching the foot of the animal. If there's no reflex, proceed the surgery.
7. Make an incision at xiphoid process. Make large transverse abdominal incision with large scissors and rat tooth forceps.
8. Clamp the xiphoid process with straight hemostats, retract, exposing abdominal contents. Tape the hemostats' handle.
9. The falciform ligament will be visible immediately inferior and deep to the xiphoid process. Wet two cotton tips in the solution reservoir. With a wet cotton bud and small scissors, divide this ligament without severing the superior vena cava or the diaphragm.
10. Inject 20 U heparin through the diaphragm into the heart to prevent blood coagulation.
11. Cut the ligaments between each lobe (*see* **Note 4**).
12. Free inferior caudate lobe (ICL) by gently rolling up the stomach and making an incision to the lesser omentum along the junction of the lobe and the posterior side of stomach. Using wet cotton tipped swabs and a gentle rolling motion, move the lobe superiorly to the stomach through the hole produced by these incisions (*see* **Note 5**).
13. Reveal the portal vein (PV) by tucking the stomach around the ribcage to the right and shifting the spleen, and the intestines downwards to the right side out of the body of the animal.
14. Identify the PV and the bile duct. Put a damped gauze on the small intestine and pull lightly aside to provide downward tension on the PV.
15. Prepare two sutures around the PV for cannulation: The proximal position is prepared superior to a small branch extending from the right side of the PV. With a wet cotton bud in the left hand and a pair of curved forceps in the right, create a window on each side of the aforementioned areas (*see* **Note 6**). Prepare the distal suture by repeating the same procedure in a position inferior to the branch of the PV. Insert the curved forceps underlying the PV with a clockwise rolling motion. Feed the forceps with the 3.0 or 5.0 sutures and gently pull the sutures in a counterclockwise rolling motion through the proximal and distal windows that were just created. Tie a loose ligature with each.
16. Insert an 18-gauge catheter. Retract the needle and simultaneously advance the catheter. Advance the catheter until it lies between the two sutures (*see* **Note 7**).
17. Connect the catheter with the 30 ml syringe filled with PBS.

18. Tie the proximal suture first by tightening the knot and securing with two more. Repeat with the distal suture.
19. Cut the inferior vena cava (IVC). Perfuse the liver by slowly pushing 20 ml PBS to clear the blood within the liver (*see* **Note 8**).
20. Excise the liver from the body cavity by ligating the superior hepatic vena cava (SHVC), bile duct, and the distal part of the PV, and freeing the ligaments around the liver (*see* **Note 9**).
21. Remove the liver into the petri dish containing PBS, and flush the liver with 10 ml PBS that left in the syringe. Place the liver at −80 °C prior to decellularization.

3.2 Perfusion Decellularization

1. Thaw harvested rat livers before perfusion while setting up the perfusion system (Fig. 1a).
2. Fill the perfusion system with PBS and keep it running for 10 min. Fill a 10-cm petri dish with PBS.
3. Reduce the flow rate to 1.2 ml/min.
4. Carefully transfer the thawed liver to the PBS filled 10-cm petri dish.
5. Remove any air bubbles in the tubing and the catheters. Make sure there is no air inside the catheters (*see* **Note 10**).
6. Carefully connect the liver to the perfusion system through the portal vein catheter while PBS is flowing through the tubing.
7. Record date and time. Continue with PBS perfusion overnight.
8. Stop the flow and transfer the tubing from PBS containing reservoir to 0.01% SDS containing reservoir. Resume the flow and perfuse with 0.01% SDS for 5 min.
9. Stop the flow and transfer the tubing from 0.01% SDS containing reservoir to PBS containing reservoir. Resume the flow and perfuse with PBS for 1 h.
10. Repeat SDS (0.01%) and PBS perfusion steps (steps 9 and 10) three more times increasing SDS perfusion time to 10, 15, and 20 min.
11. Continue perfusion with 0.01% SDS for 24 h.
12. Start perfusion with 0.1% SDS for 24 h (*see* **Note 11**).
13. Perfuse with 0.2% SDS for 3 h. Go to step 16 if the effluent is clear and the liver is transparent (*see* **Note 12**).
14. Perfuse with 0.5% SDS for 3 h. Go to step 16 if the effluent is clear and the liver is transparent (*see* **Note 13**).
15. Perfuse with dH_2O for 15 min.
16. Perfuse with 1% Triton X-100 for 30 min.
17. Perfuse with PBS for 1–2 h.

18. Store in a clean and sealed petri dish soaked in PBS at 4 °C until ready to use (*see* **Note 14**).
19. Sterilization: (a) Using a 30-ml syringe, flush the decellularized liver with sterile PBS containing 0.1% (v/v) peracetic acid and 4% (v/v) ethanol and incubate for 3 h at 4 °C in a sealed petri dish. (b) Transfer the liver into a new petri dish that contain sterile PBS and flush the decellularized liver twice with sterile PBS. (c) Wash with sterile PBS containing 2% penicillin-streptomycin, 10 μg/ml gentamicin, and 2.5 μg/ml amphotericin B. Store the decellularized liver in the same solution at 4 °C until ready to use for recellularization experiments.

3.3 Characterization of Decellularized Liver Matrix

3.3.1 Computed Tomography (CT) Scanning of Decellularized Livers

1. Calibrate CT scanner before each application.
2. Place liver in a sterile ziplock bag, take measurement of liver's dimensions to ensure full scan.
3. Perform non-contrast CT scan through the decellularization process at 0 h, at the end of PBS/0.01% SDS alteration stage, 0.01% SDS perfusion stage, 0.1% SDS perfusion stage, and at the end of decellularization.
4. Use CereTom® CT scanner software to produce and construct Dicom 3 compliant images of the scans.
5. Determine Hounsfield units using AMIDE open source software.

3.3.2 Perfusate Collection (for Nondestructive Decellularization Analysis Method)

1. When setting up the decellularization, place a 600 ml or 1000 ml beaker under each liver graft ensuring perfusate is collected in the beaker.
2. Collect 50 or 100 ml of perfusate at appropriate time points for analysis.
3. Store at −80 °C until analysis is performed.

3.3.3 Immunoassay and Biochemical Assays

Whole liver weights prior to decellularization and after decellularization should be noted. Biopsies and perfusate samples for biochemical assays should be taken prior to starting decellularization [e.g., $t = 0$, 24, 48, 66 and 76 h (end)].

Total Protein Assay

1. Crush the tissue using the mortar and pestle set (*see* **Note 15**).
2. Once the tissue is crushed, add RIPA buffer and digest on a shaker at 4 °C for 2.5 h. Pipet the tissue digestion up and down several times after 2 h digestion.

3. Place the tissue digestion in centrifuge, spin at 3000 × *g* for 20 min and then 4000 × *g* for another 10 min.
4. Collected 1 ml of supernatant, ensuring no solid was left in the sample.
5. Proceed the Pierce BCA protein assay by following the manufacturer's recommendations.

Western Blot

1. After determining the protein concentration with Pierce BCA assay, dilute the samples to the same concentration in PBS.
2. Add 50 μl of β-mercaptoethanol to 950 μl Laemmli, then take diluted samples and add 1:1 ratio of Laemmli buffer.
3. Boil mixed sample in dH_2O for 5 min, centrifuge for 1 min at 13,000 × *g*.
4. Load sample into wells of the gel along with protein ladder.
5. Set-up the gel running system and run the samples on gel (*see* **Note 16**).
6. Transfer the protein bars from gel to film (*see* **Note 17**).
7. Incubate with primary antibody, followed by secondary antibody incubation (*see* **Note 18**).
8. Develop the film with Odyssey scanner (*see* **Note 19**).

DNA Assay (See ***Note 20****)*

1. Weigh 20 mg liver biopsy sample.
2. Follow the manufacturer's instructions of Purelink Genomic DNA kit for DNA extraction and purification.
3. Use Picogreen Assay kit to determine the extracted DNA concentration.

GAG Assay

The GAG content of the DLM is determined according to Farndale et al. [12] (*see* **Note 21**).

1. Cut DLM samples into 100 mg small pieces.
2. Lyophilize the DLM samples.
3. Hydrolyze each sample in 0.25–0.5 ml of 6 N HCl at 95 °C for 20 h.
4. Cool the samples to room temperature.
5. Mix 10 μl sample and 250 μl dimethylene blue and immediately measure the absorbance at 525 nm.

Collagen Assay

Quantify collagen content by applying Quickzyme Total Collagen assay kit. Follow the manufacturer's recommendations.

3.3.4 Corrosion Casting

1. During the surgery, prepare the graft with cannulation of the IVC, portal vein, and common bile duct.
2. After decellularization, follow the manufacturer's recommendations to proceed corrosion cast.
3. Inject 4 ml polymer mixture through the cannula.
4. Place the DLM in an ice bath for 2–3 h.
5. Macerate the tissue in 1 N KOH solution for 2 h at 50 °C or 24 h at room temperature.

3.3.5 Histological Analysis

1. Fix the DLM by immersing in 10% formalin for 24 h.
2. Transfer the DLM to 70% ethanol.
3. Proceed further histology steps by embedding the tissue in paraffin, sectioning, and staining with hematoxylin and eosin.

3.4 Perfusion Recellularization

1. Set-up the seeding perfusion system that is equipped with a peristaltic pump, bubble trap, and a perfusion chamber (Fig. 1b). Fill the perfusion system by pipetting 200 ml of seeding medium into the perfusion chamber.
2. Place the DLM into the perfusion chamber and connect it to the perfusion system through the portal vein cannula while the pump is running at 5 ml/min to avoid formation of any air bubbles. Allow the medium to perfuse the DLM for 30 min.
3. Stop the flow in the perfusion system and slowly inject 10–30 million hepatocytes through a 3-way stopcock situated before the bubble trap.
4. Restart the flow at 10 ml/min and recirculate the medium for 10 min.
5. Stop the flow of medium again, inject an additional 10–30 million hepatocytes. Resume the flow again to 10 ml/min. Repeat this procedure two more times until a total of 40–120 million hepatocytes have been introduced into the DLM.
6. Once addition of cells into the DLM is complete, cut the portal vein catheter at 45° angle to free the recellularized liver graft and transfer the graft to the in vitro perfusion chamber (*see* **Note 22**).
7. Aseptically, set-up sterile perfusion system (Fig. 1c).
8. Fill the perfusion system by pipetting 25 ml of culture media into the perfusion chamber. Start the pump at 10 ml/min and remove any air from the tubing.
9. Place the recellularized liver graft into the perfusion chamber and connect the DLM to the perfusion system through the portal vein cannula while the pump is running at 5 ml/min to avoid introducing any air bubbles into the recellularized liver graft.

10. Close the perfusion chamber and seal tightly to avoid any leakage during the culture.
11. Fill the chamber with additional 15–55 ml culture media using a sterile syringe through the 3-way/4-way stopcock till the total culture media reaches 40–80 ml.
12. Incubate the perfusion system at 37 °C with 10% CO_2 and keep the perfusion flow rate at 8–10 ml/min.
13. Connect the oxygenator to a 95% O_2 and 5% CO_2 gas mixture tank and set the gas flow rate to 0.5 l/min.
14. Continue the culture with daily changes of culture medium (*see* **Note 23**).

4 Notes

1. The in vitro perfusion chamber consists of two silicone sheets pressed together by plastic frames. The chamber has one inlet and one outlet port, each equipped with size 14 silicone tubing and 3-way/4-way stopcocks for medium change and sampling purposes. At the inlet, a 20G catheter that has been cut at a 45° angle at ½ in. length is placed to connect the recellularized graft.
2. Turn on the 100% O_2 tank and make sure the regulator is at 50–55 psi. Turn up the flow rate to 0.8–1 LPM. Turn on the vaporizer to 5. When the breathing of the animal is slow and regular, and there is no muscle tone, turn off the vaporizer and flush the chamber by pressing the oxygen flush valve button for 10 s to ensure a safe working environment.
3. Before placing the animal in the nose cone, open the valve to the nose cone circuit, close the valve to the chamber circuit, and turn on the vaporizer to 5 again.
4. With two wet cotton buds move the medium lobe and the left lateral lobe (LLL) up in a gentle rolling motion against the diaphragm. Cut the hepatogastric ligament between LLL and superior caudate lobe (SCL). Roll SCL up against LLL and cut the ligament posterior to the SCL.
5. The ICL lies posterior to the stomach. This lobe is nestled in the lesser curvature of the body of the stomach and inferior to the esophagogastric junction. It is anchored to the stomach and the spleen by the lesser omentum, which must be severed to free this lobe. Be careful not to sever any of the vasculature of the lesser omentum.
6. To create windows, use a pair of forceps making initial incisions and conduct an open-close motion along the vessel with the

forceps. When necessary, hold a cotton bud on the left to create a surface where the forceps can apply pressure on.

7. Grasp the 18G catheter in the right hand with thumb and middle finger. Create traction with left hand by gently pressing the most distal end of PV. Insert the cannula from as distal to the sutures as possible. Retract the needle by pushing the retraction button with index finger. Blood should flow out. The beveled edge of the catheter should lie directly superior to the proximal suture.
8. Keep the syringe at the same position by gently taping it.
9. Roll the anterior liver lobes back to the physiological position. Gently retract the liver away from the diaphragm and identify the connection with the SHVC. Sever the SHVC and continue to carefully sever all the ligaments that anchor the liver to the body and surrounding tissue. Cut the diaphragm. Cut the ligaments connecting the liver and the esophagus as well as the diaphragm. Grasp the bile duct and sever. Grasp the portal vein and cannula and cut away from the vascular system without dislodging from the liver. Using the cotton tips, lift the inferior right lobe (IRL) that lies directly superior to the IVC. Completely sever the IVC from it. Further lift to sever the ligaments that anchor the liver deep to this lobe.
10. If there's any bubble in the catheter, fill the catheter with PBS using a syringe and a needle.
11. Alternative: in the shortened protocol [13], instead of perfusion for 24 h, perfuse 4 h with 0.1% SDS.
12. Alternative: in the shortened protocol [13], instead of perfusion for 3 h, perfuse 1 h with 0.2% SDS.
13. Alternative: in the shortened protocol [13], instead of perfusion for 3 h, perfuse 1 h with 0.5% SDS.
14. Optional: depending on the experimental purpose and utility, ligate and resect all lobes except for the median lobe.
15. To crush the tissue: fill the reservoir underneath the mortar with enough liquid nitrogen. The darker area in the metal cup is the coolest point in the cup. Crush the tissue in this exact area to prevent tissue from sticking to the cup. Be careful with the first initial crush to prevent loss of tissue. Hold spatula and sample tube in the liquid nitrogen to ensure the low temperature. Scrape crushed tissue and tap into sample tube which is on dry ice. Rinse spatula, pestle, and metal cup with water, repeat for next sample.
16. Put voltage lid on, making sure to match black to black, red to red. Bubbles should be observed once the power supply is turned on. Fill entire tank with running buffer and make sure

buffer fills space between empty glass plates. Run at 200 V until marker reaches bottom of plate.

17. Make 1× transfer buffer and add methanol, place in −80 °C to chill before use. To set-up the transfer system: Remove gel plate with green spatula. Soak sponge in transfer buffer, put on black side of cassette. Soak filter paper: place on sponge then place gel on top. Soak membrane: cut to size first, soak, lay on top of gel. Soak another filter paper, place on membrane, and remove any bubbles. Soak another sponge. Close up cassette and place in chamber. Black side should face black side of chamber. Place chamber, add transfer buffer, magnetic stir bar, icepack, and then place the lid on top. Run for 90 min at 100 V.
18. Cut the film as needed. Indicate the reaction side. Wash the membrane with TBST for two times on orbital shaker. Block the membrane with blocking buffer for 1 h at room temperature on the shaker. Wash again with TBST after the blocking. Incubate with primary antibody (in this case, β-actin at 0.2 μg/ml) overnight at 4 °C on shaker. Wash three times with TBST after the primary antibody incubation. Apply secondary antibody at an appropriate concentration. Incubate at dark on shaker for 1 h. Wash the film with TBST for three times.
19. Wipe the surface of the scanner before reading. Lay the membrane facing down on the scanner. Determine the membrane size in the program per the ruler on the scanner. Start scanning.
20. The quantification of DNA, GAG, and collagen content should be normalized accordingly. In our study, we weighed the sample, the whole liver weight of pre- and post-decellularization. The results were normalized to wet tissue weight and corrected by the ratio of post- to pre-decellularization whole liver weights at each time point. The normalization calculation is as follows:

$$\text{Target content per mg initial wet weight} = \frac{\text{Target content concentration } (\mu g/\text{ml}) \times \text{Sample volume (ml)}}{\text{Tissue sample wet weight } (\mu g)} \times \frac{\text{Whole liver post weight (g)}}{\text{Whole liver pre weight (g)}}$$

21. Plot standard curve by applying serial dilution of 0.5 mg/ml chondroitin sulfate A in PBS. The preparation of standard samples is the same as the DLM samples.
22. Consequently, collect the perfusate into four 50-ml centrifuge tubes and centrifuge at 600 rpm for 10 min. Discard the supernatants and combine the pellets into a single tube. Determine

the number of cells and their viability via trypan blue exclusion to establish the seeding efficiency.

23. In ideal situation, the culture can be kept for up to 10 days. Sample the medium daily for measurement of metabolites in the perfusate such as albumin, urea, and total bile acid. At the end of the culture period, sample the recellularized liver graft for molecular and histological analysis.

References

1. Badylak SF, Taylor D, Uygun K (2011) Whole-organ tissue engineering: decellularization and recellularization of three-dimensional matrix scaffolds. Annu Rev Biomed Eng 13:27–53
2. Gilbert TW, Sellaro TL, Badylak SF (2006) Decellularization of tissues and organs. Biomaterials 27:3675–3683
3. Macchiarini P, Jungebluth P, Go T et al (2008) Clinical transplantation of a tissue-engineered airway. Lancet 372:2023–2030
4. Ott HC, Matthiesen TS, Goh S-K et al (2008) Perfusion-decellularized matrix: using nature's platform to engineer a bioartificial heart. Nat Med 14:213–221
5. Petersen TH, Calle EA, Zhao L et al (2010) Tissue-engineered lungs for in vivo implantation. Science 329:538–541
6. Ott HC, Clippinger B, Conrad C et al (2010) Regeneration and orthotopic transplantation of a bioartificial lung. Nat Med 16:927–933
7. Sullivan DC, Mirmalek-Sani S-H, Deegan DB et al (2012) Decellularization methods of porcine kidneys for whole organ engineering using a high-throughput system. Biomaterials 33:7756–7764
8. Bolland F, Korossis S, Wilshaw S-P et al (2007) Development and characterisation of a full-thickness acellular porcine bladder matrix for tissue engineering. Biomaterials 28:1061–1070
9. Uygun BE, Soto-Gutierrez A, Yagi H et al (2010) Organ reengineering through development of a transplantable recellularized liver graft using decellularized liver matrix. Nat Med 16:814–820
10. Uygun BE, Izamis M-L, Jaramillo M et al (2016) Discarded livers find a new life: engineered liver grafts using hepatocytes recovered from marginal livers. Artif Organs 13:27
11. Dunn JC, Tompkins RG, Yarmush ML (1991) Long-term in vitro function of adult hepatocytes in a collagen sandwich configuration. Biotechnol Prog 7:237–245
12. Farndale RW, Buttle DJ, Barrett AJ (1986) Improved quantitation and discrimination of sulphated glycosaminoglycans by use of dimethylmethylene blue. Biochim Biophys Acta 883:173–177
13. Geerts S, Ozer S, Jaramillo M et al (2016) Nondestructive methods for monitoring cell removal during rat liver decellularization. Tissue Eng Part C Meth 22:671–678

Methods in Molecular Biology (2018) 1577: 271–281
DOI 10.1007/7651_2017_63

Published online: 15 August 2017

Decellularization of Liver and Organogenesis in Rats

Nana Shirakigawa and Hiroyuki Ijima

Abstract

Recently, organ construction has been attempted using decellularized organs. In this study, we used decellularized rat liver to construct liver tissue by recellularization. The right lobe of the rat liver was decellularized with 4% Triton X-100 solution, recellularized with 10^7 rat hepatocytes, and albumin synthesis in the recellularized right lobe was observed. Therefore, we introduce a method of decellularizing rat liver, which retains its fine vascular structure after removal of all the cells, perform organogenesis using the decellularized liver, and evaluate the structural and functional properties of the products.

Keywords Decellularization, Liver construction, Organ engineering, Organogenesis, Recellularization, Tissue engineering, Vascular structure

1 Introduction

Recently, organ construction has been studied using decellularized organs [1]. As the decellularization process differs between organ types and species, a suitable decellularization method should be developed. We attempted to construct a liver based on using decellularized liver to resolve the donor shortage for liver transplantation. Some methods have been reported for liver decellularization [1–3], but suitable decellularization methods may differ at the recellularization stage (see **Note 9**). Here, we introduce a decellularization and recellularization method for rat liver construction. We developed a decellularization method in which the decellularized liver retained its fine vascular structure even when all the cells were removed (Figs. 1 and 2) [4]. Next, we developed a recellularization method that reconstructs blood vessels with the aid of the vascular structure of the decellularized liver (Fig. 3) [5]. Liver cells were seeded from sides of the decellularized liver using a needle attached to a syringe containing collagen sol suspension, which formed a gel after injection. In this chapter, we describe these methods in detail.

Rat liver consists of a left lobe, middle lobe, caudate lobe, and right lobe and can regenerate if more than 30% of its original volume is retained in both rats and humans [6, 7]. Therefore,

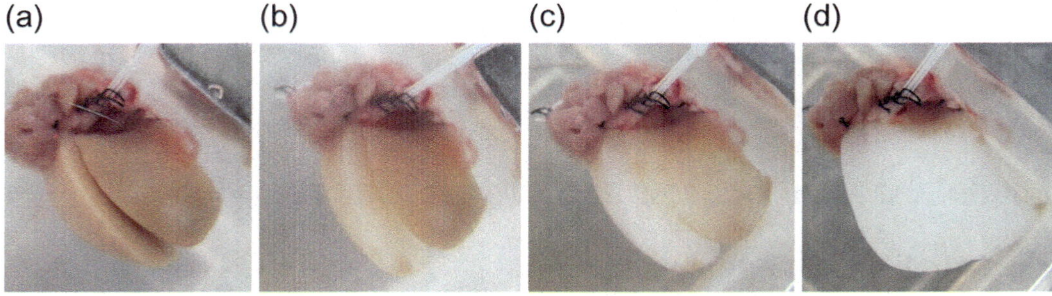

Fig. 1 Appearance of the right lobe of the rat liver during decellularization. Immediately after starting the flow of detergent solution (**a**), a few minutes later (**b**, **c**) and after flow is complete (**d**)

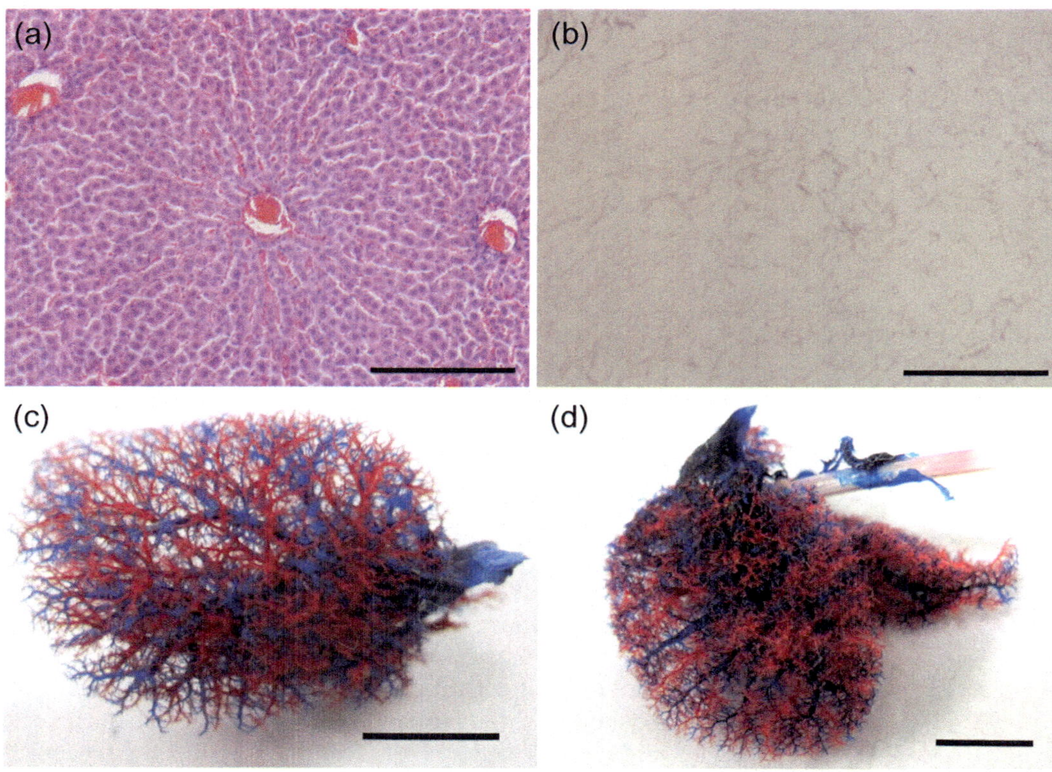

Fig. 2 Histological analysis by hematoxylin and eosin staining (**a**, **b**) and molded resin of vascular structure (**c**, **d**). Native liver (**a**, **c**) and decellularized liver (**b**, **d**). Scale bars indicate 100 μm (**a**, **b**) and 5 mm (**c**, **d**), respectively (reproduced from ref. [5] with permission from Elsevier)

although the whole liver was used in the primary study, only the right lobe, which accounts for 24% of the whole liver volume [8], can be used; this size is better for heterotopic transplantation.

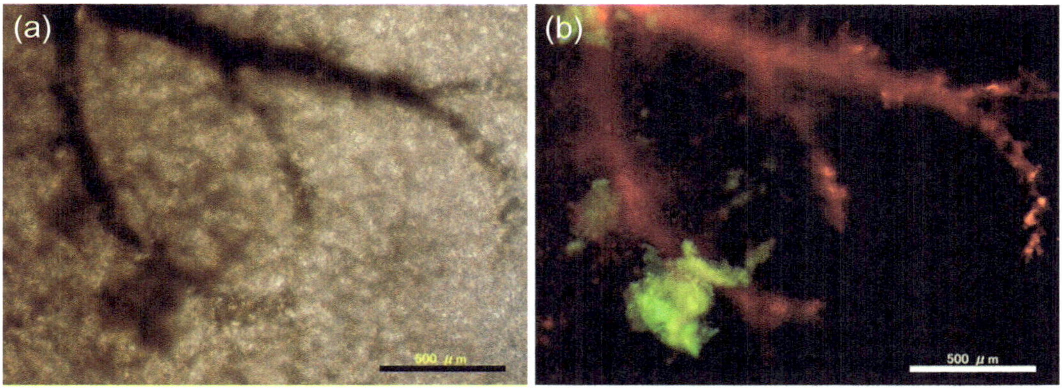

Fig. 3 Pase-contrast (**a**) and fluorescence (**b**) micrograph of the edge of recellularized liver. HUVECs are *red* and HepG2 cells are *green*. Scale bars indicate 500 μm (reproduced from ref. [5] with permission from Elsevier)

2 Materials

Basic dissecting instruments for small animals, such as scissors and forceps, are needed. These instruments are sterilized by autoclaving or spraying with 70% ethanol and drying.

2.1 Solutions

1. Calcium-magnesium-free phosphate-buffered saline (CMF-PBS). (CMF-PBS is autoclaved before use. If this solution is used without a clean bench for decellularization, add 100 mg/L streptomycin and 60 mg/L penicillin.)
2. 0.19 g/L GEDTA in 150 mL CMF-PBS (solution for removing blood).
3. 4% Triton X-100 in CMF-PBS (solution for decellularization).
4. 0.5 mg/mL DNase and 0.5 mg/mL RNase in CMF-PBS (see **Note 1**).
5. 5% Glutaraldehyde solution (product of 25% glutaraldehyde is diluted fivefold with CMF-PBS).
6. Culture medium for seeded cells.

2.2 Steps for Removing Blood from the Liver and Culture of Recellularized Liver

1. A 1.2-m silicon tube (inner diameter 3 mm, outer diameter 5 mm) and 2-m silicon tube (inner diameter 1 mm, outer diameter 2 mm) are prepared. The 1-mL syringe cut at 0.25 mL is attached to the tip of each tube. The wider tube is used to remove the blood and the decellularization circuit. The narrow tube is used for DNase treatment and circulation culture of recellularized liver.
2. A roller pump is used for the feeding solution.

3. The catheter of 18G Insyte (BD, Franklin Lakes, NJ, USA) was used to connect the syringe attached to the tube and the portal vein. Otherwise, the tube for extending the blood circuit (Kawasumi Laboratories, Inc., Tokyo, Japan) is cut to 1 cm and attached to an 18G needle on which the tip is cut and filed. This tube was also used for the connection.
4. An air vent can be added before the liver to prevent the flow of air into the liver (see **Note 4**).
5. A sampling port can be included during culture, if needed. The air vent may also function as a sampling port.
6. The case for the harvested rat right lobe for decellularization: 4 × 5 × 5 cm box composed of an acrylic board with a 1-cm depth slit on the vertical sides.
7. The case for culture of recellularized rat right lobe: size depends on the purpose of the culture.

3 Methods

3.1 Obtaining Decellularized Liver

All procedures are performed at room temperature (approximately 25 °C) in an experiment room unless specified.

1. A 6–8-week-old male rat is anesthetized (see **Note 2**) and laid on its back, and if possible, the hands and legs are anchored to the bed. For sterilization, 70% ethanol is sprayed onto the rat, which can prevent its hair from moving as well. First, the skin is cut over the median line from the lower abdomen to the xiphoid process. Second, the muscle layer is also cut along the same line. During this step, the liver is peeled off of the muscle layer with other attached organs to prevent the liver from being injured. The xiphoid process should not be cut because if it is, it may stick to the liver. The muscle layer is cut over the xiphoid process, and the ligament around the xiphoid process is cut from the side. Third, the skin and muscle layer are cut from the lower abdomen to each side (in a reverse T) and opened to create a wide viewing angle.
2. The bowels are moved to the right side with a cotton swab and the portal vein is exposed. Two sutures are brought under the portal vein using forceps (see **Note 3**).
3. (a) Attaching a catheter to the portal vein.

   A small amount of solution is injected into the portal vein through a catheter to remove the blood. The only external cylinder is pushed into the portal vein up to near side of the junction of the portal vein (Fig. 4). Next, the external cylinder and portal vein are bound by banding the preset sutures. The needle is removed, and the tip of the tube is reconnected to the external cylinder. Removal of the needle causes the blood to flow backwards if the external cylinder is fixed in the portal vein. When reconnecting the tube, use caution to prevent air from flowing into the liver (see **Note 4**).

(b) Attaching tube to portal vein.

The portal vein is cut in half and the cut tube is carefully injected (Fig. 5) to prevent air from flowing into the blood vessel. This step may be easier if performed after blood flows out from the inferior vena cava or hepatic vein.

4. The solution for removing blood is introduced via the fixed tube (see **Note 5**).
5. The heart and hepatic vein are exposed by cutting around the xiphoid process to the rib. The heart is cut two or three times to outflow the solution to remove the blood. Next, the bone and skin are cut along the rib bone and the chest is opened. The solution can be overflowed from the right chest side.
6. If necessary, the hepatic vein is cut into half and the tube is inserted and fixed (see **Note 6**).
7. The joint of the left lobe is tied in two parts, and the left lobe is cut off. The joint of the middle lobe is tied and cut off of the middle lobe. One caudate lobe is positioned on the dorsal side of the stomach. The ligaments between the caudate lobe and stomach must be cut, and the caudate lobe is moved to a position above the stomach with a cotton swab. Next, the joint of the caudate lobe is tied, and the caudate lobe is cut. Additionally, all blood vessels connected to the right lobe are tied and cut except for the portal vein and hepatic vein. The common bile duct is also tied and cut. During cutting, the tied suture should remain on the side of the right lobe. The position of each blood vessel can be studied in the orthotopic liver transplantation procedure [9] (see **Note 7**).
8. The right lobe of the liver is resected from the rat's body and moved in a circuit for decellularization (see **Note 8**), being careful to prevent air flow into the circuit. Before moving the liver to the decellularization circuit, the case used to store the harvested right lobe is filled with CMF-PBS. Next, 500 mL of 4% Triton X-100 solution (see **Note 9**) is flowed into the liver for 3 h for decellularization (see **Note 10**).
9. CMF-PBS is flowed for 30 min to wash out the decellularization solution.
10. The pump and tube are adjusted to a low flow rate and DNase and RNase solution is circulated at 0.5 mL/min and 37 °C for 6 h.

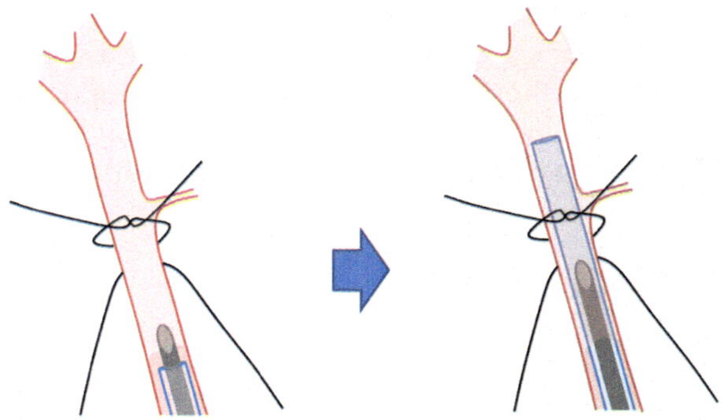

Fig. 4 Catheter fixed into the portal vein. The position of the tip of the catheter is important

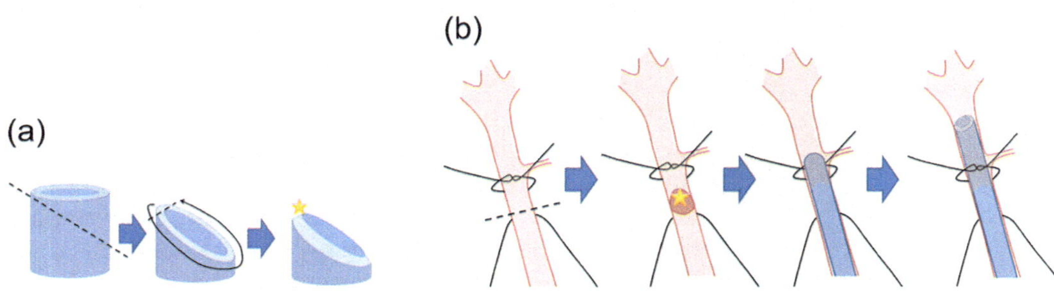

Fig. 5 Process of cutting the tip of the tube (**a**) and inserting the tube into the portal vein (**b**). Star position of (**a**) is matched to the star position of (**b**) when inserted

11. CMF-PBS is introduced inside for washing, and cross-linking is performed by flowing 5% glutaraldehyde solution for 30 min to decrease antigenicity and increase strength [10, 11].
12. CMF-PBS is introduced inside for washing, and the decellularized right lobe of the rat liver is obtained.

3.2 Evaluation of Decellularized Liver

The decellularized liver should be evaluated at three points: (1) complete cell removal, (2) retained original three-dimensional structure, and (3) retained original extracellular-matrix components. Here, we conduct typical evaluations.

1. Complete cell removal.

The obtained decellularized liver is embedded in paraffin. Histological evaluation of paraffin sections by methods such as hematoxylin and eosin staining is performed.

The DNA in the decellularized liver is extracted using the QuickGene SP kit DNA tissue (SP-DT) (Kurabo Industries Ltd, Osaka, Japan) and quantified using the Fluorescent DNA Quantitation Kit (Bio-Rad, Hercules, CA, USA) (see **Note 11**).

2. Retained original three-dimensional structure (see **Note 12**).

Resin (0.4 mL) is injected via the portal vein and hepatic vein and incubated overnight for hardening. Next, the surrounding tissue is removed using 33% KOH solution (see **Notes 13** and **14**) to obtain the resin casting of the vascular structure.

3. Retained original extracellular-matrix components.

The decellularized liver is solubilized using an enzyme such as protein kinase K. Glycosaminoglycan in the solubilized solution can be quantified using the quantitation kit.

A frozen section of decellularized liver is prepared and immunostained to observe components such as collagen I, collagen IV, fibronectin, and laminin.

3.3 Recellularization of Decellularized Liver (See Note 16)

1. The obtained decellularized liver is sterilized by flowing 70% ethanol via the portal vein for 30 min (see **Note 17**).
2. The sterilized decellularized liver is shifted to a clean bench from the experiment room. Next, the decellularized liver is connected to the sterilized circuit; sterilization can be conducted using an autoclave. CMF-PBS is flowed through the portal vein to remove the ethanol. Next, the culture medium is introduced through the portal vein into the decellularized liver to replace PBS.
3. Next, 1×10^7 rat hepatocytes (see **Note 18**) suspended in 1 mL collagen sol are injected using a 26G needle attached to a syringe into the decellularized liver at 20 different places. The liver is incubated for 30 min at 37 °C, and the injected sol forms a gel (see **Note 19**).
4. A suspension containing 2×10^7 human umbilical vein endothelial cells (HUVECs) in culture medium is seeded through the portal vein, and the liver is incubated for 6 h to allow for attachment of HUVECs.
5. Circulation culture is initiated.

3.4 Evaluation of Recellularized Liver

Both the structure and functions of the recellularized liver should be evaluated. The structure can be evaluated by histological analysis or immunostaining of the recellularized liver. Cells stained with any cell tracker can be observed easily. Otherwise, the distribution of blood vessels and liver cells can be determined by multi-chromatic immunostaining of frozen sections.

Additionally, protein synthetic ability, drug metabolism, and ammonia metabolism are typically analyzed to determine liver function. The evaluations and functionality levels differ depending on the purpose of constructing the recellularized liver. These purposes are roughly classified into two categories: construction of an in vitro liver model and patient treatment for recovery from a disease. Liver function analysis during circulation culture can be used for recellularized liver as an in vivo model of the liver. For use in patient treatment, additional analysis for the ability of extracorporeal blood circulation to rats with liver failure or for transplantation treatment is needed (see **Note 20**). Here, as the most basic and simple analysis of liver function during circulation culture, the albumin synthetic ability is introduced as an index of protein synthetic ability.

1. 100–200 μL culture medium is collected at different time points during circulation culture. The collected samples are stored in the freezer.
2. Collected samples are melted and analyzed using an enzyme-linked immunosorbent assay.

4 Notes

1. Each enzyme is dissolved in deionized water at 100 mg/mL and stored at 1 mL/tube in a freezer. Before use, the enzymes are melted and added to 18 mL CMF-PBS.
2. Although anesthetization can be conducted by inhalation or intraperitoneal injection, liver injury must be prevented.
3. Use caution to avoid injuring the liver and blood vessels with the forceps.
4. Avoid air flow into the liver during all procedures. If there is flow of air into any part of the liver, the solutions used for blood removal, decellularization, or the cell suspension cannot be introduced from the point of air contamination to the tip of the liver. Therefore, the contamination with air affects decellularization and recellularization. Air that has entered the liver cannot be removed, and an air vent can be used before the liver in the circuit, where necessary.
5. Flow the solution into the entire liver immediately to avoid blood clotting in the liver or tube connection.
6. Use caution while inserting the tube to the proper depth. When it is unsuitable, the solution from the portal vein cannot overflow from the hepatic vein. Additionally, the diaphragm must be cut off as much as possible while avoiding injury to the vein as the remaining diaphragm affects the linearity hepatic vein and connected tube.
7. Here, only the portal vein and hepatic vein are retained. The hepatic artery, common bile duct, and vena cava inferior connecting the right lobe are ligated. However, to construct a transplantable graft, these elements should be retained.

8. The solution used for decellularization will affect the proteins on the liver surface. The overflowed solution should be removed from the decellularized liver as soon as possible. It contains large amounts of cell debris and should be discarded, particularly after the first stage of decellularization.
9. The gel, which is thought to contain DNA, can be analyzed if a strong solution is used for decellularization such as 1% sodium dodecyl sulfate. Therefore, the decellularization method should be optimized for the species and target organ. Various detergents and enzyme combinations such as DNase and RNase can be optimized. The remaining three-dimensional structure should be optimized for the overall recellularization process. To construct fine blood vessels using the vascular structure of the decellularized liver, the vascular structure should be maintained in the decellularized liver. Optimization should be performed by evaluating the remaining fine structure after removal of the cells [4]. An imperfect vascular structure wall allows the movement of cells from the structure into the decellularized liver [5, 12–15].
10. If some parts are not decellularized completely, the liver can be turned by turning the connected tube. In the body, there is abdominal pressure and support by other organs. Thus, blood vessel linearity is suitable and blood can flow to every part of the liver. The condition of the resected liver is very different from that of the liver in the body. Therefore, these conditions must be controlled in vitro.
11. The decellullarized liver can be solubilized by using protein kinase K. Next, DNA can be quantified using a kit with Hoechst, and any leftover detergent will react with Hoechst. The decellularized liver should then be washed thoroughly before analysis. Additionally, DNA content should be carefully normalized. Although some reports describe normalization by dry weight, the dry weight will be different before and after decellularization [1].
12. Although computerized tomography can be used to easily observe the vascular structure in decellularized liver by flowing contrast medium through the portal vein, the contrast medium will disperse from the vascular structure wall. Decellularized liver must be kept in the solution to maintain the three-dimensional structure. Observation by computed tomography may be difficult for decellularized liver covered with solution.
13. The resin may shrink during hardening and should be cooled with ice to prevent this. When the surrounding tissue is removed with KOH solution, the solution is poured carefully and should cover the resin to avoid crushing the fine branches during melting of the surrounding tissue.

14. The molded resins are coated with contrast medium or gold and observed by three-dimensional computed tomography for quantitative analysis [16].
15. Detergent absorption to the extracellular matrix is an important problem in this research field. How detergents affect decellularized liver intended for clinical use requires further investigation.
16. The bile duct may also be reconstructed by recellularization of the decellularized bile duct.
17. Sterilizing solution may denature the extracellular matrix of the decellularized liver. It may be better to perform all procedures under aseptic conditions without the sterilizing procedures.
18. Here, rat hepatocytes are used. However, to construct a miniature human liver, HepG2 cells can be useful for recellularization. The characteristics of cells such as the size or growth ability differ by cell type. The seeding of cell types and number of cells seeded should be optimized.
19. Methods of cell seeding show some patterns [5]. Cells were seeded through the portal vein in some reports. The cells can be seeded at different times, and the time of seeding requires optimization. Additionally, the flow rate of circulation culture must be optimized.
20. The strength of the recellularized liver must be increased to endure abdominal pressure. Development of a suturing method between recellularized liver and the recipient's blood vessels is needed for transplantation.

References

1. Shirakigawa N, Ijima H (2017) Decellularized tissue engineering. In: Tripathi A, Melo JS (eds) Advances in biomaterials for biomedical applications, advanced structured materials 66. Springer Nature, Berlin, Germany
2. Crapo PM, Gilbert TW, Badylak SF (2011) An overview of tissue and whole organ decellularization processes. Biomaterials 32:3233–3243
3. Caralt M, Velasco E, Lanas A et al (2014) Liver bioengineering: from the stage of liver decellularized matrix to the multiple cellular actors and bioreactor special effects. Organogenesis 10:250–259
4. Shirakigawa N, Ijima H, Takei T (2012) Decellularized liver as a practical scaffold with a vascular network template for liver tissue engineering. J Biosci Bioeng 114:546–551
5. Shirakigawa N, Takei T, Ijima H (2013) Base structure consisting of an endothelialized vascular-tree network and hepatocytes for whole liver engineering. J Biosci Bioeng 116:740–745
6. Meier M, Andersen KJ, Knudsen AR et al (2016) Liver regeneration is dependent on the extent of hepatectomy. J Surg Res 205:76–84
7. Helling TS (2006) Liver failure following partial hepatectomy. HPB (Oxford) 8:165–174
8. Suh KS, Lilja H, Kamohara Y et al (1999) Bioartificial liver treatment in rats with fulminant hepatic failure: effect on DNA-binding activity of liver-enriched and growth-associated transcription factors. J Surg Res 85:243–250
9. Oldani G, Lacotte S, Morel P et al (2012) Orthotopic liver transplantation in rats. J Vis Exp 65:e4143
10. Wang F, Guan X, Wu T et al (2016) Acellular endocardium as a novel biomaterial for the intima of tissue-engineered small-caliber vascular grafts. Artif Organs 40:E253–E265

11. Mathapati S, Bishi DK, Guhathakurta S et al (2013) Biomimetic acellular detoxified glutaraldehyde cross-linked bovine pericardium for tissue engineering. Mater Sci Eng C Mater Biol Appl 33:1561–1572
12. Uygun BE, Soto-Gutierrez A, Yagi H et al (2010) Organ reengineering through development of a transplantable recellularized liver graft using decellularized liver matrix. Nat Med 16:814–820
13. Baptista PM, Siddiqui MM, Lozier G et al (2011) The use of whole organ decellularization for the generation of a vascularized liver organoid. Hepatology 53:604–617
14. Zhou P, Lessa N, Estrada D et al (2011) Decellularized liver matrix as a carrier for the transplantation of human fetal and primary hepatocytes in mice. Liver Transpl 17:418–427
15. Barakat O, Abbasi S, Rodriguez G et al (2012) Use of decellularized porcine liver for engineering humanized liver organ. J Surg Res 173: e11–e15
16. Shirakigawa N, Hara Y, Sakamoto H et al (2016) Tissue-engineered bioreactors with flow channels molded by polypod particles. Adv Biomed Eng 5:105–110

Methods in Molecular Biology (2018) 1577: 283–292
DOI 10.1007/7651_2017_97

Published online: 04 November 2017

Simple and Quick Method to Obtain a Decellularized, Functional Liver Bioscaffold

Matteo Ghiringhelli, Alessandro Zenobi, Stefano Brizzola, Fulvio Gandolfi, Valentino Bontempo, Sandro Rossi, Tiziana A.L. Brevini, and Fabio Acocella

Abstract

The development of new approaches for organ transplantation has become crucial in the last years. In particular, organ engineering, involving the preparation of acellular matrices that provide a natural habitat for reseeding with an appropriate population of cells, is an attractive although technically demanding approach. We here describe a method that allows for the derivation of functional in vitro hepatic organoids and that does not require a previous selection of all the parenchymal hepatocytes and non-parenchymal cells, namely, Kupffer cells, liver endothelial cells, and hepatic stellate cells. The procedure also replaces the costly standard collagenase perfusion step with a trypsin-based enzymatic digestion that results in high-yield decellularization. A combination of physical and chemical treatments through deep immersion and intra-luminal infusion of two different consecutive solutions is used: (1) deionized water (DI) and (2) DI + Triton X 1% + ammonium hydroxide (NH_4OH) 0.1%. This ensures the isolation of the hepatic constructs that reliably maintain original architecture and ECM components while completely removing cellular DNA and RNA. The procedure is fast, simple, and cheap and warrants an optimal organoid functionality that may find applications in both toxicological and transplantation studies.

Keywords Hepatocyte, Scaffold, Decellularization, Organoid, Liver bioengineering

1 Introduction

The treatment of end-stage liver diseases sees transplantation as the definitive therapy. However, poor availability of donor organs causes a significant increase in the rate of morbidity and mortality, with 1 million deaths in 2010 for liver cirrhosis all over the world [1]. Organ transplantation also involves problems related to rejection, organ damage, and neoplasms caused by immunosuppressive therapy. In this context, tissue engineering may represent a valuable alternative to rebuild entire organs in vitro that may be implanted in the patient. The approach involves the removal of the cellular component or tissues from allogeneic or xenogenic donors to obtain three-dimensional biological scaffolds, constituted by the extracellular matrix (ECM). The advantage is remarkable since it offers the possibility of reconstructing genetically compatible organs in vitro that may subsequently be cultured with cellular populations taken from the patient [2]. The type of matrix generated contains the optimal structure and components to host primary cells, providing an ideal environment for surface cell receptors [3]. Many reports in the literature describe the regeneration of in vitro organs, and numerous decellularization protocols have been developed. However, most of the researches are limited to the rat, which is a relatively meaningful model for regenerative medicine. Furthermore, many protocols involve costly or time-consuming procedures with detergents and chemical or enzymatic agents, with specific methods that include agitation in solution, thermal shock (i.e., freeze-thaw cycles), ultrasound, hydrostatic pressure, convective flow, and manual disruption [4]. Very promising and consistent techniques that can be applied to any cadaveric solid organ have been reported [5, 6]. However, there exists a great variability in currently used perfusion-decellularization protocols, as most of them have been developed from anecdotal experience. Different types and combinations of decellularizing agents (e.g., detergents, osmotically active solutions, nucleases), different exposure times, as well as varying perfusion pressure have been described, which, in most cases, make direct comparisons between methods difficult [7–9].

We here describe a new simple, fast, and cheap strategy that allows the decellularization and consequent recellularization to obtain a functional 3D bioengineering liver. ECM is obtained with a combined application of physical, chemical, and enzymatic methods and provides the optimal homing environment for the engrafting of different hepatic cells types. The scaffolds generated with this methodology show a complete decellularization with preservation of micro-architecture, including the basement membranes and vascular network and a low amount of persisting DNA and RNA. The isolated ECM can be repopulated with a single cell-type or a co-culture population that may include immortalized cell lines or progenitors and displays optimal homing results. Cells used for scaffold repopulation are isolated with an enzymatic and mechanic method consisting in trypsinization of the liver specimens, in order to avoid the long and expensive collagenase P perfusion digestion solution. They colonize the entire ECM in three-dimensional structures, exhibit self-organization ability, and produce albumin and factor VII. The whole protocol is developed in the rabbit which represents one of the key animal model species for liver regenerative medicine [10].

2 Materials

2.1 Animals and Liver Explantation

1. Two adult New England white rabbit not eviscerated carcasses collected from a local slaughterhouse.
2. Topical skin chlorhexidine for surgical scrub with sterile gauzes (*see* **note 1**).
3. Sterile surgical instruments composed of four curved hemostat clamps, #3 handle with #5 blade, one straight Mayo scissor, one regular Metzenbaum scissor, one Potts scissor, and one DeBakey forceps (*see* **note 2**).
4. Cotton pad.
5. Sheath introducer 8.5 Fr (*see* **note 3**).
6. 3-0 silk sutures.
7. 2 ml sodium heparin (5000 U/ml).
8. 50 ml Dulbecco's phosphate-buffered saline (PBS): 2.7 mM KCl, 1.5 mM KH_2PO_4, 136.9 mM NaCl, and 8.9 mM Na_2HPO_4 (anhydrous).

2.2 Liver Decellularization Circuit

1. Gear pump (ISMATEC 405 MCP-Z Standard).
2. Disposable Luer lock infusion set.
3. High-density polyethylene basin (30 × 30 × 10 cm).
4. Polyethylene net.
5. Deionized water.
6. 1% Triton X-100.
7. Ammonium hydroxide (NH_4OH).
8. Peracetic acid 20%.
9. NaCl 0.9% solution.

2.3 Cell Isolation and Cell Culture

1. Hepatocyte culture medium: 77% (v/v) Dulbecco's modified Eagle medium/F12 (DMEM) high glucose, 20% (v/v) fetal bovine serum (FBS), 1% (v/v) L-glutamine solution, 2% (v/v) antibiotic-antimycotic solution, 0.0002% hepatocyte growth factor (HGF) (*see* **note 4**).
2. #3 handle with #15 blade.
3. 100 mm Petri dish.
4. Falcon 50 ml conical centrifuge tubes.
5. Trypsin-EDTA solution (1×): 0.5 g/l porcine trypsin, 0.2 g/l EDTA·4Na in Hank's Balanced Salt Solution with phenol red.
6. Cell strainer 100 μm.
7. T-75 flask.

8. Porcine gelatin 0.1%: dissolve 0.1 g of porcine gelatin in 100 ml of water. Sterilize solution with autoclave.
9. Laboratory centrifuge.
10. Counting chamber.
11. Cell culture incubator (Sanyo MCO-18AIC).

3 Methods

All the procedures described below must be performed following the normal surgical antisepsis rules. Every instrumentation touching or in connection to the liver has to be sterilized. All the cell isolation and recellularization steps must be achieved under laminar flow hood in sterile conditions. Make sure that all culture procedures are carried out on thermostatically controlled stages and the recellularized scaffolds are maintained at 37 °C throughout their handling.

3.1 Liver Explantation

1. Collect two adult female New England white rabbit 3 kg not eviscerated carcasses from a local slaughterhouse, carrying it with a certified ice container.
2. Before the incision, treat the abdominal wall muscles with topical chlorhexidine skin disinfectant alternating with chlorhexidine-soaked gauze and dry gauze.
3. Perform a ventral midline laparotomy extending from the pubis to the xyphoid combined with a rooftop incision using the #3 handle with #15 blade (*see* **note 5**).
4. Dissect the falciform ligament with a Metzenbaum scissor. Use a wet gauze bandage to hold the medial and left liver lobes cranially under the dome of the diaphragm.
5. Mobilize the upper omental liver lobe and then move the stomach to the left.
6. Fix the stomach with a clamp, and dissect the minor omentum with Metzenbaum scissors and DeBakey forceps.
7. Dissect the bile duct from the adipose and pancreatic tissue.
8. Cut the bile duct 1.5 cm from the bile duct bifurcation.
9. Dissect the portal vein from the surrounding tissue.
10. Dissect the common hepatic artery from the surrounding tissue.
11. Free the hepatic artery from the surrounding tissue, and tie it twice with silk 3-0, with the ties distant from each other.
12. Dissect the artery between the two ligatures.
13. Circumscribe the inferior vena cava between the right kidney and the right lateral liver lobe.

14. Dissect the vein from the retroperitoneal space, and inject 1000 I.E. heparin in 1 ml 0.9% saline solution into the inferior vena cava. Retract the cannula and close the incision with a cotton pad. Wait 1–2 min for the effect of heparin.
15. Ligate the distal end of the portal vein, and then cannulate the vein with a sheath introducer 8.5 Fr cutting the vein near the knots (*see* **note 6**) with a Potts scissor and a vein pick catheter introducer (*see* **note 7**).
16. Fix the decellularization cannula with 3-0 silk sutures.
17. Inject a total of 2 ml heparin sodium (100 U/ml) through the vein to prevent coagulation.
18. Transect infrahepatic inferior vena cava to allow the outflow of the perfusate.
19. Perfuse a total of 50 ml phosphate-buffered saline (PBS) slowly through the PV clear blood from the liver.
20. Free the suprahepatic inferior vena cava, the hepatic artery, and the common bile duct.
21. Isolate and transfer the whole liver to the decellularization circuit (*see* Fig. 2).
22. To collect the liver for the hepatocyte isolation protocol, cut the organ with Mayo scissor at the level of the grove liver for vena cava and its peritoneal ligaments.
23. Complete the liver isolation cutting the common bile duct, hepatic artery, and portal vein.
24. Transfer the whole liver to a 100 mm Petri dish.

3.2 Liver Decellularization

1. 1% Triton X-100: add 10 ml to a 1000 ml bottle. Add 990 ml deionized water and shake the bottle twice. Filter the 1% Triton X-100 solution through a sterile filter.
2. Pre-inject through the portal vein 400 ml NaCl 0.9% and 800 UI heparin (*see* Fig. 1).
3. Prime the circuit with the deionized water.
4. Place the liver in the high-density polyethylene basin.
5. Attach the liver to the pump (ISMATEC 405 MCP-Z Standard) (*see* Fig. 2).
6. Place the polyethylene net over the liver to fix the organ (*see* **note 8**).
7. Perform the decellularization steps as follows: deionize (DI) water for 24 h with a flow rate of 7 ml/min; Triton X 1% + ammonium hydroxide (NH_4OH) 0.1% for 48 h with a flow rate of 10 ml/min; DI for 24 h 10 ml/min; Triton X 1% + ammonium hydroxide (NH_4OH) 0.1% for 24 h; DI for 24 h 10 ml/min.

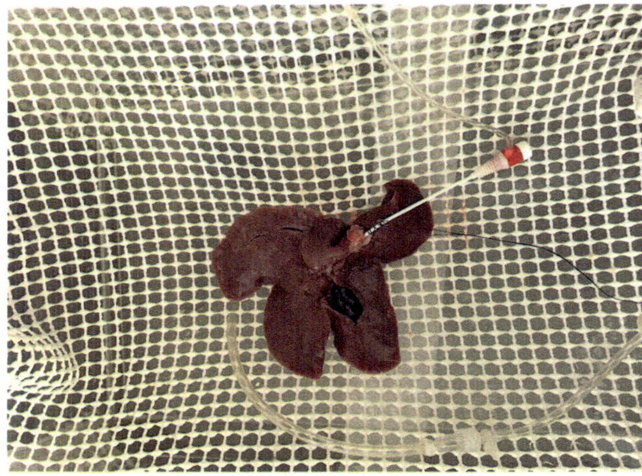

Fig. 1 The liver is washed with a sterile solution containing NaCl and heparin to clean and destroy all the little clots present in the vascular net

8. Disinfect the scaffold by perfusing with peracetic acid for 30 min at 12 ml/min.
9. Wash the scaffold with sterile NaCl 0.9% for 2 h.
10. Use a light source to transilluminate the decellularized scaffold and check the macroscopic appearance of the parenchyma, its homogeneity, and preservation of the vascular tree (*see* Fig. 3).
11. Store in sterile PBS for further repopulation experiments.

3.3 Cell Isolation and Cell Culture

1. Prepare T-75 flask adding 1.5 ml of sterile 0.1% porcine gelatin. Wait 2 h to coat the surface at room temperature.
2. Wash liver biopsies with new PBS, supplemented with 2% antibiotic-antimycotic solution.
3. Place biopsies in a 100 mm Petri dish, and cut into approximately 2 mm^3 fragments with sterile scalpels (*see* **note 9**).
4. Place fragments in tube containing 30 ml of Trypsin solution. Gently shake from 10 to 30 min at 37 °C (*see* **note 10**).
5. At the end of 30 min, filter digested tissue with cell strainer 100 μm, and collect cells in a new 50 ml tube.
6. Centrifuge filtered cells at 1250 × *g*/min for 5 min at room temperature.
7. Remove supernatant and resuspend the cells in 11 ml of culture medium.
8. The total volume obtained is 15 ml (11 ml medium + 4 ml pellet).
9. Plate cells in gelatin pre-coated T-75 flasks (*see* Sect. 3.3, step 1) and culture for 6 h.

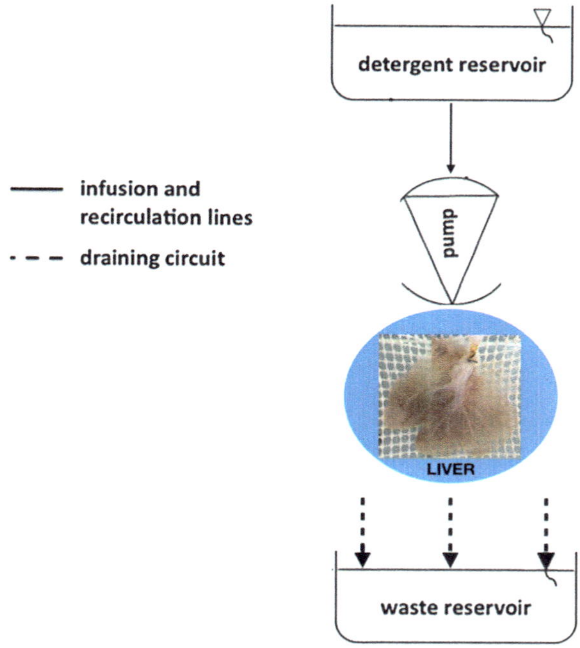

Fig. 2 Schematics of the dynamic decellularization system. Lines represent the hydraulic circuit, which includes infusion and recirculation subsections (*solid lines*) driven by a peristaltic pump and draining subsection (*dashed lines*) driven by atmospheric pressure

10. After 6 h centrifuge at 1250 g/min for 5 min at room temperature, and resuspend the pellet to obtain 1×10^6 cell concentration in 500 μl.
11. Place every single decellularized liver scaffolds in a 35 mm Petri dish.
12. Seed the cells directly onto scaffold using a 1000 μl pipette.
13. Insert the 35 mm Petri dishes containing the ECM and cells in a 100 mm Petri dish with another 35 mm Petri dish containing deionized water to prevent exsiccation.
14. Transfer the scaffold to the incubator under aseptic conditions with (5%) CO_2 at 37 °C, to encourage cell engrafting and ECM repopulation (*see* Fig. 4).

4 Notes

1. It is crucial to use sterile surgical instrument to avoid the microbiological overgrowth before starting the decellularization perfusion. If there is an infection and bacterial colonization of the ECM, the decellularization process could fail, and

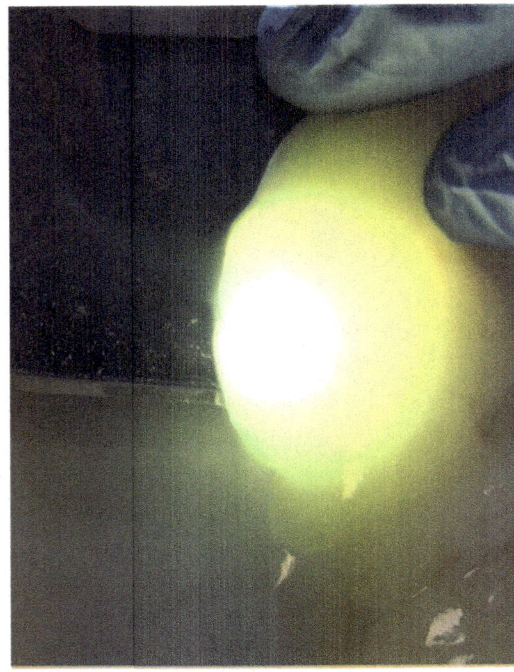

Fig. 3 Transillumination of the decellularized left liver lobe. This technique shows the translucent color homogeneity after a complete cell removal from the liver parenchyma

constituents of the ECM could degenerate before the end of the perfusion.

2. A microsurgical set could help to be more precise and delicate to the tissue, if not you can use pediatric surgical instruments.
3. The sheath introducer is important to safely manage the portal vein avoiding the leak of the cannula out of the venostomy and the collapse of the vessel.
4. Dissolve 5 μg of HGF in 2.5 ml of 0.1% BSA in PBS. Aliquot and store at −20 °C.
5. This surgical incision of the abdominal wall muscles let to leave the normal abdominal topography and give a good exposure of the surgical field.
6. Avoid making the venostomy near the liver hilum, but insert the cannula distally.
7. After venostomy utilize the vein pick catheter introducer, and eject PBS or saline solution to help the introduction of the portal vein catheter.
8. The polyethylene net maintaining the liver in the decellularization fluid avoids Glisson's capsule damages due to their prolonged air exposure.

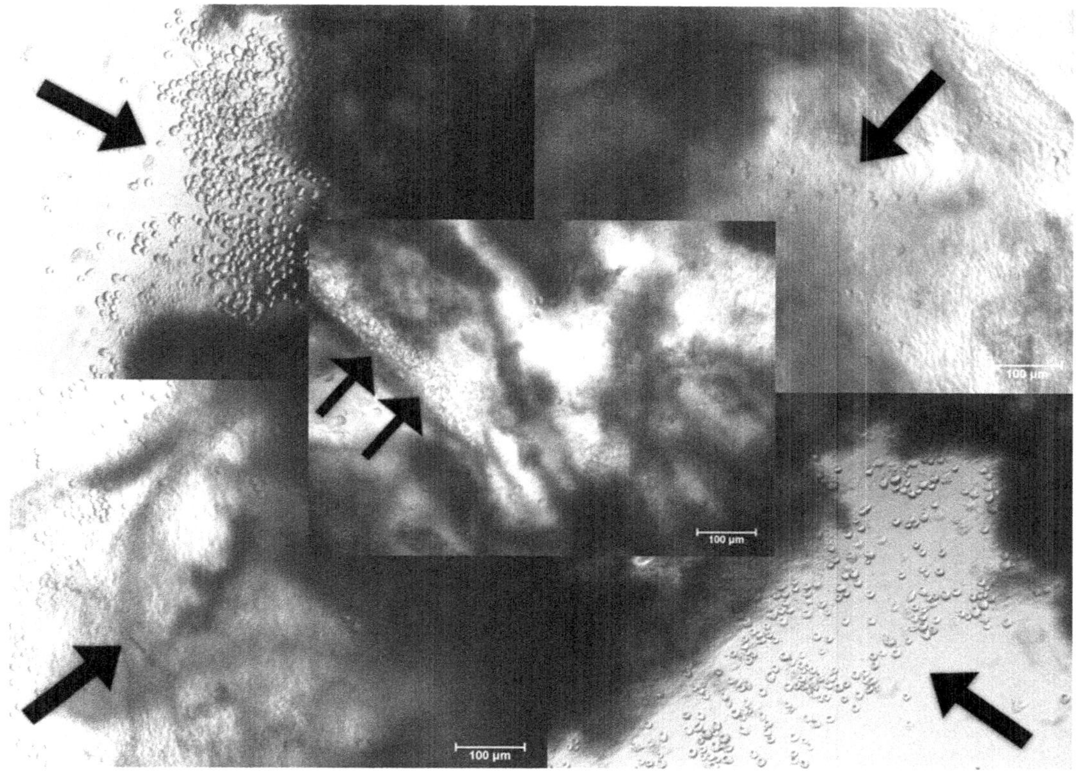

Fig. 4 Cells colonize the scaffold (*arrows*) and form clusters bordering and entering the recellularizing matrix

9. The appropriate shaking time is 20 min (gentle cycle). The liquid must be homogeneous and the color similar to white coffee.
10. The digested and predigested liver tissue is gluey and gelatinous. All the digestion procedures take about 40 min.

Acknowledgement

This work was supported by Carraresi Foundation and by Fondazione CMT Cura Mini-invasiva Tumori ONLUS. T.A.L.B., F.G., M.G., and A.Z. are members of the COST Actions CA16119.

References

1. Mokdad AA, Lopez AD, Shahraz S, Lozano R, Mokdad AH, Stanaway J, Murray JLC, Naghavi M (2014) Liver cirrhosis mortality in 187 countries between 1980 and 2010: a systematic analysis. BMC Med 12:145. https://doi.org/10.1186/s12916-014-0145-y
2. Badylak SF, Taylor D, Uygun K (2011) Whole-organ tissue engineering: Decellularization and recellularization of three-dimensional matrix scaffolds. Annu Rev Biomed Eng 13:27–53. https://doi.org/10.1146/annurev-bioeng-071910-124743

3. Gilbert TW, Sellaro TL, Badylak TF (2006) Decellularization of tissues and organs. Biomaterials 27(19):3675–3683. https://doi.org/10.1016/j.biomaterials.2006.02.014
4. Gilbert TW (2012) Strategies for tissue and organ decellularization. J Cell Biochem 113 (7):2217–2222. https://doi.org/10.1002/jcb.24130
5. Ott HC, Matthiesen TS, Goh SK, Black LD, Kren SM, Netoff TI, Taylor DA (2008) Perfusion-decellularized matrix: using nature's platform to engineer a bioartificial heart. Nat Med 14(2):213–221. https://doi.org/10.1038/nm1684
6. Tapias LF, Ott HC (2014) Decellularized scaffolds as a platform for bioengineered organs. Curr Opin Organ Transplant 19(2):145–152. https://doi.org/10.1097/MOT.0000000000000051
7. Uygun BE, Soto-Gutierrez A, Yagi H, Izamis ML, Guzzardi MA, Shulman C, Milwid J, Kobayashi N, Tilles A, Berthiaume F, Herti M, Nahmias Y, Yarmush ML, Uygun K (2010) Organ reengineering through development of a transplantable recellularized liver graft using decellularized liver matrix. Nat Med 16(7):814–820. https://doi.org/10.1038/nm.2170
8. Shupe T, Williams M, Brown A, Willenberg B, Petersen BE (2010) Method for the decellularization of intact rat liver. Organogenesis 6 (2):134–136. PMCID: PMC2901817
9. Barakat O, Abbasi S, Rodriguez G, Rios J, Wood RP, Ozaki C et al (2012) Use of decellularized porcine liver for engineering humanized liver organ. J Surg Res 173(1):e11–e25. https://doi.org/10.1016/j.jss
10. Rahman TM, Hodgson HFJ (2000) Animal models of acute hepatic failure. Int J Exp Pathol 81(2):145–157. https://doi.org/10.1046/j.1365-2613.2000.00144.x

Methods in Molecular Biology (2018) 1577: 293–305
DOI 10.1007/7651_2017_98

Published online: 04 November 2017

Liver Bioengineering Using Decellularized Whole-Liver Scaffolds

Iris Pla-Palacín, Pilar Sainz-Arnal, Sara Morini, Manuel Almeida, and Pedro M. Baptista

Abstract

Currently, due to the progress made in the field of regenerative medicine, whole-organ bioengineering is becoming a valid alternative to cope with the shortages of organs for transplantation. In this chapter, we describe the main techniques carried out for pig liver bioengineering, which serves as an essential model for future human liver bioengineering. These include porcine whole-liver decellularization, endothelial and mesenchymal stem cell isolation, porcine ES-derived hepatoblasts, and scaffold recellularization using a bioreactor perfusion system.

Keywords Decellularization, Liver scaffold, Organ bioengineering, Perfusion bioreactor system, Recellularization

1 Introduction

Due to its role in the production of several proteins, vitamins, lipids, and carbohydrates, the synthesis of substances necessary for homeostasis and digestion, detoxification of various metabolites, and glucose level regulation, the liver is one of the body's vital organs in metabolism and homeostasis.

The high mortality of end-stage liver disease is a global public health issue, to which liver transplantation is the only definitive treatment to improve the survival and quality of life of these patients. However, there is still a significant gap between organ supply and demand.

Whole-organ bioengineering and regenerative medicine are promising new technologies that can help reduce liver shortage by increasing the number of organs available for transplantation. In this perspective, decellularization is an attractive technique in regenerative medicine to prepare scaffolds. This consist of the removal of cells from a tissue or organ using detergent perfusion through the vasculature, leaving behind an intact structure of the extracellular material, which is fundamental to regulate cell differentiation and function. The perfusion decellularization method represents a useful procedure to create whole-organ scaffolds ready to be recellularized with freshly isolated cells and maintained in bioreactors for cell expansion, differentiation, and function [1, 2]. In this protocol, we describe how to prepare porcine liver scaffolds and isolate and expand endothelial, mesenchymal stem cells and iPS-derived hepatoblasts for posterior recellularization and maintenance of this bioengineered liver in a bioreactor perfusion system.

2 Materials

2.1 Liver Harvesting and Cannulation

1. Scalpel.
2. Forceps (with and without teeth).
3. Scissors.
4. Silk suture 4-0.
5. Cannula 18G.
6. Straight fitting 6.4 mm diameter.

2.2 Decellularization and Scaffold Preparation for Bioreactor

1. Distilled water.
2. Peristaltic pump with 14G and 17G tubing (Masterflex L/S with Masterflex L/S easy load pump head, Cole Parmer).
3. Silicone tubing (silicone tubing size 14G and 17G, Cole Parmer).
4. Pulse dampener.
5. Pressure sensor (APT Pressure Transduced, Panlab).
6. Pressure controller (Panlab).
7. Male Luer.
8. Four-way stopcock.
9. Detergent solution: 1% Triton X-100 with 0.1% ammonium hydroxide in distilled water.
10. X-ray source.
11. 150 cm culture dish.

2.3 Cell Isolation and Culture

1. Piglet umbilical cord.
2. PBS− (without calcium and magnesium).
3. PBS+ (with calcium and magnesium).
4. 0.05% trypsin/EDTA.
5. Neutralization medium: DMEM/F12 with 10% FBS.
6. 18G cannula.
7. Endothelial growth medium 2.
8. Fibronectin (5 μg/cm^2).

9. 6-multiwell cell culture plate.
10. 150 cm cell culture dish.
11. 50 mL conical centrifuge tubes.
12. Pig femur.
13. Surgical saw.
14. 18G needle.
15. Scalpel.
16. Scissors.
17. Forceps.
18. Bovine gelatin (0.02%).
19. 100 and 40 μm cell strainers.
20. Histopaque-1077.
21. Centrifuge.
22. Incubator at 37 °C and 5% CO_2.
23. MEF medium: 81% DMEM high glucose with 15% FBS, 2% P/S, 1% L-Glut, 1% DMEM nonessential amino acids.
24. Matrigel (Geltrex hESC-qualified reduced growth factor basement membrane) – diluted to 2 mg/mL in DMEM/F12 medium.
25. TrypLE.
26. 293FT cells.
27. Gelatin.
28. 100 mm tissue culture dish pre-coated with 0.1% gelatin.
29. DMEM high glucose.
30. 293FT medium: DMEM high glucose with 10% FBS.
31. FuGENE HD Transfection Reagent.
32. rtTA plasmid.
33. STEMCCA (OKSM) Lentivirus Reprogramming Kit.
34. D8.9/psPAX2 plasmid.
35. VSV-G plasmid.
36. Microcentrifuge.
37. Sterile microcentrifuge tubes (1.5 mL).
38. 0.45 μm filters.
39. Beckman Coulter Optima L-90K ultracentrifuge with SW-32 rotor.
40. 0.25% Trypsin-0.53 mM EDTA.

41. Polybrene.
42. Doxycycline.
43. Hemocytometer.
44. Pluripotent stem cell medium: mTeSR1.
45. RPMI cell differentiation medium: RPMI 1640, HEPES medium with 1% nonessential amino acids, 1% P/S.
46. B27 without insulin.
47. B27 with insulin.
48. Activin A.
49. Bone morphogenetic protein 4.
50. Fibroblast growth factor 2.
51. Hepatocyte growth factor.
52. EGM2 bullet kit (Lonza).
53. MCDB 131 culture medium (ThermoFisher Scientific).
54. DMEM/F12 culture medium (ThermoFisher Scientific).

2.4 Bioreactor System

1. Bioreactor vessel (Glass Ball Spinner, 250 mL, Bellco Biotechnology Inc., Vineland, NJ, USA).
2. Peristaltic pump (same description as in Sect. 2.2, item 2).
3. Silicone tubing (same description as in Sect. 2.2, item 3).
4. Pressure sensor (same description as in Sect. 2.2, item 5).
5. Pressure controller.
6. Magneto.
7. Magnetic stirrer.
8. Pulse dampeners (Cole Palmer).
9. Four-way stopcocks (Cole Palmer).
10. Smart site connections (Cole Parmer).
11. Luer-lock syringe 20 mL.
12. 20 μm filter.
13. Empty (sterile and clean) culture media bottle (as a secondary reservoir).
14. Bioreactor culture medium: 50% DMEM/F12 + 50% MCDB 131 containing 5% FBS, 1% L-glutamine, 1% penicillin/streptomycin, 5 μg/mL insulin, 10 μg/mL transferrin, 50 μg/mL VEGF, 40 μg/mL EGF, 40 μg/mL FGF-2, and 40 μg/mL IGF-1.

3 Methods

3.1 Pig Cadaveric Liver Harvesting

1. The suprahepatic vena cava is dissected as close to the atrium as possible.
2. The common bile duct is dissected as close to the duodenum as possible.
3. The portal vein is carefully dissected from the surrounding tissue, to visualize it and its branches. Lateral branches are ligated with 4-0 silk suture and cut as close to the intestines as possible.
4. The hepatic artery is carefully dissected from the surrounding tissue, to visualize it and its branches. Lateral branches are ligated with silk suture 4-0 and cut as close to the stomach as possible.
5. Infra-hepatic vena cava is carefully dissected and cut without damaging the lobe. The cut should be performed before the renal bifurcation (see Note 1).
6. Liver attachments are cut to remove the intact organ.
7. The diaphragm is carefully dissected around the esophagus.
8. The portal vein is cannulated with a straight fitting of 6, 4 mm diameter. The hepatic artery should be cannulated with a 20G cannula.
9. The gallbladder is emptied and cleaned.

3.2 Liver Decellularization

Decellularization is usually performed at room temperature, after one freezing/thawing cycle. Scaffolds are kept sterile at 4 °C until use. Components are assembled as shown in Fig. 1, and results are seen in Fig. 2.

1. Cannulas are attached to a peristaltic pump by using 17G tubing for the portal vein and 14G for the hepatic artery.
2. 2 L of distilled water is perfused through the portal vein at a pressure of 25 mmHg for both the hepatic artery and the portal vein.
3. 10 L of decellularization solution is perfused afterward (see Note 2).
4. 20 L of distilled water is perfused to remove the detergent from the tissue.

3.3 pUVEC Iisolation

1. Umbilical cords are obtained from piglets that had undergone spontaneous abortions in farms of animal production.
2. Locate umbilical cord vein and inject 20 mL of PBS with a 18G cannula.

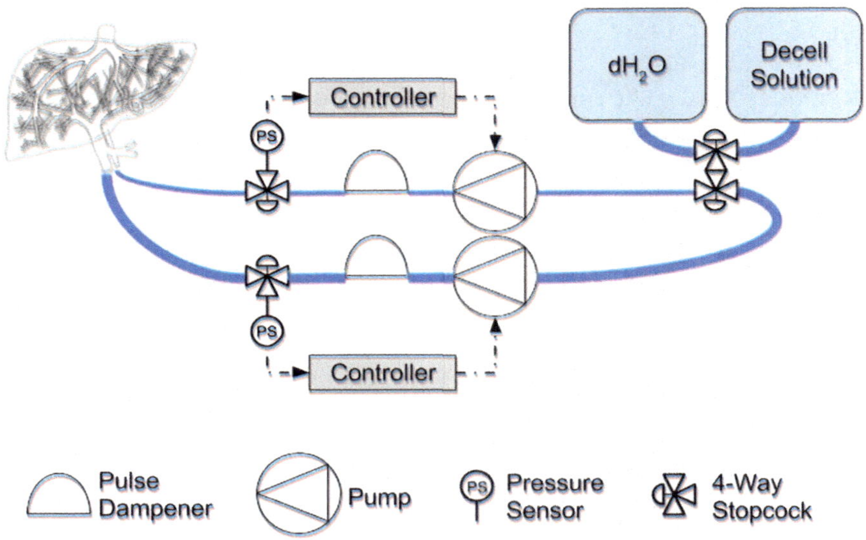

Fig. 1 Schematic diagram of the decellularization setup. The fine line represents the hepatic artery line; the thicker one represents the portal vein line. The liver is first washed with distilled water and then perfused with the decellularization solution detergent. Finally it is washed again to remove the detergent from the liver

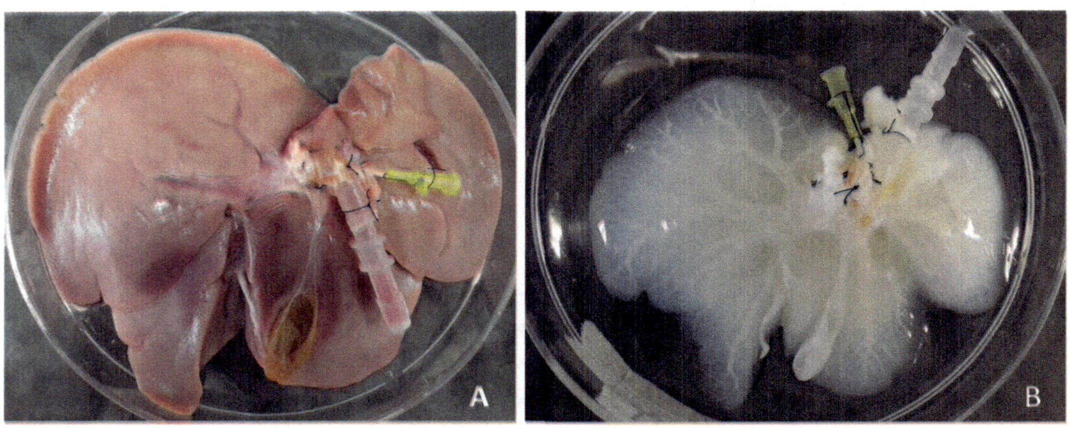

Fig. 2 Piglet liver before (**a**) and after decellularization (**b**)

3. Inject 20 mL of pre-warmed trypsin. Collect and inject again. Repeat this step for 7 min.
4. Collect all the trypsin and neutralize with 20 mL of DMEM with 10% FBS and centrifuge 400 × g for 5 min.
5. Aspirate the supernatant and resuspend the pellet in 10 mL of EGM-2 with 20% FBS.
6. Transfer the cells to a 6-multiwell dish coated with fibronectin (0.03 mg/mL), and place it into the incubator at 37 °C.
7. Allow the cells to grow for 3–4 days. Change media at day 4.

8. Once some colonies are detected, change media to EGM-2 with 10% FBS. Change media every 3 days.
9. Trypsinize the culture when it reaches 80% confluence (see Note 3).

3.4 pMSC Isolation

1. Pig bone marrows are isolated from cadaveric piglets slaughtered by farms of animal production due to malformations or trauma.
2. Femurs are carefully dissected and cleaned from muscle/ligament tissue.
3. Cut the epiphysis and flush the bone marrow with a 18G needle with 5 mL of DMEM and 1% P/S.
4. Collect this cell suspension and centrifuge at 300 × *g* for 5 min.
5. Reconstitute the pellet with 10 mL of DMEM +1% P/S, and filter sequentially through 100 and 40 μm cell strainers.
6. Separate with a histopaque density gradient centrifugation.
7. Resuspend the pellet with 10 mL of DMEM/F12 with 10% FBS and 1% P/S, and plate the cells in a 10 cm culture dish previously coated with a bovine gelatin solution (0.2%). Place it in the incubator.
8. Allow the cells to grow for 2 days. Change media at day 2.
9. Trypsinize the culture when it reaches 70–80% confluence.

3.5 Generation of iPS-Derived Hepatoblasts

Cultured porcine mesenchymal stem cells with less than four passages were reprogrammed to generate iPS cells by transduction with four human reprogramming factors: Sox2, Klf4, Oct4, and c-Myc. We followed the protocol published by Rajarajan et al [3]. Once the iPS cells were obtained and characterized properly, they were differentiated to hepatoblast-like cells.

3.5.1 Lentiviral Production

1. Culture 293FT cells onto gelatinized 100 mm dish in 293FT media until cells reach 80–90% confluency.
2. Mix in microcentrifuge tube 770 μL DMEM high glucose with 50 μL FuGENE Reagent per 10 cm dish, and incubate for 5 min at room temperature.
3. Add 5.5 μL VSV-G and 8.25 μL D8.9 and mix well.
4. Add a total of 11 μg vector DNA (STEMCCA Lentivirus Reprogramming Vector pluripotent transcription factors at 11 μg/infection). Mix gently and incubate for 30 min at room temperature.
5. During incubation, add 10 mL of fresh 293FT medium to 293FT cells.
6. Add entire FuGENE/DNA complex to the 100 mm dish drop by drop, and rotate the plate to mix the contents. Incubate for 12–24 h.

7. Change 293FT medium 12–24 h later.
8. After 24 h, collect all media and store at 4 °C. Add 10 mL of fresh 293FT medium to the dish. Repeat every subsequent 24 h for three collections.
9. Filter collected viral medium through a 0.45 μm filter, and centrifuge at 50,000 × *g* for 1.5 h at 4 °C.
10. Decant supernatant and add 200 μL of serum-free DMEM medium to the pellet and let it stand overnight at 4 °C.
11. Resuspend, aliquot, and store the virus at −80 °C.

3.5.2 Infection of Fibroblasts Using Pluripotency Transcription Factors

1. Culture pig MSC in a gelatin-coated single well of a 6-well plate until 90% confluency.
2. Trypsinize the cells in culture, count the number of cells, and adjust the cell concentration to 10^4 cells/mL.
3. Add polybrene at a final concentration of 8 μg/mL to the cell-containing medium. Mix by pipetting up and down.
4. Thaw the four viral constructs Oct-3/4, Sox2, Klf4, and c-Myc, and combine them into a cocktail. Add into the cell mixture. Mix gently pipetting up and down (see Note 4).
5. Plate 1 mL of the cell/virus mixture onto a single well of a gelatin-coated 6-well plate, and incubate for 24 h.
6. After this time, wash with warm MEF medium, replenish with 2 mL of fresh MEF medium, and continue incubation.
7. Leave the culture until iPS colonies appear. Once they do (around day 7), pick colonies with a 200 μL pipet.

After post-viral transduction, the cells are grown on X-ray radiation inactivated mouse embryonic fibroblast (MEF) feeder layer with mTeSR1. After that, cells were adapted to grow onto Matrigel.

3.5.3 Hepatoblast-Like Cell Production

Differentiation of iPS cells into hepatoblast-like cells is induced as previously described [4].

1. Pig pluripotent stem cells should be maintained on Matrigel-coated 100 mm tissue culture dishes using mTeSR1 (see Note 5).
2. After 50% of confluence, wash the cells with sterile PBS−, and incubate the cells for 2 min with 3 mL of PBS−/0.02% EDTA at room temperature. As soon as the cells begin to detach, remove the PBS−/0.02% EDTA solution, and wash the plate with 6 mL of mTeSR1.
3. Pipet the cell solution to obtain small clusters, centrifuge cells at 200 × *g* for 5 min, and suspend in mTeSR1.

4. Transfer the cell suspension to a suitable number of wells of a 6-well matrigel-coated culture plate. Culture the cells overnight (see Note 6).
5. Change the culture medium to RPMI 1640 medium supplemented with 2% B27 (without insulin), 100 ng/mL Activin A, 10 ng/mL BMP4, and 20 ng/mL FGF2, and culture for 2 days with daily medium changes.
6. Substitute the culture medium again to RPMI 1640 with 2% B27 (without insulin), 100 ng/mL Activin A for 3 days with daily medium changes.
7. Culture the definitive endoderm cells with RPMI 1640 containing 2% B27 (with insulin), 20 ng/mL BMP4, and 10 ng/mL FGF-2 for 5 days with daily medium changes.
8. Culture the hepatic progenitor cells with RPMI 1640 containing 2% B27 (with insulin) and 20 ng/mL HGF for 5 days with daily medium changes.

By day 16 of the differentiation protocol, the cells should display a morphology that resembles hepatoblasts, with 80–90% of the cells expressing AFP and, quite commonly, some lipid droplets within the cytoplasm of the cells.

3.6 Scaffold Preparation and Sterilization for Bioreactor

1. After liver decellularization, the left and central lobes are removed, to decrease the size of the scaffold and, consequently, the total cells needed to recellularize it.
2. A 4-0 silk suture is passed through the vascular structure of the left lobe to ligate it. The vessels are then cut, and the lobe is removed. This same procedure is performed to remove the central lobe.
3. The diaphragm should be trimmed as much as possible.
4. The remnant right lobe scaffold is then put in a 150 cm culture dish with 20 mL of distilled water.
5. The scaffold is then sterilized with X-ray at 160 kV, 6.3 mA, for 99.9 min (see Note 7).

3.7 Bioreactor Assembly and Recellularization

For liver recellularization, it is necessary to assemble the setup described in Fig. 3. The process consists of three phases: the first is the priming, the second is cell seeding, and the third is the maintenance of the recellularized scaffold (for 7 days).

1. All the components used for the bioreactor setup are sterilized (see Note 8).
2. All the components are assembled in a culture biosafety cabinet, using sterile gloves, sterile lab coat, and a mask, to reduce contamination.

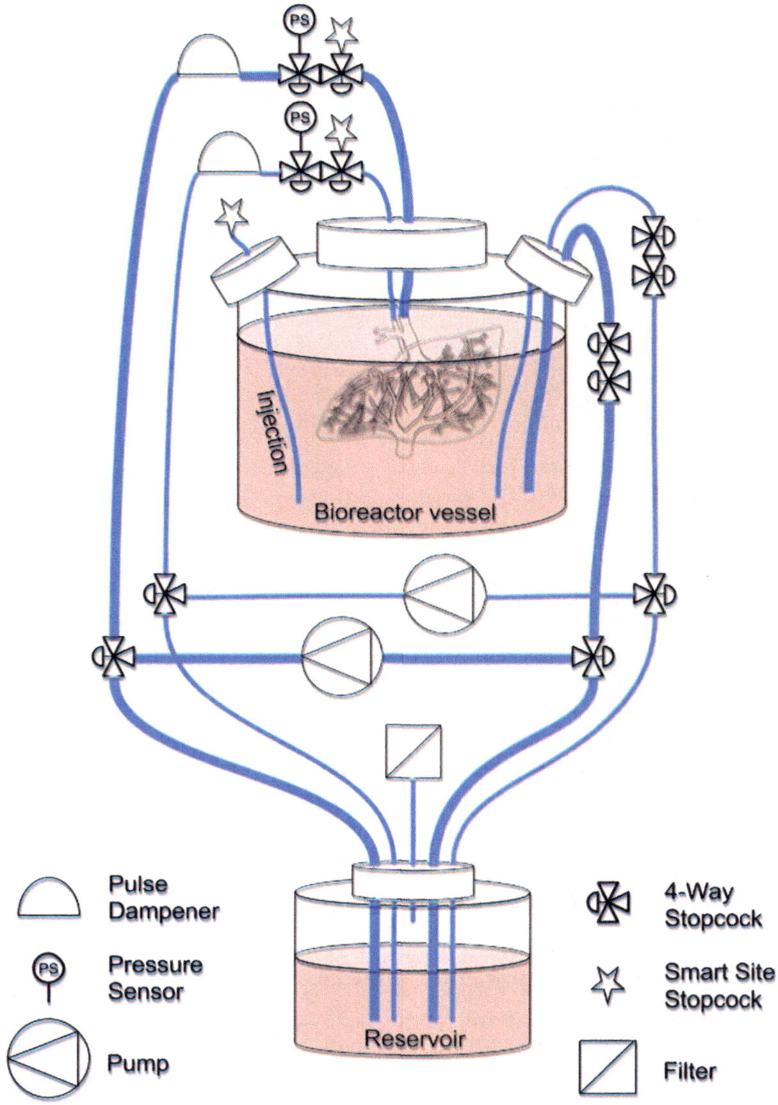

Fig. 3 Schematic diagram of the bioreactor system. Cells are injected through the injection tube to the media within the spinner flask. Then, they are perfused by the peristaltic pumps all across the circuit to the liver, entering the organ through the portal vein and the hepatic artery

3. The tubing is connected to the four-way stopcocks and smart site stopcocks using male and female Luer locks. The pulse dampeners are linked to the tubing.
4. Once the tubing is connected to the bioreactor vessel and the reservoir and the circuit is closed, it is primed with culture medium using a 20 mL Luer-lock syringe, to remove the air from the system.

5. The liver is attached to the tubing through the portal fitting and the hepatic cannula and is suspended in the culture medium within the spinner flask.
6. The bioreactor system is transferred to an incubator at 37 °C/ 5% CO_2 and attached to the peristaltic pumps.
7. The liver is then perfused at low-pressure conditions overnight before the cell seeding: 25 mmHg in both lines.

3.8 pUVEC Preparation for Bioreactor Seeding

1. Aspirate the EGM-2 media from the 150 cm culture dish, and wash with 10 mL of PBS without calcium and magnesium.
2. Add 7 mL of trypsin/EDTA and incubate for 4 min at 37 °C.
3. Collect the trypsin with the detached cells, and transfer to a 50 mL conical tube.
4. Rinse the dish with 7 mL of DMEM +10% FBS + 1% P/S for any leftover cells, and transfer to the 50 mL conical tube.
5. Centrifuge at 300 × *g* for 5 min. Aspirate supernatant and reconstitute the pellet with 10 mL of bioreactor media.
6. Cell counting.
7. Keep the cells on ice until ready to be injected in the bioreactor system.

3.9 pMSC Preparation for Bioreactor Seeding

1. Aspirate the DMEM/F12 media from the 150 cm culture dish, and wash with 10 mL of PBS without calcium and magnesium.
2. Add 7 mL of trypsin/EDTA and incubate for 4 min at 37 °C.
3. Collect the trypsin with the detached cells and transfer to a 50 mL conical tube.
4. Rinse the dish with 7 mL of DMEM with 10% FBS and 1% P/S for any leftover cells, and transfer to the 50 mL conical tube.
5. Centrifuge at 300 × *g* for 5 min. Aspirate supernatant and reconstitute the pellet with 10 mL of bioreactor media.
6. Cell counting.
7. Keep the cells on ice until ready to be injected in the bioreactor system.

3.10 iPS-Derived Hepatoblast Preparation for Bioreactor Seeding

1. Aspirate the media.
2. Add 3 mL of TrypLE for 10 min at room temperature.
3. Collect the TrypLE with the detached cells, and transfer to a 50 mL of conical tube.
4. Inactivate with DMEM/F12 with 10% FBS and 1% P/S.
5. Centrifuge 200 × *g* at room temperature, and reconstitute the pellet with 10 mL of bioreactor media.
6. Cell counting.

7. Keep the cells on ice until ready to be injected in the bioreactor system.

Once all cell types are detached and counted, they should be centrifuged again. Pellets should be mixed with 10 mL of bioreactor media for scaffold seeding and supernatants discarded.

3.11 Scaffold Seeding

1. An ammount of approximately 60 million hepatoblasts, 45 million pUVECs and 12 million pMSCs (ratio 10:7:2) are co-seeded through the portal vein and the hepatic artery of the piglet liver scaffold by perfusion at 60 mmHg in the hepatic artery and 25 mmHg in the portal vein.
2. All types of cells used are put together in 10 mL of bioreactor media (see Note 9) and injected into the bioreactor through the injection tube. Peristaltic pumps will recirculate the cells to the portal vein and hepatic artery.
3. Cells are seeded in two steps, separated by 4 h.
4. After the second seeding, a maintenance pressure is set up for 1 week, to maintain a specific fluid flow inside the scaffold that allows the attached cells not to detach. A set point of 25 mmHg will be used for both lines.
5. The scaffold is left in the bioreactor system for 1 week, changing media every 2 days.
6. After 1 week, the reseeded scaffold needs to be perfused with 4% paraformaldehyde for 15 min, after which small pieces are taken for tissue processing.

4 Notes

1. Porcine infra-hepatic vena cava penetrates the liver right lobe. Some care is needed not to disrupt this hepatic lobe.
2. Depending on each liver, it might be necessary to prepare more or less detergent.
3. Colonies usually appear between 5 and 21 days after pUVECs isolation.
4. A four-in-one mono-cassette virus is recommended. 1 μg/mL of doxycycline should be added in every medium change to induce gene expression. The amount of virus added is calculated using a multiplicity of infection (MOI) of at least 20. To determine the desired amount of virus: virus volume (μL) required = [(# pMSCs seeded for infection)/virus titer (U/mL)] × [(desired MOI)/1 mL] × 1000 μL.
5. It is essential to start the differentiation protocol of iPCs cells into hepatoblast-like cells without minimal morphological evidence of differentiation.

6. The cell density at the initiation of differentiation usually has a dramatic effect on differentiation efficiency, so the best density should be determined empirically.
7. It is desirable to use gamma irradiation for scaffold sterilization, but this procedure is not always available.
8. Tubing, spinner flask, and pulse dampener are sterilized by steam at 121 °C. Some other components cannot be autoclaved: four-way stopcocks and smart connectors are sterilized with ethylene oxide; the pressure sensor is cleaned with a bactericide solution for 15 min and then washed with sterile PBS. The outer part is cleaned with ethanol.
9. Before loading the cell suspension into a Luer-lock syringe, it is vital to resuspend the cells carefully to reduce cell aggregation which could obstruct the scaffold vascular network.
   In this chapter, we describe a seeding method in which cells are seeded with continuous perfusion. Some other authors have used multistep infusion systems, where cells are delivered directly inside the scaffold.

Acknowledgments

This work was supported by Instituto de Salud Carlos III, through a predoctoral fellowship i-PFIS IFI15/00158 (I. P-P), Gobierno de Aragón and Fondo Social Europeo through a predoctoral fellowship DGA C066/2014 (P. S-A), and by a predoctoral fellowship from Fundação para Ciência e Tecnologia (PD/BD/114057/2015), Portugal (S.M.). PMB was supported by an H2020-MSCA-IF-2014 from the European Research Agency and by the PI15/00563 Research Project from Instituto de Salud Carlos III, Madrid, Spain.

References

1. Baptista PM et al (2011) The use of whole organ decellularization for the generation of a vascularized liver organoid. Hepatology 53:604–617
2. Soto-Gutierrez A et al (2011) A whole-organ regenerative medicine approach for liver replacement. Tissue Eng Part C Methods 17:677–686
3. Rajarajan K, Engels MC, Wu SM (2012) Reprogramming of mouse, rat, pig, and human fibroblasts into iPS cells. Curr Protoc Mol Biol. Chapter 23, Unit 23 15
4. Mallanna SK, Duncan SA (2013) Differentiation of hepatocytes from pluripotent stem cells. Curr Protoc Stem Cell Biol 26, Unit 1G 4

Methods in Molecular Biology (2018) 1577: 307–315
DOI 10.1007/7651_2017_32

Published online: 28 June 2017

Decellularization of Intact Lung Tissue Through Vasculature and Airways Using Negative and Positive Pressure

Steven Skolasinski and Angela Panoskaltsis-Mortari

Abstract

Decellularization allows the production of extracellular matrix (ECM) scaffolds. Here we describe the use of combined positive pressure and negative pressure to drive decellularization reagents into the vasculature and airways, respectively, of structurally intact lungs in order to remove cells and cellular material leaving an intact ECM scaffold.

Keywords Decellularization, Lung, Negative pressure bioreactor, Organ scaffold

1 Introduction

Organ decellularization involves the removal of cells and cellular material from an organ leaving the extracellular matrix (ECM) scaffold. The resulting scaffold is composed of many different molecules including collagen, elastin, fibronectin, and proteoglycans [1]. Such scaffolds can be used to study the sub-structure of the organ or can be recellularized for a variety of investigational or clinical applications [2, 3].

The lung is a heterogeneous organ at both the structural and cellular level. It contains over 60 different cell types and structurally can be grossly divided into parenchyma, blood vessels, conducting airways, and alveolar airways [4]. The alveolar tissue, which makes up approximately 65% of the total lung tissue area, is highly cellular with only 12–15% being ECM [5]. In contrast, the non-alveolar tissue is much less cellular, with approximately 50% of the tissue being ECM.

Several methods of lung decellularization have been described [1, 6, 7]. The various methods differ in which reagent combinations are used, which portal of entry is used to deliver the reagents (airway, vasculature, or both), and how the reagents are applied (positive vs. negative pressure). We have found that utilizing both airways and vasculature, under negative and positive pressure, respectively, allows for the best penetration of the decellularization reagents. Different reagent combinations provide different degrees of decellularization and have different effects on the ECM [8]. An intact ECM is important to the functional structure of the lung, and some components of the ECM such as glycosaminoglycans also seem to play a role in directing cell differentiation and are therefore vitally important when attempting to recellularize a scaffold [1]. This protocol prepares lung tissue for recellularization and thus is designed to optimize removal of cells and functional cellular material while minimizing effects on the ECM. Collagen, the most abundant molecule in lung ECM, remains relatively intact after decellularization with this protocol [9]. Other components of the ECM such as laminin, elastin, and glycosaminoglycan's remain structurally intact, but at lower levels than non-decellularized lungs due to some of these molecules being removed by the decellularization process [1, 9]. DNA is completely eliminated by this process, at least as determined using conventional detection methods.

2 Materials

2.1 Reagents

Prepare reagents in 5–10 L carboys with spouts. Prepare all reagents with deionized water. Sterilely filter all reagents through 0.22 μm filter after mixing.

1. Deionized water (DI H_2O).
2. Triton: 0.1% v/v Triton X-100.
3. Sodium deoxycholate: 2% w/v (*See* **Note 1**).
4. Sodium chloride: 1 M NaCl.
5. DNase: 30 μg/mL DNase, 1.3 mM $MgSO_4$, 2 mM $CaCl_2$ (*See* **Note 2**).
6. Phosphate buffered saline (PBS).

2.2 Pumps and Controller

1. Pump controller device: Device which can be programmed to control rate and direction of flow based on pressure measurements and a time schedule. Controllers are available from several manufactures (we use Harvard Apparatus Catalog # 132500).
2. Two peristaltic pumps: Capable of pumping approximately 5–100 mL/min (we use Harvard Apparatus Catalog # 132400).
3. Pressure transducer (working range between 0 and 500 mmHg).
4. Four low resistance check valves (one-way valves) with barb connectors to attach tubing.

2.3 Bioreactor

1. Air-tight chamber with tubing attachments in the lid. Sufficient size to contain lung without touching walls of chamber. This can be custom built (See **Note 3**).

2.4 Ventilator Reservoir

1. Erlenmeyer flask with spout, 1 L capacity. Attach tube from interior of spout to bottom of the flask (Fig. 1).

2.5 Principles of Bioreactor System Function

2.5.1 Vascular

The vascular pump uses positive pressure to pump reagent from the reagent carboy into the pulmonary artery. The reagent passes through the vasculature and drains out of the pulmonary vein.

2.5.2 Airway

The airway is infused with reagent which is drawn into the trachea from the reagent carboy by negative pressure as fluid is removed from the bioreactor by the Ventilator Pump. The reagent is then pushed into the Ventilator Reservoir by positive pressure as fluid is returned to the Bioreactor from the Ventilator Reservoir. A Check

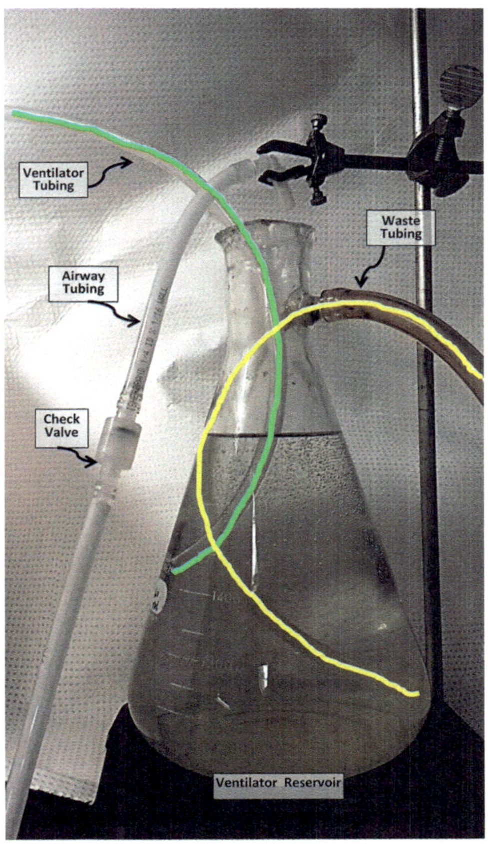

Fig. 1 Photo of ventilator reservoir. *Lines* are drawn over tubing in this image to allow the path to be followed inside the flask

Valve between the reagent carboy and the trachea ensures that no reagent refluxes from the airway back to the reagent carboy. Likewise, a Check Valve between the trachea and the Ventilator Reservoir ensures no air or waste reagent is drawn back into the lung. The distal end of the Airway Tubing is suspended above the Ventilator Reservoir to allow visualization of fluid flow.

2.5.3 Ventilator System

The Ventilator Pump alternately moves fluid back and forth between the Bioreactor and the Ventilator Reservoir. Check valves at the proximal end of the Ventilator Tubing help to remove debris from the lower portion of the bioreactor and return fluid to the upper portion. The distal end of the Ventilator Tubing is positioned halfway between the spout and the bottom of the Ventilator Reservoir. This ensures that the Ventilator Tubing will be deep enough to remain submerged when fluid is drawn from the Ventilator Reservoir, but not so deep that debris is drawn in from the bottom. The Waste Tubing allows excess fluid to drain from the Ventilation Reservoir. The interior portion of the Waste Tubing should extend to the bottom of the Ventilator Reservoir to allow debris to flow to the Waste Container.

3 Methods

1. Attach tubing to the decellularization reagent carboy and run pumps to prime tubing with reagent and expel all air.
2. Fill Ventilator Reservoir with DI H_2O. Ensure that the end of the Ventilator Tubing is properly positioned in the Ventilator Reservoir (*See* **Note 4**).
3. Insert a barb connector into the trachea and place a zip-tie over the tissue and barb connector to secure lung to the connector (Fig. 2).
4. Insert a barb connector into pulmonary artery and place a zip-tie over the tissue and barb connector to secure lung to the connector (Fig. 2).
5. Attach lung to bioreactor tubing using the barb connectors (Figs. 3 and 4).
6. Fill bioreactor with DI H_2O. Fill bioreactor to the brim, minimizing airspace as much as possible.
7. Close bioreactor. Ensure lid is air-tight.
8. Loosen the lid of the reagent carboy to prevent a vacuum from forming when fluid is removed.
9. Zero the pressure transducer.
10. Set the *vascular* pump to maintain a continuous pressure of 15 mmHg with a minimum flow of 5 mL/min.

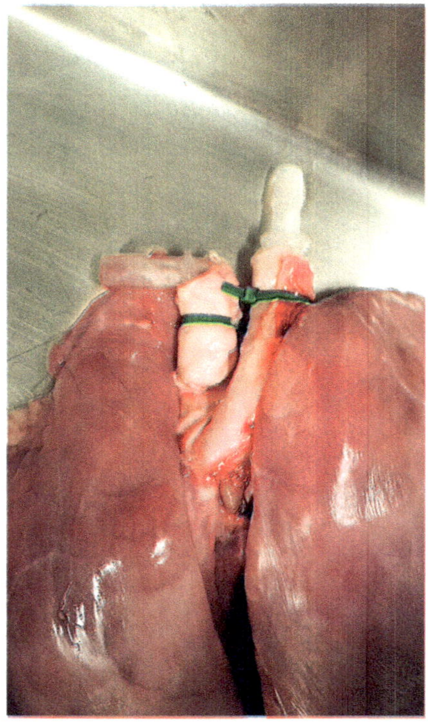

Fig. 2 Lung with barb connectors in the pulmonary artery (*left*) and trachea (*right*)

11. Set *ventilator* pump as follows:
    (a) Expel approximately 100 mL (*See* **Note 5**) of fluid from the bioreactor over 1–2 min. This will generate negative pressure in the bioreactor and draw fluid from the carboy into the airway.
    (b) Pause for 10 min.
    (c) Pump 70 mL (*See* **Note 6**) into the bioreactor over 1–2 min.
    (d) Pause for 1 min.
    (e) Repeat starting from step a.
12. Allow the bioreactor to run for the duration listed in Table 1.
13. After the time has elapsed for a particular decellularization reagent, pause the pumps.
14. Disconnect the tubing from the reagent carboy and reconnect to the next reagent carboy.
15. Restart the pumps. Continue to transition from one carboy to the next, referencing Table 1 for run durations, until the lung has been treated with all of the decellularization reagents (*See* **Note 7**).

Table 1
Decellularization reagent run times

Decellularization reagent	Run times (in hours)
DI H_2O	2
Triton	6
Sodium deoxycholate	6
NaCl	2
DNase	2
PBS	5

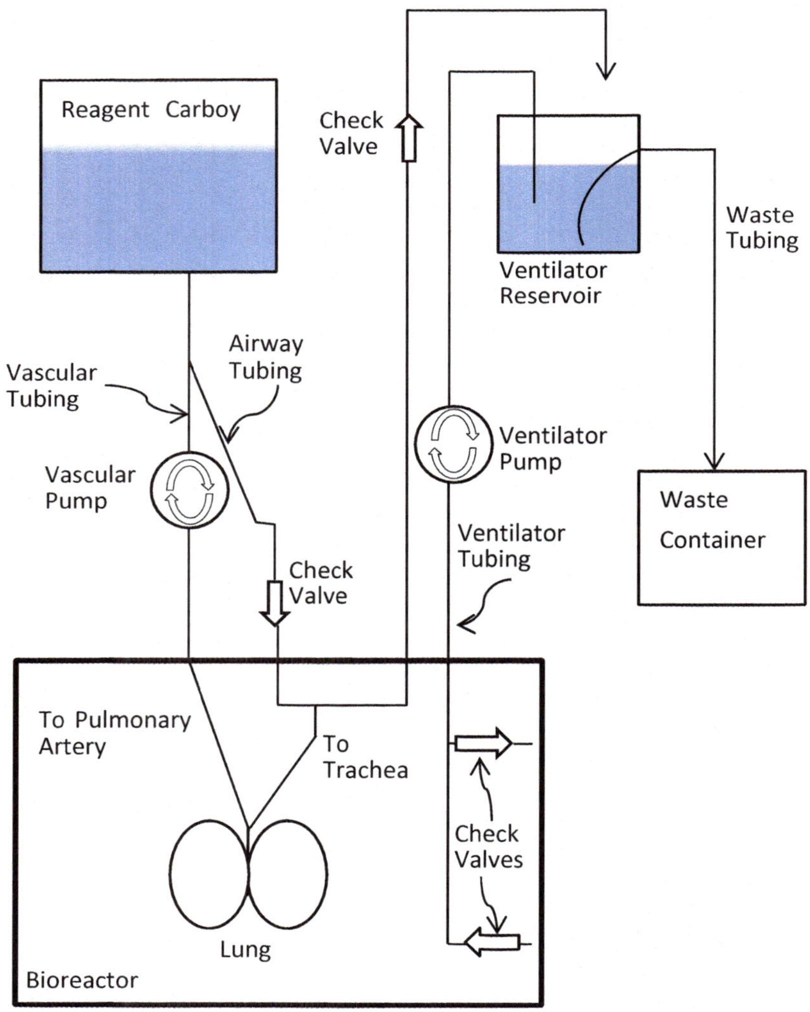

Fig. 3 Bioreactor assembly

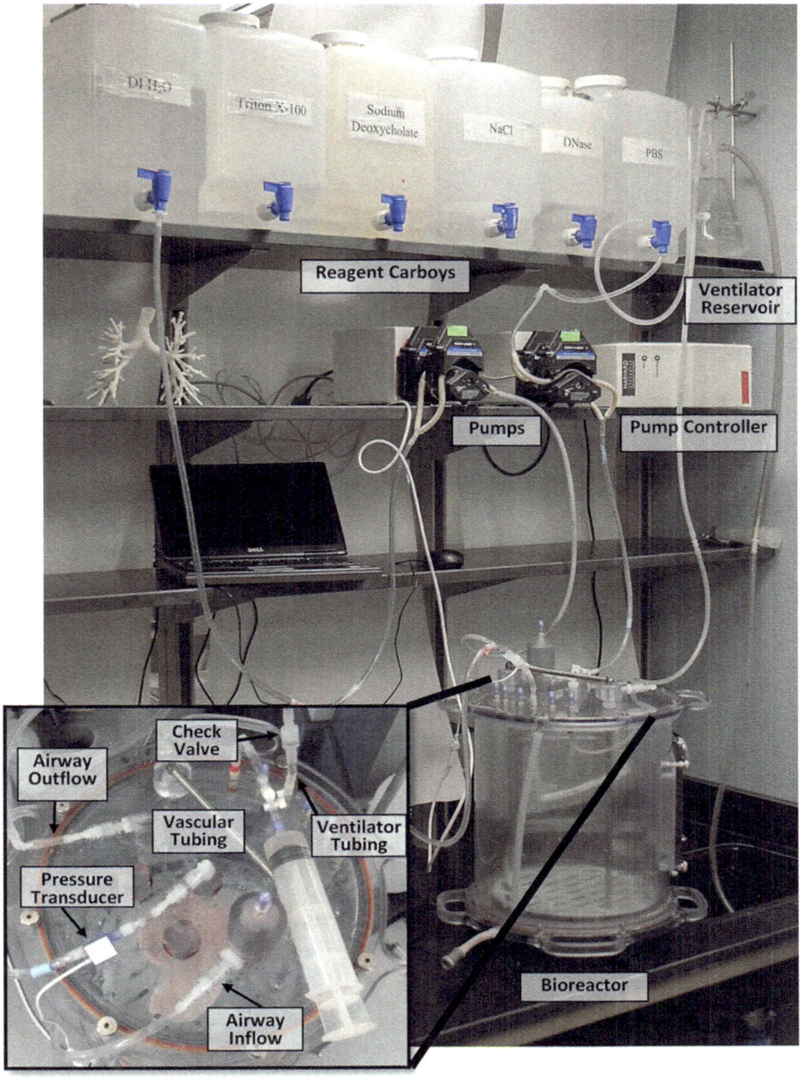

Fig. 4 Photo of decellularization system with close-up of the top of the bioreactor showing the connectors

16. The process can be further automated by the use of a computer-controlled manifold to automatically switch between the decellularization reagents at the appropriate times without manually disconnecting the tubing from one reagent carboy and reconnecting to another.
17. Store the decellularized lung in an isotonic solution such as PBS or NaCl, with 50–100 I.U./mL penicillin and 50–100 μg/mL streptomycin.

4 Notes

1. Sodium deoxycholate is a lung irritant. Wear a mask when weighing sodium deoxycholate and avoid generating dust.

   The fine powder can be difficult to dissolve. We recommend preparing a 4–8% stock solution on a heated stir-plate which can then be diluted in the carboy as needed.

2. Do not reconstitute DNase until a few hours prior to using. DNase begins to degrade within a few hours of reconstitution. Once reconstituted, degradation can be slowed by storing between 2 and 8 °C until ready to use.
3. Our bioreactors are cylinders constructed of translucent acrylic plastic. We use a 24 cm diameter × 27 cm tall bioreactor for lungs from pigs under approximately 20 kg, or individual lung lobes. We use a 34 cm diameter × 36 cm tall bioreactor for lungs from larger pigs.
4. The distal end of the Ventilator Tubing is positioned halfway between the spout and the bottom of the Ventilator Reservoir. This ensures that the Ventilator Tubing will be deep enough to remain submerged when fluid is drawn from the Ventilator Reservoir, but not so deep that debris is drawn in from the bottom.
5. The volume expelled is the tidal volume of the lungs and may need to be adjusted based on the size of the lungs.
6. The volume returned to the bioreactor is less than the volume expelled to account for the volume of the reagent being continuously added through the vasculature.
7. If the decellularization process needs to be separated into multiple days, remove the lung from the bioreactor and store at between 2 and 8 °C in an isotonic solution such as PBS or NaCl, with 50–100 I.U./mL penicillin and 50–100 μg/mL streptomycin. Once the lung is in the storage container, use a damp paper towel to cover any portions of the lung exposed to air to prevent those areas from drying out.

References

1. Price AP, England KA, Matson AM et al (2010) Development of a decellularized lung bioreactor system for bioengineering the lung: the matrix reloaded. Tissue Eng Part A 16:2581–2591. doi:10.1089/ten.TEA.2009.0659
2. Crabbé A, Liu Y, Sarker SF et al (2015) Recellularization of decellularized lung scaffolds is enhanced by dynamic suspension culture. PLoS One 10:1–17. doi:10.1371/journal.pone.0126846
3. Nichols JE, La Francesca S, Vega SP et al (2016) Giving new life to old lungs: methods to produce and assess whole human paediatric bioengineered lungs. J Tissue Eng Regen Med n/a-n/a. Doi: 10. 1002/term.2113

4. Franks TJ, Colby TV, Travis WD et al (2008) Resident cellular components of the human lung: current knowledge and goals for research on cell phenotyping and function. Proc Am Thorac Soc 5:763–766. doi:10.1513/pats.200803-025HR
5. Stone KC, Mercer RR, Freeman BA et al (1992) Distribution of lung cell numbers and volumes between alveolar and nonalveolar tissue. Am Rev Respir Dis 146:454–456. doi:10.1164/ajrccm/146.2.454
6. Tsuchiya T, Mendez J, Calle EA et al (2016) Ventilation-based decellularization system of the lung. Biores Open Access 5:118–126. doi:10.1089/biores.2016.0012
7. Raredon MSB, Rocco KA, Gheorghe CP et al (2016) Biomimetic culture reactor for whole-lung engineering. Biores Open Access 5:72–83. doi:10.1089/biores.2016.0006
8. He M, Callanan A (2013) Comparison of methods for whole-organ decellularization in tissue engineering of bioartificial organs. Tissue Eng Part B Rev 19:194–208. doi:10.1089/ten.teb.2012.0340
9. Price AP, Godin LM, Domek A et al (2015) Automated decellularization of intact, human-sized lungs for tissue engineering. Tissue Eng Part C Methods 21:94–103. doi:10.1089/ten.TEC.2013.0756

Methods in Molecular Biology (2018) 1577: 317–325
DOI 10.1007/7651_2017_45

Published online: 25 May 2017

Alveolar Rhabdomyosarcoma Decellularization

Michela Pozzobon, Mattia Saggioro, Stefania D'Agostino, Gianni Bisogno, Maurizio Muraca, and Piergiorgio Gamba

Abstract

In cancer research, it is an urgent need in the obtainment of a simple and reproducible model that mimics in all the complexity the pathological microenvironment. Specifically, the will to improve the overall survival of young patients affected by rhabdomyosarcoma compels researchers to develop new models resembling the multifaceted environment of the pathology to deeply study the disease under novel and different aspects. To this end, we developed a decellularization protocol for alveolar rhabdomyosarcoma (ARMS) able to maintain the three-dimensional structure. The attained extracellular matrix (ECM) can be used as 3D in vitro model suitable to both recapitulate the in vivo cancer microenvironment, and also for drug testing. Here, we first describe a detergent-enzymatic method and then we analyze the decellularization efficiency and the scaffold proteins.

Keywords: Decellularization, Extracellular matrix, Rhabdomyosarcoma, 3D in vitro model

1 Introduction

Rhabdomyosarcoma is a soft tissue sarcoma that arises mainly in or near the muscles. Two subtypes can be distinguished, the embryonal with a more favorable prognosis and the alveolar with worse outcome. Current standard treatment for alveolar rhabdomyosarcoma (ARMS) consists in a multimodal therapy involving chemical/radio treatment and surgery but the survival rate of 70 % should be improved [1, 2]. The tumor aggressiveness is related to the high propensity to rapidly grow and widely disseminate and the role of the microenvironment appears of paramount importance. Indeed, cancer progresses within a dynamically evolving extracellular matrix (ECM) which is a network of secreted factors, basement (collagen, fibronectin, and laminin), and interstitial (proteoglycans) matrix proteins that provide not only mechanical and structural support to the cells and the tissues but also spatial coordination of signaling processes [3]. Specifically, ECM influences cell recruitment from blood, growth, differentiation, and survival of the cells. Therefore, dysregulation or mutation of ECM components, as it happens in cancer, entails in a broad range of pathological conditions [4, 5]. For all these reasons, in the last decades, the ECM has been considered a promising therapeutic target [6, 7].

Recently, researchers focused on studying cancer microenvironment drawing the attention toward simple in vitro models to investigate the complex interaction between the matrix and the cells. To this end, commercially available substrates mimicking the natural proteins have been used [8]. However, synthetic matrices do not recreate the composition of tissue-derived ECM, mainly because: (1) both cytokines/growth factors and the typical 3D structure are absent; and (2) murine matrices (i.e., Matrigel®) are derived from mouse tumors and thus are not representative of human tissues. Moreover, a major limitation is represented by the fact that cells are grown as monolayers on flat solid surface lacking cell–cell and cell–matrix interactions characterizing the native tumors. It is also demonstrated that cells in petri dish undergo cytoskeletal rearrangements acquiring artificial polarity, which in turn causes aberrant gene and protein expression [9]. Recently, 3D models such as tumor tissue explant, tumor on a chip, organoids with growth factor cocktails mimicking the various organ stem cell niches have been developed but the lack of tight contact between cells and tissue architecture makes all these samples limited for the clinical translation of the basic research results [9, 10]. In this scenario, tumor ECM obtained after decellularization process has been recently appointed as the best system to model native tissue, recapitulating fundamental aspects of the natural structure and soluble factors [11]. Able to preserve the original shape and biochemical properties but deprived of the genetic material, ECM can create a cross talk with seeded cells [12, 13]. Our group already developed a method for muscle decellularization [14]; here, we set an efficient and reproducible strategy to decellularize ARMS tissue obtaining ECM with maintained structural proteins and three-dimensional organization. This 3D in vitro model could represent a new tool to recapitulate ARMS complexity after cell seeding and to perform drug screening.

2 Materials

2.1 Equipment and Supplies

1. Biologic safety cabinet
2. Syringe Insulin 0.3 mL (Terumo)
3. Serological pipettes (5, 10, and 25 mL)
4. 20, 200, and 1,000 μL micropipettes
5. 20, 200, and 1,000 μL micropipette tips
6. 1.5 mL microcentrifuge tubes
7. Polystyrene conical tubes (15 and 50 mL).
8. Nanodrop 2000 spectrophotometer (Thermo Scientific) or comparable for DNA quantification

9. Electrophoresis chamber and power supply (Bio-Rad) for agarose gel electrophoresis
10. Microscope slides (frosted end)
11. Cryostat (Leica CM1520) or comparable for tissue slide preparation
12. Stove for immunostaining
13. Inverted light microscope (Olympus IX71) or comparable for histological analyses
14. Inverted fluorescent microscope (Leica DMI6000B) or comparable for immunofluorescent analyses
15. Spectrophotometer (SpectraMax Plus) or comparable for protein quantification

2.2 Rhabdomyosarcoma Production

1. Matrigel basement membrane matrix, phenol red-free (BD)
2. Commercially available RH30 cell line from bone marrow of 16-year-old male with untreated ARMS, expressing translocation t (2;13)
3. RAG $2^{-/-}\gamma c^{-/-}$ mice

2.3 Decellularization Reagents

1. Sterile phosphate-buffered saline (PBS)
2. Penicillin/streptomycin (pen/strep) (GIBCO, Thermo Fisher Scientific)
3. 1 % sodium dodecyl sulfate (SDS) (Sigma-Aldrich)
4. 2,000 kU Deoxyribonuclease 1 (DNase 1) (Sigma-Aldrich) in 1 M NaCl solution
5. 1 % Triton X-100 (Bio-Rad)

2.4 DNA Extraction and Quantification

1. DNeasy Blood & Tissue kit (Qiagen) for DNA extraction
2. 1 % agarose gel (in 1× TBE) for electrophoresis. Running buffer is 1× TBE: 0.5 M Tris base, 0.5 M Boric acid, and 0.01 M EDTA (disodium salt)
3. SYBR® Safe DNA gel stain (Thermo Fisher Scientific) for visualization of DNA in agarose gel

2.5 Samples Fixing and Sectioning

1. 4 % paraformaldehyde (PFA) for fixation
2. 10, 15, and 30 % sucrose solutions for dehydration
3. Optimal cutting temperature (O.C.T., Kaltek) for embedding
4. Sectioning samples in 7–9 μm slices using a cryostat (Leica CM1520)

2.6 Immunofluorescence Analyses

1. 0.5 % Triton X-100 in 1× PBS for permeabilization (30 min)
2. 10 % horse serum in 1× PBS for saturation (12 min)

3. Primary antibody:
   - Polyclonal rabbit anti-mouse laminin (Sigma-Aldrich), dilution 1:200. Use 1 % bovine serum albumin (BSA) in 1× PBS for dilution (1 h at 37 °C)
   - Polyclonal rabbit anti-mouse Collagen I A1 (Life Technologies), dilution 1:80 in 1 % BSA (overnight at 4 °C)
   - Monoclonal mouse anti-mouse Fibronectin (Life Technologies), dilution 1:100 in 1 % BSA (overnight at 4 °C)
   - Monoclonal mouse anti-mouse α-Smooth Muscle Actin (Abcam), dilution 1:100 in 1 % BSA (1 h at 37 °C)
   - Monoclonal mouse anti human Vimentin (Abcam), dilution 1:100 in 1 % BSA (1h at 37 °C)
   - Monoclonal mouse anti-human Myod (DAKO), dilution 1:50 in 1 % BSA (overnight at 4 °C)
   - Monoclonal mouse anti-human Hunu (Millipore), dilution 1:200 in 1 % BSA (1h at 37 °C)
4. Secondary antibody:
   - Alexa Fluor chicken anti-rabbit 488 (Life Technologies), dilution 1:200 in 1 % BSA
   - Alexa Fluor goat anti-mouse 594 (Life Technologies), dilution 1:200 in 1 % BSA
5. Fluorescent mounting medium plus 100 ng/mL 4′,6-diamidino-2-phenylindole (DAPI, Sigma) for nuclei counterstaining

2.7 Histological Analyses

1. Hematoxylin & Eosin (H&E) kit for frozen sections (Bio-Optica)
2. Alcian Blue kit for frozen sections (Bio-Optica)

2.8 Extracellular Matrix Component Quantification

1. Sircol™ Soluble Collagen Assay (Biocolor) for collagen quantification

3 Methods

3.1 Rhabdomyosarcoma Tumor Mass Induction and Characterization

1. For one single injection, resuspend 2×10^6 RH30 cells in 30 μL of PBS (Protocol number 304/2017 approved by the Ministry of Italian Health).
2. Add 30 μL of phenol free Matrigel and mix gently without forming bubbles.
3. Load the insulin syringe with cell suspension in Matrigel and inject it in flanks of 16-week-old RAG $2^{-/-}\gamma c^{-/-}$ mice.

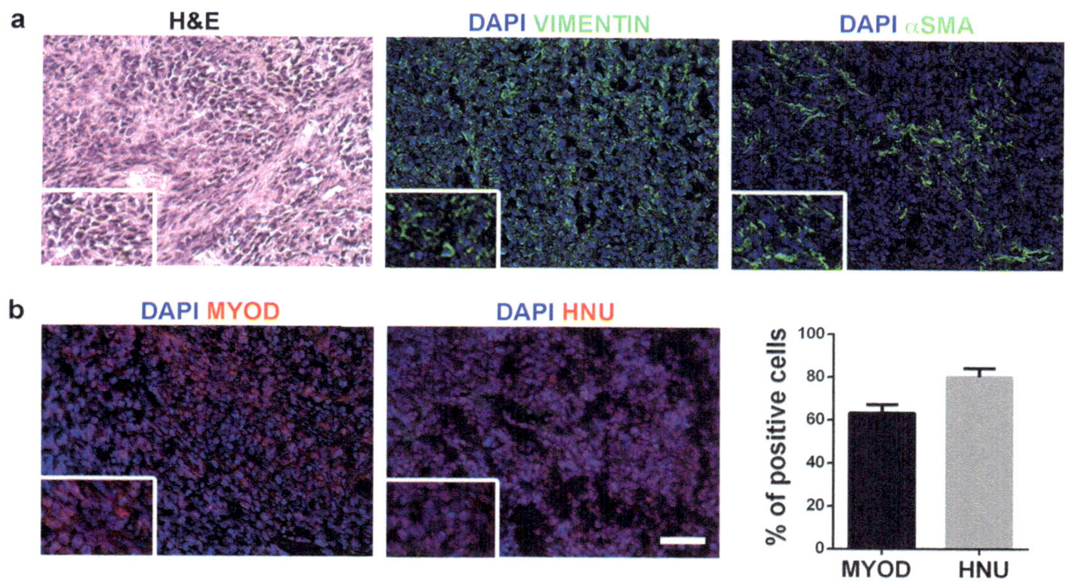

Fig. 1 Xenogenic alveolar rhabdomyosarcoma (ARMS) characterization. (**a**) Hematoxylin Eosin underlines the typical cell distribution of human ARMS. IF of the most used protein markers of human ARMS: Vimentin, alpha smooth muscle actin (αSMA). (**b**) Human nuclei staining and MYOD detection were present above 70 %. Scale bar: 50 μm

4. Sacrifice mice after 4 weeks from injection and extract the tumor mass.
5. Process tumor sample and perform immunofluorescence to verify the expression of the characteristic proteins (VIMENTIN, a SMA). Evaluate the percentage of human nuclei and muscle cells (Fig. 1a, b) (*see* **Note 1**).

3.2 Decellularization Procedure

1. Collect ARMS and carry them in sterile PBS to the laboratory processing room (*see* **Note 2**).
2. Quickly rinse the sample in PBS with 3 % pen/strep.
3. Transfer the samples on a petri dish using sterile disposable plastic pincers and clean the tumor mass by removing residual muscle fibers and the capsule that envelopes the ARMS.
4. Cut the sample into pieces in a range of 250–300 mg (*see* **Note 3**).
5. Put the samples in 40 mL of sterile and deionized water with 3 % pen/strep at 4 °C, for 24 h (*see* **Note 4**).
6. Start cycle of Detergent-Enzymatic Treatment (DET) (Fig. 2a):
   - Put samples in 40 mL of 1 % SDS for 4 h at room temperature (RT) and in gentle agitation (*see* **Note 5**). SDS is used to disrupt cell membranes.

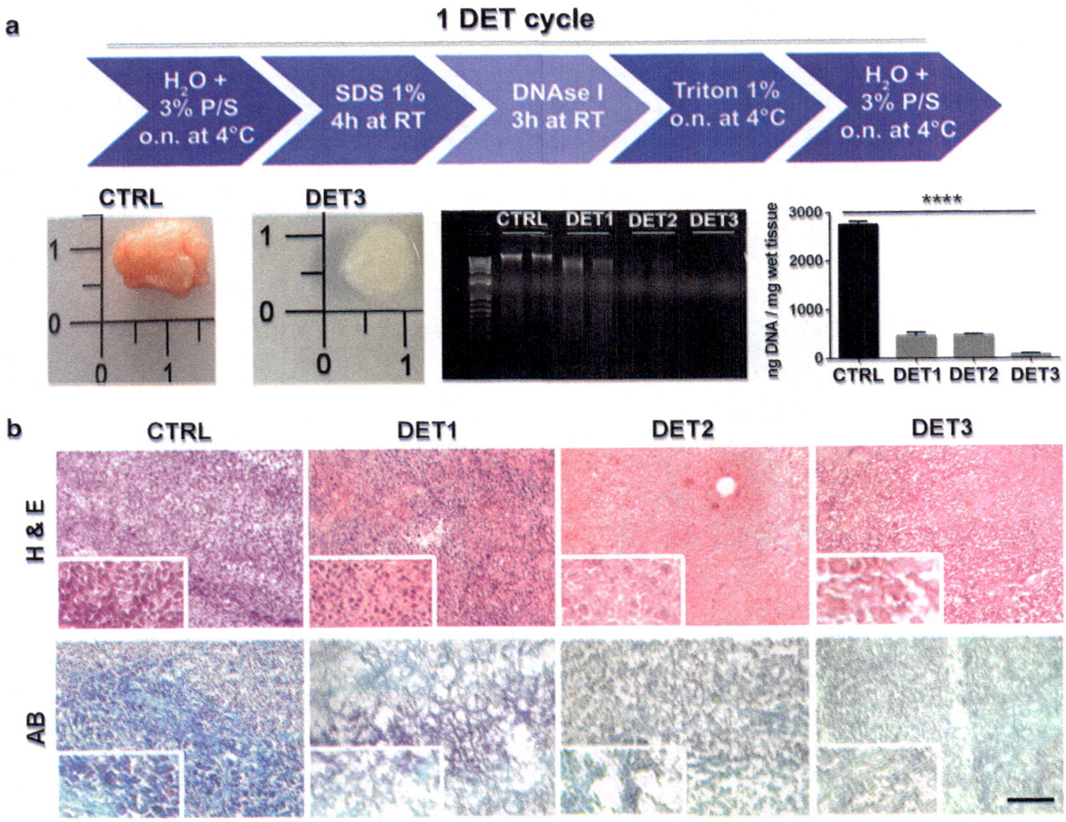

Fig. 2 Decellularization protocol and efficiency of the procedure. (**a**) Different phases of the protocol. Gross appearance of fresh and decellularized sample. DNA quantification and agarose gel. (**b**) *Upper row*: Hematoxylin Eosin; *lower row*: Alcian blue (AB) at different cycles. Nuclei are depleted but tissue shape is maintained at different cycles. Scale bar: 50 μm ****$p<0.0001$

- Wash samples from three to five times with sterile and deionized water.
- Put samples in 40 mL of 2,000 kU DNase 1 in 1 M NaCl for 3 h at RT and in gentle agitation in order to breakdown DNA.
- Wash samples with sterile and deionized water.
- Put samples in 40 mL of 1 % Triton X-100 overnight at 4 °C and in gentle agitation. Triton X-100 is used to remove cell debris.

End of one cycle of DET.

7. Repeat the steps from point 5.1 to point 5.5 for three times.
8. Wash samples for 5 days in sterile 1× PBS and 3 % pen/strep (*see* **Note 6**).

9. Freeze in liquid nitrogen (snap frozen) the samples dedicated for DNA and ECM components quantification or fix in 4 % PFA, dehydrate in increasing sucrose gradient, embed in O.C. T., and then section for histological, immunofluorescence analyses.

3.3 DNA Quantification

DNA quantification: proceed with DNA isolation protocol following manufacturer instruction of item 1 of Sect. 2.4 dissolving about 15 mg of tissue. After decellularization, remaining genetic material should be less than 50 ng DNA/mg tissue and if DNA fragments are present, they should have a maximum length of 200 bp [14]. Use items 8 and 9 of Sect. 2.1 and items 2 and 3 of Sect. 2.4 to quantify DNA amount and fragments length (Fig. 2a).

3.4 Extracellular Matrix Structural Protein Analysis and Collagen Quantification

1. *ECM three-dimensional structure analysis*: following section 2.5, after cryostat cutting sections, proceed with the immunostaining protocol using reagents in Sect. 2.7 and the antibodies described in the items 3, 4, and 5 of Sect. 2.6.

   Through items 1 and 2 of Sect. 2.7, determine morphological aspect and evaluate the loss of weakly sulfured mucins, hyaluronic acid, sialomucins, and sulfured mucins (Figs. 2b and 3a, b).

2. *ECM components quantification*: using item 1 of Sect. 2.8 to quantify collagen in order to verify the ECM integrity not only from the morphological point of view but also from the quality of retained proteins (Fig. 3a).

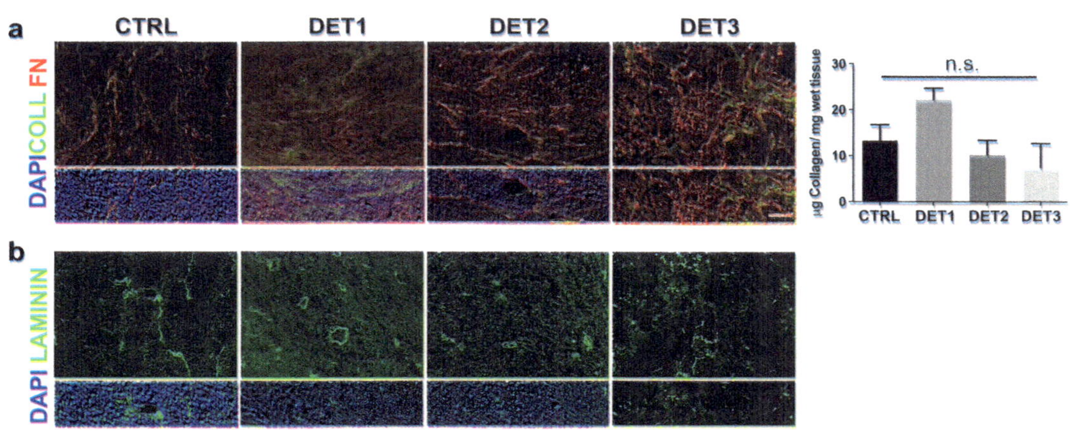

Fig. 3 Extracellular matrix (ECM) preservation. Basal protein ECM immunofluorescence highlights the good maintenance of fibronectin and collagen, also after quantification (**a**) and laminin (**b**) after three detergent-enzymatic treatment (DET) cycles. Scale bar: 50 μm

4 Notes

1. After 4 weeks, we found that more than 70 % of cells were of human origin and more than 50 % expressed MYOD protein. The xenogenic tissue morphology closely recapitulated the human counterpart [1].
2. If it is not possible to decellularize the sample immediately, it can be stored in sterile deionized water with 3 % pen/strep at 4 °C not longer than 48–72 h.
3. Tumor mass after 4 weeks from injection weights between 0.8 and 1.2 g. For each decellularization batch, reserve enough fresh ARMS as a control for: DNA quantification, immunostaining protocols, and ECM component quantification. This protocol has been optimized using a piece of ARMS weighting around 300 mg. After three DET cycles, the loss of weight is around 80 % of the initial mass since the tumor is full of cells, lost after the detergent-enzymatic procedure.
4. Sterility is of paramount importance when ECM matrices are repopulated with cells or used in vivo. It is necessary to work in aseptic conditions from the specimen collection (i.e., remove the hairs and the skin before murine biopsy) to the decellularization process (i.e., sterile reagents and supplies) in order to preserve sterility.
5. To obtain an efficient decellularization, do not use more than six samples in a volume of 40 mL of SDS or 40 mL of DNase 1.
6. Washing step is mandatory to obtain a suitable sample both for in vitro and in vivo tests. If necessary, extend the washing period by changing the solution twice a day.

Acknowledgments

Saggioro M and D'Agostino S are supported by Istituto di Ricerca Pediatrica Città della Speranza. Pozzobon M is supported by University of Padova, Grant number GRIC15AIPF, Assegno di ricerca Senior. Pozzobon M is a coinventor of the Italian Patent N. 0001422436 entitled "Matrice acellulare per ricostruzione *in vivo* di muscolo scheletrico."

References

1. Kashi VP, Hatley ME, Galindo RL (2015) Probing for a deeper understanding of rhabdomyosarcoma: insights from complementary model systems. Nat Rev Cancer 15:426–439. doi:10.1038/nrc3961
2. Hettmer S, Li Z, Billin AN et al (2014) Rhabdomyosarcoma: current challenges and their implications for developing therapies. Cold Spring Harb Perspect Med. doi:10.1101/cshperspect.a025650

3. Pickup MW, Mouw JK, Weaver VM (2014) The extracellular matrix modulates the hallmarks of cancer. EMBO Rep 15:1243–1253. doi:10.15252/embr.201439246
4. Harisi R, Jeney A (2015) Extracellular matrix as target for antitumor therapy. Onco Targets Ther 8:1387–1398. doi:10.2147/OTT.S48883
5. Lambert AW, Pattabiraman DR, Weinberg RA (2016) Review emerging biological principles of metastasis. Cell 168:670–691. doi:10.1016/j.cell.2016.11.037
6. Mouw JK, Ou G, Weaver VM (2014) Extracellular matrix assembly: a multiscale deconstruction. Nat Rev Mol Cell Biol 15:771–785. doi:10.1038/nrm3902
7. Onisto M, Slongo ML, Gregnanin L et al (2005) Expression and activity of vascular endothelial growth factor and metalloproteinases in alveolar and embryonal rhabdomyosarcoma cell lines. Int J Oncol 27:791–798
8. Bissell MJ, Radisky D (2001) Putting tumours in context. Nat Rev Cancer 1:46–54. doi:10.1038/35094059
9. Clevers H (2016) Modeling development and disease with organoids. Cell 165:1586–1597. doi:10.1016/j.cell.2016.05.082
10. Nath S, Devi GR (2016) Three-dimensional culture systems in cancer research: focus on tumor spheroid model. Pharmacol Ther 163:94–108. doi:10.1016/j.pharmthera.2016.03.013
11. Gilbert TW, Sellaro TL, Badylak SF (2006) Decellularization of tissues and organs. Biomaterials 27:3675–3683. doi:10.1016/j.biomaterials.2006.02.014
12. Chen HJ, Wei Z, Sun J et al (2016) A recellularized human colon model identifies cancer driver genes. Nat Biotechnol 34(8):845–851. doi:10.1038/nbt.3586
13. Crapo PM, Gilbert TW, Badylak SF (2011) An overview of tissue and whole organ decellularization processes. Biomaterials 32:3233–3243. doi:10.1016/j.biomaterials.2011.01.057
14. Piccoli M, Urbani L, Alvarez-Fallas ME et al (2016) Improvement of diaphragmatic performance through orthotopic application of decellularized extracellular matrix patch. Biomaterials 74:245–255. doi:10.1016/j.biomaterials.2015.10.005

Methods in Molecular Biology (2018) 1577: 327–335
DOI 10.1007/7651_2018_195

Published online: 06 September 2018

Human-Scale Liver Harvest and Decellularization for Preclinical Research

Kazuki Tajima, Hiroshi Yagi, and Yuko Kitagawa

Abstract

Decellularized scaffolds are promising for organ regeneration; to facilitate analyses of decellularization and preclinical research, experiments using large animals are indispensable. An efficient protocol using large animals requires the modification of current methods to account for differences in anatomy, size, stiffness, and animal handling. This chapter describes a method for the decellularization of livers from large animals (especially porcine livers), including anesthesia, organ harvest, organ decellularization, and storage, with a summary of potential pitfalls. This methodology can hypothetically be applied to large animals for preclinical studies and subsequent clinical applications.

Keywords Acellular scaffold, Bioengineering, Decellularization, Extracellular matrix, Human scale, Porcine liver, Preclinical trial, Regeneration therapy

1 Introduction

Decellularization is a unique regenerative approach in which all viable cells are eliminated while preserving native extracellular matrix components, such as collagen, laminin, and fibronectin [1]. Substantial evidence indicates that organ-specific extracellular matrix components contribute to cell migration, proliferation, and differentiation [2–5]. We have reported the efficacy of decellularized liver scaffold as a regenerative organ and showed that the organ-specific scaffold could support cell function and viability [6–8]. The decellularized scaffold also preserves the 3D architecture of the original vessel structure. Three-dimensional organs require the deep diffusion of nutrients and oxygen; thus, the decellularized 3D vessel structure is important for the survival and long-term function of reseeded cells [6].

Despite studies of the utility of decellularized organs in small animals, such as rats [6, 8], experiments with large animals are indispensable for preclinical research. Our group has demonstrated that porcine whole livers could be decellularized by the portal vein perfusion method [9]. Subsequently, other groups have reported various decellularization techniques using porcine livers; however, a standard method has not been established [10–12]. Therefore, we developed an efficient method based on extensive experiments since our initial report. We hope that this method will enhance preclinical experiments and enable clinical applications.

2 Materials

2.1 Porcine Anesthesia

1. Intramuscular injection set: a 21G needle (0.80 × 38 mm), 10-mL syringe, and medical extension tube (75 cm, 0.8 mL, TOP Corporation, Tokyo, Japan) are connected.
2. Sedatives: medetomidine (Nippon Zenyaku Kogyo Co., Ltd., Fukushima, Japan) and midazolam (Astellas Pharma Inc., Tokyo, Japan) are mixed for intramuscular injection. For additional dosage, only medetomidine is administered by intravenous injection.
3. Inhalation anesthesia: an anesthetic vaporization instrument (SN-487, Shinano MFG. Co., Ltd., Tokyo, Japan) and artificial respirator (SN-480-3, Shinano MFG. Co., Ltd.) are used to maintain the inhalation of anesthesia (isoflurane inhalation solution, Pfizer Japan Inc., Tokyo, Japan).
4. Tracheal intubation: endotracheal tube (Lo-contour oral/nasal tracheal tube cuffed reinforced 5.0 mm, Covidien Japan Co., Ltd., Tokyo, Japan), lidocaine spray (Xylocaine spray 8%, Aspen Japan K.K., Tokyo, Japan), and a laryngoscope (Macintosh improved blade KN-306-D1 280 mm, Natsume MFG. Co., Ltd., Tokyo, Japan) are used (*see* **Note 1**).
5. Intraoperative infusion: a 22G indwelling needle (Surflo 0.85 × 25 mm, Terumo Corporation, Tokyo, Japan), administration set, and sterile saline solution are used.

2.2 Porcine Liver Harvest

All surgical instruments are sterilized by autoclaving before use.

1. Disinfection: povidone-iodine solution (popiyodon solution 10%, Yoshida Pharmaceutical, Saitama, Japan).
2. Surgical instruments: surgical sterile gown and globes, surgical cap and mask, sterile drape, surgical scalpel, tissue forceps (180 mm), dressing forceps (180 mm, fine), Balfour abdominal retractor, Metzen balm scissors, Kelly forceps, Kocher and Pean forceps, sterile gauze, and surgical silk sutures (1-0, 2-0, and 3-0).
3. Electrocautery system (SHAPPER Dx, Senko Medical Instrument MFG. Co., Ltd., Tokyo, Japan) and disposable switch pencil (Senko Medical Instrument MFG. Co., Ltd., Tokyo, Japan).
4. Anticoagulant: heparin (heparin sodium injection AY, AY Pharmaceuticals Co., Ltd., Tokyo, Japan).

5. Perfusion and cannulating tube: sterilized saline, 60-mm sterilized Masterflex tube (L/S18, Yamato Scientific Co., Ltd., Tokyo, Japan), and administration set are used.
6. Storage: a sealed container (Screw-top Keeper 1200 mL, Iwasaki Industry Inc., Nara, Japan) sterilized by autoclaving and an ultralow temperature freezer (REVCO UXF60086, Asahi Life Science Co., Ltd., Saitama, Japan) are used.

2.3 Porcine Liver Decellularization

All solutions are made with ultrapure water (milli-Q Integral MT, Merck Ltd., Tokyo, Japan).

1. Wash solution: phosphate-buffered saline (PBS, Wako Pure Chemical Industries, Osaka, Japan).
2. Decellularization reagent 1: sodium dodecyl sulfate (Wako Pure Chemical Industries, Osaka, Japan) dissolved to 0.5% in ultrapure water.
3. Decellularization reagent 2: 1% Triton X-100 (Sigma-Aldrich Japan, Tokyo, Japan), 0.05% ethylene glycol bis (2-aminoethyl ether)-N,N,N',N'-tetraacetic acid (EGTA, Tokyo Chemical Industry Co., Tokyo, Japan), and 4 mM CHAPS (Dojindo Laboratories, Kumamoto, Japan) are dissolved in wash solution.
4. Tubing pump and tube: Masterflex L/S with Masterflex L/S Easy Load Pump Head III (Yamato Scientific Co., Ltd., Tokyo, Japan) and Masterflex tube (L/S18).

2.4 Sterilization and Storage of the Scaffold

1. Antimicrobial reagent: PBS containing 1% penicillin-streptomycin (Sigma-Aldrich Japan, Tokyo, Japan), 0.25 μg/mL amphotericin B, and 1% gentamycin (Thermo Fisher Scientific, Yokohama, Japan) solution.
2. Leak check reagent: DMEM (Sigma-Aldrich Japan).
3. γ-Ray irradiation facility: Radia Industry Co., Ltd. (Gunma, Japan).
4. Sealed container: Screw-top Keeper 1200 mL (Iwasaki Industry Inc.) is sterilized by autoclaving.
5. Refrigerator.

3 Methods

3.1 Porcine Liver Harvest

Although liver decellularization can be performed using cadaver organs, we describe the harvest method for a living donor, considering subsequent recellularization and transplantation. Therefore, careful attention is needed to avoid damaging the liver, including the capsule, and to maintain a sufficient length of the portal vein and suprahepatic vena cava (both vessels can reach 70 mm or more).

1. Sedate a pig with a medetomidine (0.02 mg/kg) and midazolam (0.5 mg/kg) cocktail by intramuscular injection (*see* **Notes 2** and **3**). After 5–10 min, while the pig is lying down, cannulate the auricular vein using an indwelling needle, and administer an additional dose of medetomidine (0.5 mg/kg) to ensure complete sedation. Infuse the pig with saline (5–10 mL/kg/h) via the auricular vein during the operation.
2. Anesthetize the larynx with lidocaine spray and perform tracheal intubation using a laryngoscope and endotracheal tube (*see* **Note 4**). After intubation, maintain anesthesia by the inhalation of isoflurane (2–2.5%) using a vaporization instrument and respirator.
3. After the stabilization of anesthesia, disinfect the surgical field with povidone-iodine solution, wearing a surgical cap, mask, gown, and gloves. Then, fix a sterile drape on the abdomen aseptically.
4. Make a midline skin incision using a surgical scalpel. Cut the subcutaneous fat and the linea alba using an electrocautery pencil. Hold the cutoff section of the linea alba with Kocher forceps, lift the abdominal wall, and hold the peritoneum with tissue forceps. Make an incision with a scalpel, without injury to the intra-abdominal organs. Continue the midline incision from the xiphoid process to the bladder, and insert Balfour abdominal retractors.
5. Keep gauze on the liver, and make an incision at the falciform ligament of the liver to near the suprahepatic vena cava. Bring the left lobe of the liver to the right and make an incision at the left triangular ligament.
6. Open the hepatogastric ligament and cut the ventral hepatoduodenal ligament, taking care not to damage the artery and bile duct.
7. Perform portal vein peeling using Metzen balm scissors, tissue forceps, and Kelly forceps and ligate and dissect the posterior superior pancreaticoduodenal vein, right gastric vein, and splenic vein with 3-0 silk sutures.
8. Hold the gallbladder with Pean forceps and peel the gallbladder bed. Ligate and dissect the cystic duct and the cystic artery with 2-0 silk sutures and remove the gallbladder.
9. Peel and tape the infrahepatic vena cava with 1-0 silk sutures (*see* **Note 5**). Lift the sutures and infrahepatic vena cava and incise the coronary ligament from the infrahepatic vena cava to the diaphragm.

10. Puncture and cut the diaphragm around the suprahepatic vena cava. Ensure that the phrenic vein is appropriately ligated and dissected, taping the suprahepatic vena cava at the nearby heart.
11. Connect the liver to the body trunk only by the portal vein, infrahepatic vena cava, suprahepatic vena cava, and hepatoduodenal ligament.
12. Inject 5000 units of heparin via the auricular vein and wait 5 min.
13. Prepare saline and a tube for injection into the liver during the waiting period. Fill the administration set with saline for immediate use after portal vein cannulation.
14. Ligate and dissect the hepatoduodenal ligament with 1-0 silk sutures.
15. Cannulate the tube into the portal vein near the splenic vein and dissect the portal vein. Immediately begin the rapid perfusion of saline from the cannulated tube (Fig. 1).
16. Ligate and dissect the infrahepatic vena cava, and open the suprahepatic vena cava to release blood in the liver. Perform euthanasia by exsanguination.
17. After perfusing 1 L of saline, place the liver in a sealed container and store it in a freezer at −80 °C (*see* **Note 6**).

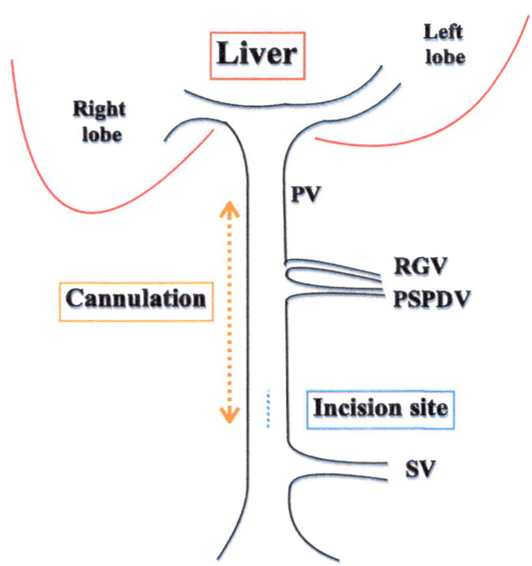

Fig. 1 Schema of the portal vein branches and incision area for cannulation. Since the trunk of the portal vein diverges immediately in the liver, the depth of cannulation is set with respect to the margin of the liver. *PV* portal vein, *PSPDV* posterior superior pancreaticoduodenal vein, *RGV* right gastric vein, *SV* splenic vein

3.2 Porcine Liver Decellularization

1. Move the frozen liver (stored at −80 °C) to a refrigerator at 4 °C and slowly thaw it over 2 days (*see* **Note 7**) (Fig. 2).
2. Connect tubes cannulated in the portal vein with tubes of the same size, and connect them to the Masterflex tubing pump (*see* **Note 8**) (Fig. 3).
3. Initiate wash buffer perfusion from the portal vein. Normally, approximately 20 L (90 mL/min) of wash buffer is necessary, and continue until blood drainage from the vena cava disappears.
4. When the turbidity of waste disappears, liver decellularization begins. Reflux decellularization reagent 1–30 L (90–200 mL/min) (*see* **Notes 9** and **10**).
5. After the whole liver becomes visibly white and decellularization is completed, wash by perfusion of 10 L of wash buffer.
6. Reflux decellularization reagent 2–20 L (150 mL/min) to wash out decellularization reagent 1, and wash by perfusion of 20 L of wash buffer (Fig. 4).
7. Store at 4 °C until sterilization (*see* **Note 11**).

3.3 Sterilization and Storage of the Scaffold

1. Perfuse 100 mL of leak check reagent from the portal vein and investigate leakage from the liver. If any rupture is detected, perform suturing or ligation (*see* **Note 12**).

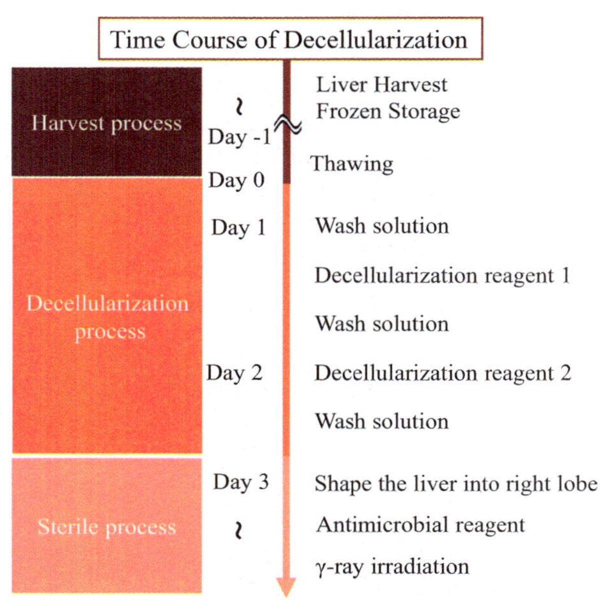

Fig. 2 Time course of porcine liver decellularization. Cryopreservation can be used for long-term storage, and thawing is necessary before use. The process takes about 7 days in total, including γ-ray sterilization

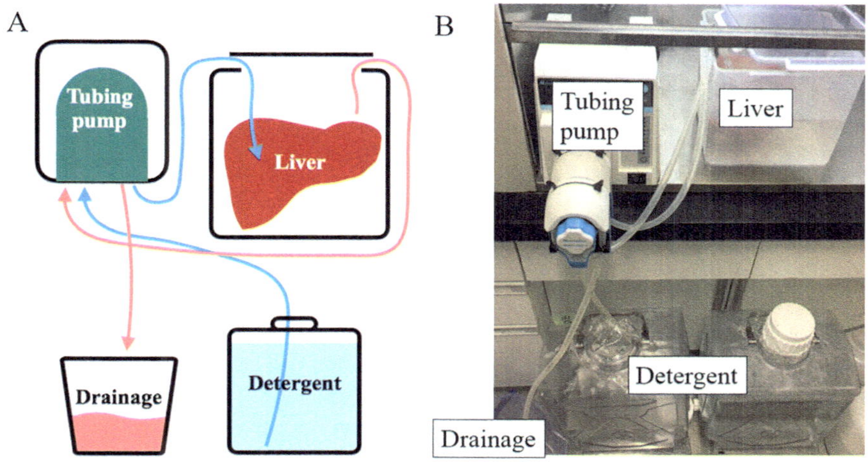

Fig. 3 Schema (**a**) and photograph (**b**) of the decellularization system. Two long tubes are used, one for perfusion of the reagent into the liver using a pump and another for drainage using the same pump

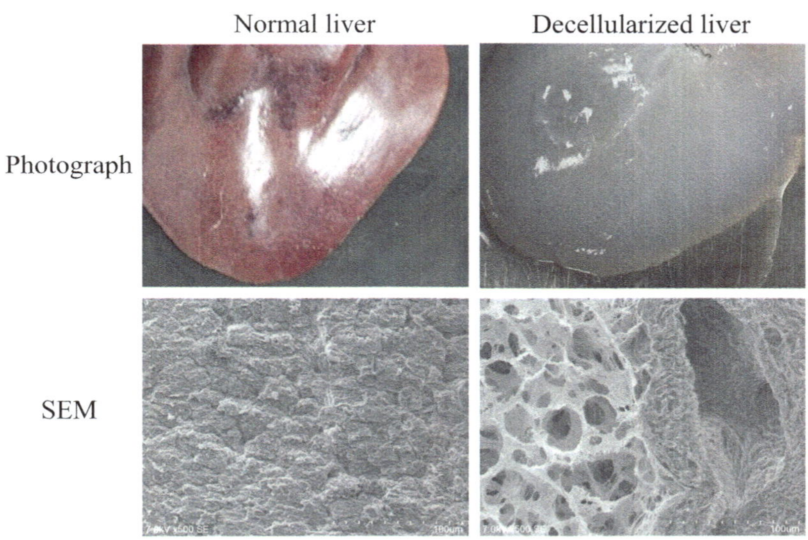

Fig. 4 Appearance of normal and decellularized porcine livers. In the decellularized liver, blood vessels appear to be transparent. Based on scanning electron microscopy (SEM) after decellularization, the structure of the extracellular matrix exhibited a mesh pattern

2. Reflux antimicrobial reagent (2 L) from the portal vein and seal it in a container.
3. Transport the sealed container to a gamma-ray sterilization facility in refrigerated conditions (4 °C), and perform γ-ray irradiation (25 kGy) (*see* **Note 13**).
4. Store at 4 °C until use.

4 Notes

1. Since pigs have a long pharynx, tracheal intubation requires a laryngoscope with a long blade (approximately 250 mm or longer for pigs of 15–20 kg).
2. Since young fattening pigs at a few months of age experience pain during intramuscular injection, reduce the dose of intramuscular injection and relieve the stress of intramuscular injection by administering additional anesthesia via the auricular vein.
3. For intramuscular injection, the extension tube is between the needle and the syringe; suck air in the syringe beforehand in order to inject all sedatives in the extension tube. Inject the whole volume quickly.
4. As pig tracheae are not straight, rotate the tracheal tube to facilitate insertion. Although tracheal intubation is possible in the recumbent position, supine position, and lying down, it is the easiest to insert the tracheal tube in animals in the supine position.
5. The right adrenal gland is located between the right kidney and the infrahepatic vena cava, and bleeding easily occurs from the adrenal vein while freeing the vena cava; it should be taped at a position away from the kidney.
6. Long-term preservation (approximately a few months) of the liver is possible by freezing at −80 °C.
7. Remove the frozen liver from the freezer, noting that the liver may crack unless it is slowly thawed in the refrigerator.
8. When connecting the portal vein tube and the perfusion tube, take care not to let air enter the tube. When air enters the tube, an air embolism occurs in the liver, and the decellularization process is incomplete.
9. Incomplete decellularization even when the decellularization solution is perfused can be explained by leakage. Check thoroughly for leaks in refluxed reagents in the whole liver.
10. As decellularization progresses, the liver tension weakens. Since the liver is not decellularized unless appropriate pressure is applied, the flow velocity is appropriately adjusted to maintain tension. An appropriate degree of pressure is necessary for complete decellularization.
11. As an indicator of decellularization, the measurement of DNA content and confirmation by hematoxylin and eosin staining are usually performed. To evaluate the extracellular matrix, azan staining and immunohistochemistry for the detection of collagen type I and IV, laminin, and fibronectin are performed.

12. Adjust the length of the portal vein and suprahepatic vena cava according to the research aims. When the decellularized scaffold is used for transplantation, leave the vessels long to account for anastomosis.
13. In our sterilization method, the decellularized scaffold exhibits sterility. Microorganisms, including viruses, have not been detected. Thus, long-term culture is acceptable. Measurements of endotoxin levels are indispensable for clinical applications.

Acknowledgments

This work was supported by grants from the Japan Agency for Medical Research and Development, Research Center Network for Realization of Regenerative Medicine, and Projects for Technological Development (to Y.K. and H.Y.).

References

1. Crapo PM, Gilbert TW, Badylak SF (2011) An overview of tissue and whole organ decellularization processes. Biomaterials 32:3233
2. Barkan D, Green JE, Chambers AF (2010) Extracellular matrix: a gatekeeper in the transition from dormancy to metastatic growth. Eur J Cancer 46:1181
3. Bissell MJ, Aggeler J (1987) Dynamic reciprocity: how do extracellular matrix and hormones direct gene expression? Prog Clin Biol Res 249:251
4. Nelson CM, Bissell MJ (2006) Of extracellular matrix, scaffolds, and signaling: tissue architecture regulates development, homeostasis, and cancer. Annu Rev Cell Dev Biol 22:287
5. Sellaro TL, Ravindra AK, Stolz DB, Badylak SF (2007) Maintenance of hepatic sinusoidal endothelial cell phenotype in vitro using organ-specific extracellular matrix scaffolds. Tissue Eng 13:2301
6. Kadota Y, Yagi H, Inomata K, Matsubara K, Hibi T, Abe Y, Kitago M, Shinoda M, Obara H, Itano O, Kitagawa Y (2014) Mesenchymal stem cells support hepatocyte function in engineered liver grafts. Organogenesis 10:268
7. Katsuki Y, Yagi H, Okitsu T, Kitago M, Tajima K, Kadota Y, Hibi T, Abe Y, Shinoda M, Itano O, Takeuchi S, Kitagawa Y (2016) Endocrine pancreas engineered using porcine islets and partial pancreatic scaffolds. Pancreatology 16:922
8. Uygun BE, Soto-Gutierrez A, Yagi H, Izamis ML, Guzzardi MA, Shulman C, Milwid J, Kobayashi N, Tilles A, Berthiaume F, Hertl M, Nahmias Y, Yarmush ML, Uygun K (2010) Organ reengineering through development of a transplantable recellularized liver graft using decellularized liver matrix. Nat Med 16:814
9. Yagi H, Fukumitsu K, Fukuda K, Kitago M, Shinoda M, Obara H, Itano O, Kawachi S, Tanabe M, Coudriet GM, Piganelli JD, Gilbert TW, Soto-Gutierrez A, Kitagawa Y (2013) Human-scale whole-organ bioengineering for liver transplantation: a regenerative medicine approach. Cell Transplant 22:231
10. Barakat O, Abbasi S, Rodriguez G, Rios J, Wood RP, Ozaki C, Holley LS, Gauthier PK (2012) Use of decellularized porcine liver for engineering humanized liver organ. J Surg Res 173:e11
11. Struecker B, Hillebrandt KH, Voitl R, Butter A, Schmuck RB, Reutzel-Selke A, Geisel D, Joehrens K, Pickerodt PA, Raschzok N, Puhl G, Neuhaus P, Pratschke J, Sauer IM (2015) Porcine liver decellularization under oscillating pressure conditions: a technical refinement to improve the homogeneity of the decellularization process. Tissue Eng Part C Methods 21:303
12. Mirmalek-Sani SH, Sullivan DC, Zimmerman C, Shupe TD, Petersen BE (2013) Immunogenicity of decellularized porcine liver for bioengineered hepatic tissue. Am J Pathol 183:558

Methods in Molecular Biology (2018) 1577: 337
DOI 10.1007/7651_2017_110

Published online: 14 January 2018

Erratum to: Decellularization Methods for Scaffold Fabrication

Sweta K. Gupta, Narayan C. Mishra, and Archna Dhasmana

Erratum to:
Methods in Molecular Biology
10.1007/7651_2017_34

The publisher regrets that an author was not mentioned in the chapter by mistake. The details of the author are provided below:

Archna Dhasmana – Department of Polymer and Process Engineering, Indian Institute of Technology, Roorkee, India

The updated original online version for this chapter can be found at
DOI 10.1007/7651_2017_34

Index

Kursad Turksen (ed.), *Decellularized Scaffolds and Organogenesis, Methods and Protocols*, Methods in Molecular Biology, https://doi.org/10.1007/978-1-4939-7309-5_1, © Springer Science+Business Media, LLC, part of Springer Nature 2018

D

I

K

L

M

U

V

W

X

Z

Printed by Printforce, the Netherlands